Springer Series in Solid-State Sciences

Volume 208

The Springer Series in Solid-State Sciences features fundamental scientific books prepared by leading and up-and-coming researchers in the field. They strive to communicate, in a systematic and comprehensive way, the basic principles as well as new developments in theoretical and experimental solid-state science.

We welcome submissions for monographs or edited volumes from scholars across this broad domain. Topics of current interest include, but are not limited to:

- Semiconductors and superconductors
- Quantum phenomena
- Spin physics
- Topological insulators
- Multiferroics
- Nano-optics and nanophotonics
- Correlated electron systems and strongly correlated materials
- Vibrational and electronic properties of solids
- Spectroscopy and magnetic resonance

Tomoe Fukamachi · Takaaki Kawamura

Resonant Dynamical X-ray Diffraction

Theory and Experiments

Springer

Tomoe Fukamachi
Saitama Institute of Technology
Fukaya, Saitama, Japan

Takaaki Kawamura
University of Yamanashi
Kofu, Yamanashi, Japan

ISSN 0171-1873 ISSN 2197-4179 (electronic)
Springer Series in Solid-State Sciences
ISBN 978-981-96-8405-2 ISBN 978-981-96-8406-9 (eBook)
https://doi.org/10.1007/978-981-96-8406-9

This Springer imprint is published by the registered company Springer Nature Singapore Pte Ltd.
The registered company address is: 152 Beach Road, #21-01/04 Gateway East, Singapore 189721, Singapore

Dedicated to the memory of Shizuo Miyake

Preface

Since the 1970s, synchrotron radiation X-rays, which are much brighter and highly collimated than X-rays from the tubes and have a continuous spectrum, have become available, allowing for not only accurate X-ray diffraction experiments but also new types of X-ray diffraction experiments. The X-ray resonant scattering described in this book is a typical example. Each atom has its own core electron level, and X-rays undergo resonant scattering accompanying the photoelectron excitation of this level. When this scattering occurs within the crystal, the X-ray scattering factor and the X-ray scattering intensity changes significantly. This scattering occurs with X-rays having energies specific to each atom in the crystal, so it provides structural information about its surrounding atoms. When this scattering occurs in the core electrons in a defect-free, so-called perfect crystal, a large change is seen in the X-ray diffraction intensity from the crystal. To describe the diffraction intensity and diffraction phenomena from a perfect crystal, a theory that incorporates multiple scattering of X-rays is required, and this is called the dynamical theory of diffraction. Resonant X-ray dynamical diffraction theory describes the diffraction intensity and diffraction phenomena from a perfect crystal when the resonant scattering occurs.

X-ray diffraction from a crystal is very useful and widely used for the crystal structure analysis. For interpreting the diffracted intensities, the single scattering theory called the kinematic theory of diffraction is usually used for this purpose. The intensities are given by the Fourier transform of the electron densities around atoms in the crystal. This in turn means that the kinematic theory is applied to analyze the electron densities, which means the atomic arrangement of the crystal. In this theory, however, the energy conservation does not hold and the calculated diffraction intensities do not always reproduce the measured intensities quantitatively. The theory cannot be applied to precise X-ray optics. The single scattering theory works only when X-ray scattering by an atom is weak or/and the size of the crystal is small compared with the penetration depth of the X-ray. Such crystals are microcrystals of several microns, and a cluster of such microcrystals called a mosaic crystal. When the orientation of the crystal grains is aligned, it is called a mosaic single crystal. If the orientations are not aligned and of powder form, it is called a powder crystal. In contrast, if three-dimensional period of atoms or molecules in a crystal is maintained on the order of centimeters, it is called a perfect crystal. In the crystal, multiple scattering of X-rays cannot be ignored. Regarding diffraction due to coherent multiple scattering by atoms in a crystal, Darwin and Ewald worked on it immediately after the discovery of X-ray diffraction. The multiple scattering theory is called the dynamical theory of diffraction (the law of energy conservation holds in this theory). Unfortunately, crystals with a high degree of perfection do not exist in nature. Researches on dynamical diffraction were stagnated for a long time. The Pendellösung fringes predicted by using the dynamical theory by Ewald were observed for the first time in X-ray diffraction from the highly pure and almost perfect artificial crystal (Si), which were required for the invention of the transistor. After this, it is recognized that researches on dynamical diffraction should be useful for evaluating the integrity of artificial crystals. Along with the subsequent remarkable development of the semiconductor industry, researches on dynamical diffraction theory have been developed.

In this book, we will first give a brief explanation about resonant dynamical X-ray diffraction (Chap. 1), and then describes diffraction by microcrystals (kinematic diffraction theory) (Chap. 2). We will explain resonant scattering (also called anomalous scattering) in atoms using the classical and quantum theories, and discuss the applications of resonant scattering in crystal structure analysis (polarity determination, phase determination) (Chap. 3). In Chap. 4, we will explain the optical properties of X-rays, such as refraction, absorption, total reflection by crystals, reflection by single-layer films, and X-ray waveguides. After deriving the basic equations of dynamical diffraction in Chap. 5, we will discuss phenomena such as Pendellösung fringes and extinction effects that are unique to perfect crystals in Chap. 6. We will explain the direction in which X-ray energy flows within the crystal, and the X-ray beam trajectory corresponding to the energy flow. We also describe the effect of angular amplification from a crystal. In Chap. 7, we develop the resonant dynamical diffraction theory and describe diffraction phenomena that depend on the resonance state (resonance circle). In Chap. 8, we will describe the interference phenomenon of the wave-packet. When the incident X-ray is to be regarded as a wave-packet and its divergence angle is about 1 s, the direction of the X-rays spreads extremely within the crystal to about several tens of degrees due to the angular amplification effect (within a Borrmann fan). The interference phenomenon of the wave-packet by this effect can be observed. In Chap. 9, the dynamical theory of diffraction is extended to apply to slightly strained crystals. In a crystal with a constant strain gradient, the trajectory of the X-ray becomes a hyperbola, and the X-ray incident on the crystal returns to the plane of incidence and is diffracted (mirage diffraction). We will discuss the interference fringes observed in the diffracted beams and the strain analysis in the crystal.

Fukaya, Japan Tomoe Fukamachi
Kofu, Japan Takaaki Kawamura

Acknowledgements The authors would like to thank thc latc Prof. Shizuo Miyake of the Institute for Solid State Physics, the University of Tokyo, for giving us the motivation to start research on resonant dynamical X-ray diffraction and for sharing his knowledge of the complex dispersion surface. The late Prof. Sukeaki Hosoya of the same institute provided us with a wealth of knowledge about the research of resonant X-ray diffraction and its applications. We would like to thank Prof. Arthur Bienenstock of Stanford University and the Director of then Stanford Synchrotron Radiation Laboratory (SSRL), who provided us with the opportunity for the research of resonant dynamical X-ray diffraction using synchrotron radiation at SSRL and subsequently conducted the collaborative research. We would also like to thank Prof. Herman Winick of SSRL, who supported us with synchrotron radiation experiments at SSRL.

We would like to thank the late Prof. Masayasu Tokonami of the University of Tokyo and the late Prof. Kazunobu Hayakawa of Hokkaido University for their valuable advice and cooperation in advancing our research on resonant dynamical diffraction.

We would like to thank Prof. Riichiro Negishi, Institute of Advanced Science, Saitama Institute of Technology (SIT), who collaborated with us on theory and experiments in the study of resonant dynamical X-ray diffraction, Prof. Dong-ying Ju of SIT, and Prof. Masayuki Shimojo of Shibaura Institute of Technology, who provided useful information in the analysis of strained crystals, and the late Assoc. Prof. Masami Yoshizawa of SIT, who worked very hard to measure the anomalous scattering factor. We would also like to thank Dr. Kenji Hirano and Dr. Jongsukswat Sukswat for their help with the experiments and data analysis.

The synchrotron radiation experiments described in this book were conducted at the Photon Factory (PF) of the High Energy Accelerator Research Organization in Japan, Spring 8 in Japan, and SSRL in the USA. In particular, many of the experiments were performed at beamline 16C at PF. We would like to thank Prof. Keiichi Hirano, the beamline director, for his cooperation.

We would like to express our gratitude to Mr. Akiyuki Tokuno and Ms. Manju Ramanathan of Springer Nature for their appropriate advice and support in publishing this book.

Competing Interests The authors have no competing interests to declare that are relevant to the content of this manuscript.

Contents

Abbreviations

Bb	Bragg-Bragg
BbL	Bragg-Bragg-Laue
BL	Bragg-Laue
CRL	Compound Refractive Lens
EXAFS	Extended X-ray Absorption Fine Structure
FWHM	Full Width at Half Maximum
IFLSD	Interference Fringes emitted from Lateral Surface in the Diffracted wave direction
IFLST	Interference Fringes emitted from Lateral Surface in the Transmitted wave direction
IFMD	Interference Fringes caused by two Mirage Diffracted waves
IFMRB	Interference Fringes between Mirage diffracted beam and Reflected Beam from the bottom surface
IFRB	Interference Fringes between once- and twice-Reflected beams from the Back surface
IFTB	Interference Fringes observed in the Transmitted Beam
LLL	Laue-Laue-Laue
MAD	Multiple Anomalous Dispersion
MBL	Multiple Bragg-Laue
MCA	Multi-Channel Analyzer
MIFMD	Modified IFMD
SC	Scintillation Counter
SR	Synchrotron Radiation
SSD	Solid State Detector
TDS	Thermal Diffuse Scattering
TRK	Thomas-Reich-Kuhn
XAFS	X-ray Absorption Fine Structure
XSW	X-ray Standing Wave

Chapter 1
Introduction

X-rays are electromagnetic waves with the wavelength of approximately 0.1 nm (=1 Å, corresponding to the energy of 12.4 keV), which is comparable to the lattice size of crystals. This causes diffraction by the crystal lattice. In classical electromagnetism, X-ray scattering is a phenomenon in which the electric field of the X-ray causes electrons to vibrate, thereby emitting X-rays in turn (Thomson scattering). While X-rays can be treated as waves, they also need to be treated as photons with a certain momentum as quanta. This chapter briefly describes the basics of X-ray diffraction and related phenomena. If the reader does not understand the explanation in this chapter, it is better to skip it and read the relevant chapter first.

1.1 X-ray Scattering

According to classical electromagnetism, Thomson scattering is the elastic scattering of X-rays by a free electron (with charge $-e$). According to quantum mechanics, an atom consists of a nucleus (consisting of protons and neutrons) and surrounding electrons. Since the mass of a proton is 1836 times that of an electron, the scattering intensity by a proton is negligible compared to that by an electron. However, core electrons are strongly bound to the atomic nucleus and vibrate with the normal mode. If the characteristic frequency of the X-ray approaches the normal mode, a resonance occurs between the X-ray and the normal mode of the electron, which is called an X-ray resonant scattering. In the X-ray region, the resonant scattering is sometimes called anomalous scattering. According to quantum mechanics, the processes of Thomson scattering and resonant scattering are different. When X-ray photons, quanta of the electromagnetic field, are scattered in an atomic system, one photon annihilates in the system and one photon is generated in Thomson scattering. In resonant scattering, one X-ray photon is absorbed when the system is excited, and another X-ray photon is emitted when the system returns to its ground state.

T. Fukamachi and T. Kawamura, *Resonant Dynamical X-ray Diffraction*, Springer Series in Solid-State Sciences 208, https://doi.org/10.1007/978-981-96-8406-9_1

Absorption and emission of photon occurs. For this reason, the effect of lifetime of the excited state appears, and the effect of the excited state in the absorption and emission process appears as well. Thomson scattering reflects the electron density distribution $\rho(\boldsymbol{r})$ ($\boldsymbol{r}$ is the position vector) around atoms in a crystal, and the crystal structure can be analyzed by measuring the diffraction intensities. The scattering factor by atoms is called the atomic scattering factor denoted as $f(\boldsymbol{s})$, with $\boldsymbol{s}$ being the scattering vector.

1.2 Optical Properties of X-ray

The optical properties of X-rays are important as a basis for understanding dynamical diffraction theory. For studying the optical properties of X-rays in an isotropic medium, the basic equation is derived from Maxwell's equation using the dielectric constant ε and X-ray susceptibility χ of the medium. In the region of visible light, the value of the refractive index n_0 is usually more than 1 for natural substances. It can be negative for some metamaterials called negative-index metamaterials [1, 2]. In the region of X-ray, on the other hand, the value of the refractive index is slightly smaller than 1 because the sign of χ is negative and its magnitude is quite small of the order from 10^{-6} to 10^{-4}. Then, total internal reflection occurs at a very small angle of incidence (glancing angle θ_i). From the measurement of this critical angle of the total internal reflection ($\theta_c = \sqrt{-\chi}$), n_0 and χ can be obtained ($n_0 = 1 + \chi/2$, the sign of χ is negative). By using Maxwell's equation and the derived basic equation, we will obtain the dispersion relation of X-rays as

$$\boldsymbol{k}_t^2 = \frac{\varepsilon}{\varepsilon_0}\boldsymbol{K}_0^2 = n_0^2\boldsymbol{K}_0^2 \tag{1.1}$$

Here, $\boldsymbol{K}_0$ is the wave vector of the incident wave in a vacuum and $\boldsymbol{k}_t$ that of the refracted wave in a medium.

Figure 1.1a, b show the relation between the incident wave and the reflected wave (of wavevector $\boldsymbol{K}_r$), and the refracted wave (of wavevector $\boldsymbol{k}_t$) (a) shows the reflection when $\theta_i > \theta_c$, and (b) shows that when $\theta_i \leq \theta_c$ and the total internal reflection occurs. Figure 1.1c shows the dispersion relations for the incident wave and refracted wave obtained from Eq. (1.1). The refractive index of the medium is given as $n_0 = |\boldsymbol{k}_t|/|\boldsymbol{K}_0|$. This figure is a cross-sectional view of two spheres centered on the origin O and with radii of $|\boldsymbol{K}_0|$ and $|\boldsymbol{k}_t|$, which are called dispersion surfaces in a vacuum and in the medium. The propagation point (tie-point) for $\boldsymbol{K}_0$ is determined from the intersection (A) of the line normal to the surface of the medium and the dispersion surface in a vacuum. The tie-point of $\boldsymbol{k}_t$ is determined from the intersection (B) of the line normal to the surface of the medium and the dispersion surface in the medium. When the refracted wave goes into the medium in Fig. 1.1a, the corresponding tie-point in a vacuum is A, and that in the medium is B. When the total internal reflection occurs as shown in Fig. 1.1b, the corresponding tie-point in

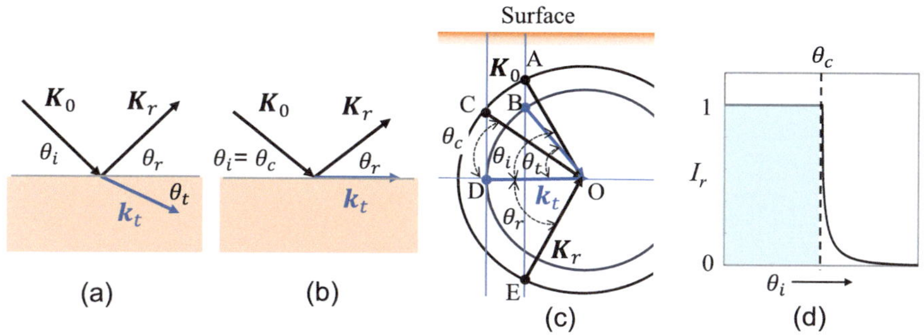

Fig. 1.1 Reflection of X-ray from a medium for $\theta_i > \theta_c$ (**a**) and $\theta_i \leq \theta_c$ (**b**). The dispersion surface (**c**) and reflectivity (**d**)

a vacuum is C, and that in the medium is D. Then the wavevector of the refracted wave is parallel to the surface. Using the dispersion surface, the law of reflection of $\theta_i = \theta_r$ and Snell's law of $|\boldsymbol{K}_0|\cos\theta_i = |\boldsymbol{k}_t|\cos\theta_t$ are derived. Here θ_r is the reflection angle from the surface and θ_t is the refraction angle.

Figure 1.1d shows the variation of reflectance as a function of the incident angle θ_i. The reflectivity (I_r) is 1 and total internal reflection occurs when $\theta_i \leq \theta_c$. It decreases rapidly when θ_i becomes larger than θ_c. In the angular region of total internal reflection ($\theta_i \leq \theta_c$), the refracted waves attenuate with depth of the medium, resulting in localization of the wave near the surface. This wave is called an evanescent wave. For X-rays, generally the electric susceptibility χ is a complex number, and the imaginary part is due to resonant scattering, which is determined by measuring the absorption rate. The optical properties described here will be explained in detail in Chap. 4.

1.3 Bragg's Law and Kinematical Theory of Diffraction

X-ray diffraction plays an extremely important role in crystal structure analysis. A crystal is a solid material in which atoms (and molecules) are arranged in a highly ordered microscopic structure. On the other hand, a solid that lacks the long-range order that is characteristic of a crystal is amorphous (amorphous or glass). In an atom, electrons are distributed approximately spherically around the nucleus and atoms in the crystal are arranged periodically in three dimensions. If we consider these atoms as points and connect them, we form a three-dimensional lattice. The minimum unit lattice is called a unit cell. Two dimensional planes linking these points are called lattice planes or crystallographic planes.

By assuming that X-rays are scattered only once by a lattice plane, we will derive the diffraction conditions. The situation is shown in Fig. 1.2. By assuming that the incident X-ray is a plane wave with its wavelength λ and the distance between adjacent lattice planes is d in the crystal, we have a relation

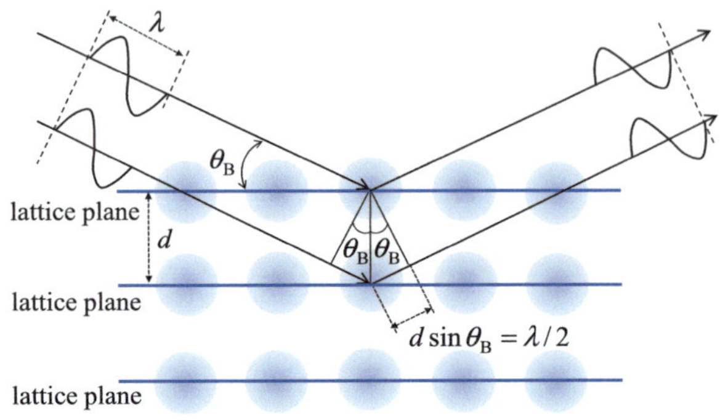

Fig. 1.2 Bragg diffraction condition for $n = 1$. The dashed lines indicate a wave front of a plane wave

$$n\lambda = 2d \sin \theta_B \; (n : \text{integer}). \tag{1.2}$$

When Eq. (1.2) holds, the optical path difference between the beam scattered by two planes adjacent with each other is $2d\sin\theta_B$ and is equal to integer times of the wavelength as shown in the figure, which results in large diffraction intensity. Equation (1.2) is known as Bragg's law of diffraction [3–5], and the incident glancing angle satisfying this equation is called the Bragg angle (θ_B). The X-rays reflected by the second lattice plane from the top should also be reflected by the first plane, but this is ignored in the single scattering approximation. By changing the direction of lattice planes of a crystal and measuring a suitable number of different Bragg angles and the diffraction intensities I_g (the subscript g denotes a reflection index), the crystal structure can be analyzed by using the Fourier transforms of these I_g's. The crystal size to which this theory can be applied is approximately less than several micrometers. This is based on single scattering theory called kinematic diffraction theory. It is assumed that the reflectance of a single lattice plane is very small and that transmitted waves do not attenuate, which breaks the law of conservation of energy. For this reason, this theory cannot be applied to high-precision X-ray optics.

1.4 Dynamical Theory of Diffraction

In the kinematical theory of X-ray diffraction, multiple scattering of X-ray by atoms in a crystal is neglected. In a crystal with its size of approximately larger 1 mm without disorder in the lattice period (called a perfect crystal) as shown in Fig. 1.3, X-rays are scattered once, twice, three times and so on by lattice planes, and multiple scattering occurs. The theory that incorporates this multiple scattering process is called the dynamical theory of diffraction. Darwin and Ewald worked on the treatment of multiple scattering, immediately after the first observation of X-ray diffraction (1912) by Max von Laue and his coworkers [6].

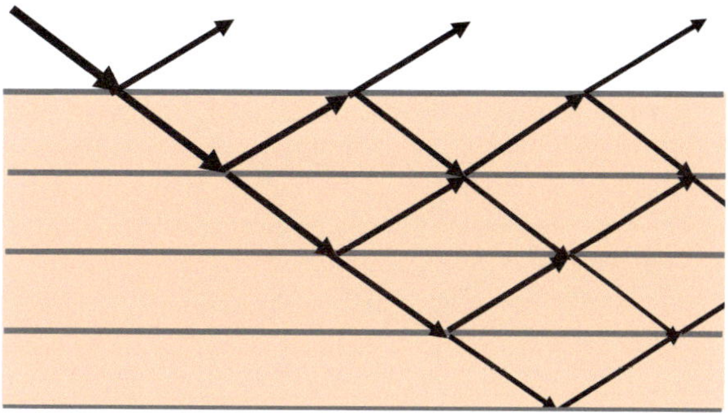

Fig. 1.3 Schematic illustration of multiple scattering by lattice planes in a perfect crystal

1.4.1 Darwin's Approach

In the Darwin's approach [7], a crystal is constructed by a stack of lattice planes as depicted in Fig. 1.4a. A plane wave of X-ray is partially transmitted through $(n-1)$-th plane and comes in the n-th plane with the amplitude T_n and partially reflected wave by the n-th plane with the amplitude R_n comes out by the n-th plane. Then we have simultaneous equations among T_n, T_{n-1}, R_n and R_{n+1}. From the equations, the reflectance from the top layer $(R_0/T_0)^2 = I_g/I_0$ (where I_0 is the intensity of the incident wave) is derived for a semi-infinite crystal. The calculated results are shown in Fig. 1.4b. θ_B is the Bragg angle in kinematical diffraction theory. The intensity curve as a function of the scattering angle (rocking curve) shows total reflection (the reflected intensity is one) with an angular width of several seconds at an angle several seconds higher than θ_B. The center of this peak is θ'_B which is slightly higher than θ_B due to the refraction effect of X-rays. The curve in Fig. 1.4b is called the Darwin curve. From this result, the penetration depth (on the order of μm) of X-rays in the total reflection region can be obtained (extinction distance). Darwin showed that actual crystals were intermediate ones between a mosaic crystal and an ideally perfect crystal. Details about Darwin's approach is given in Appendix A.

1.4.2 Ewald-Laue's Approach

Ewald [8, 9] studied the diffraction of light at very short wavelengths from the time of the discovery of X-ray diffraction, and published his work in 1916–1917. In his approach, the interaction between X-rays and matter of crystal were considered by regarding the crystal as periodic distributions of oscillating dipoles excited by the incident wave. He led interference fringes of wave fields in the crystal, so-called "Pendellösung fringes". The notion of wavefield was introduced to describe the electric field $\boldsymbol{E}_g$ corresponding to the wavevector $\boldsymbol{k}_g = \boldsymbol{k}_0 + \boldsymbol{g}$ ($\boldsymbol{g}$ is a reciprocal lattice vector) excited by the incident wave with the wavevector $\boldsymbol{k}_0$. The form of the

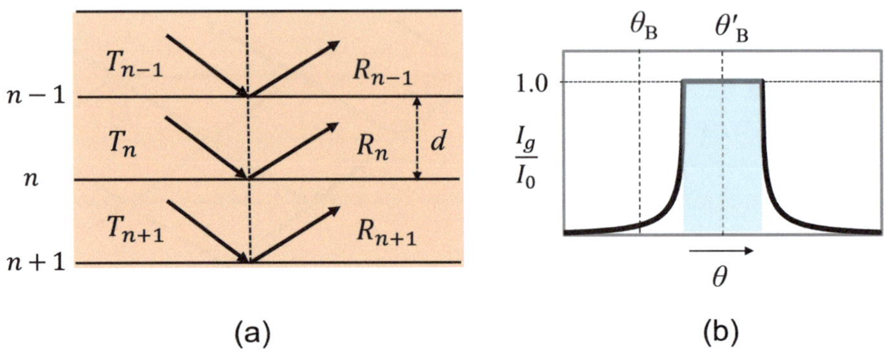

Fig. 1.4 (**a**) A schematic diagram of Darwin's model of dynamical theory. d is the spacing between two adjacent lattice planes. (**b**) Darwin curve: The diffraction intensities as a function of the scattering angle (θ). I_0 and I_g are the incident and the diffracted intensities

wavefield is known as a Bloch wave, although Ewald introduced it 15 years earlier (1913). James [10] provided a good explanation of Ewald's approach.

Ewald investigated the diffraction when there is one diffracted wave in addition to the incident wave, which is called the two-wave approximation. Figure 1.5 shows the dispersion surface in the two-wave approximation. Two dispersion surfaces: a circle (T_0) centered at point O with the radius LO corresponding to the magnitude of the incident wavevector ($n_0|\boldsymbol{K}_0|$), and a circle (T_g) centered at G (OG is the reciprocal lattice vector $\boldsymbol{g}$) with the same radius. Near the intersection L of these two circles, due to the interaction between electromagnetic waves (X-rays) and atoms in the crystal, two dispersion surfaces slightly deviated from the circular arc (thick solid line, the lower side is called branch 1 and the upper side branch 2) are formed. Once the tie-points A and A′ on this surface are given, the wave vector $\boldsymbol{k}_0$ of the incident wave and that of the diffracted wave $\boldsymbol{k}_g$ are determined. This deformed dispersion surface is given by Ewald's basic equation, and $\boldsymbol{E}_0$ and $\boldsymbol{E}_g$ can also be given. From one tie-point (A), there are two wavevectors in the crystal, one in the transmission direction and the other in the diffraction direction. As there are two tie-points (A and A′), there are four wavevectors in total. Ewald classified the boundary condition between the vacuum and the crystal into two. One is the Laue case, in which the transmitted wave and the diffracted wave come out of the bottom surface. The other is the Bragg case, in which the diffracted wave comes out of the entrance surface. In the Laue case, as shown in Fig. 1.5, waves of different branches are excited at tie-points A and A′, while in the Bragg case, waves in the same branch are excited at tie-points B and B′.

Laue [11] adopted the concept of a periodic and continuous dielectric susceptibility within a crystal and proposed a dynamical diffraction theory by assuming that the interaction between X-rays and the crystal could be described by solving Maxwell's equations in the medium of crystal. This theory is most widely used today and often referred to as Laue's dynamical theory. Since Laue's theory is based on

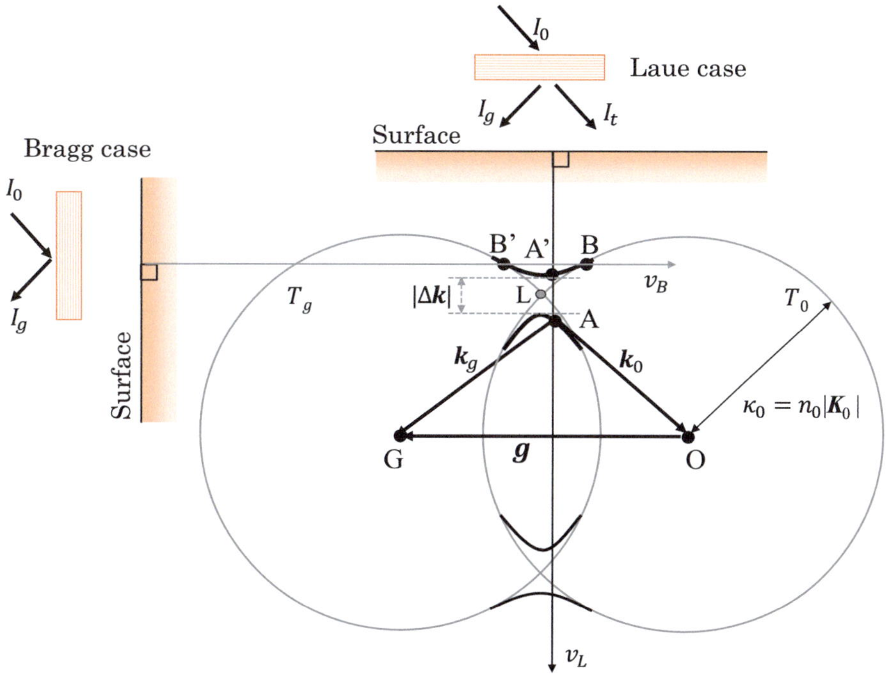

Fig. 1.5 Dispersion surface in the two-wave approximation. The radii of circles O and G are $n_0 K_0$. OG = $|\boldsymbol{g}|$ and $\angle$ OLG = $2\theta'_B$. ν_L and ν_B denote the lines normal to the surface in the Laue and Bragg cases

Ewald's theory, it will be referred to as the Ewald-Laue's dynamical theory of diffraction or dynamical diffraction theory in this book. In the analysis of dynamical theory of diffraction, there are two diffraction geometries as mentioned above: the Laue case and the Bragg case.

1.4.3 Laue Case

When the transmitted wave (with the wavevector $\boldsymbol{K}_t$) and the diffracted wave (with the wavevector $\boldsymbol{K}_g$) come out of the bottom surface, we call it as the Laue case or Laue geometry. An example is shown in Fig. 1.6. As mentioned above, transmitted waves and diffracted waves are emitted from the bottom of the crystal, as shown in Fig. 1.6a. Figure 1.6a shows the electric field in the crystal when the electric field $\boldsymbol{E}_0$ of the incident wave and the electric field $\boldsymbol{E}_g$ of one diffracted wave are excited in the case of the two-wave approximation. There are two waves $\boldsymbol{E}_0^{(1)}$ and $\boldsymbol{E}_0^{(2)}$ with different wave vectors $\boldsymbol{k}_0^{(1)}$ and $\boldsymbol{k}_0^{(2)}$ in the direction of the incident wave in the crystal. There are two waves $\boldsymbol{E}_g^{(1)}$ and $\boldsymbol{E}_g^{(2)}$ with different wave vectors $\boldsymbol{k}_g^{(1)}$ and $\boldsymbol{k}_g^{(2)}$ in the

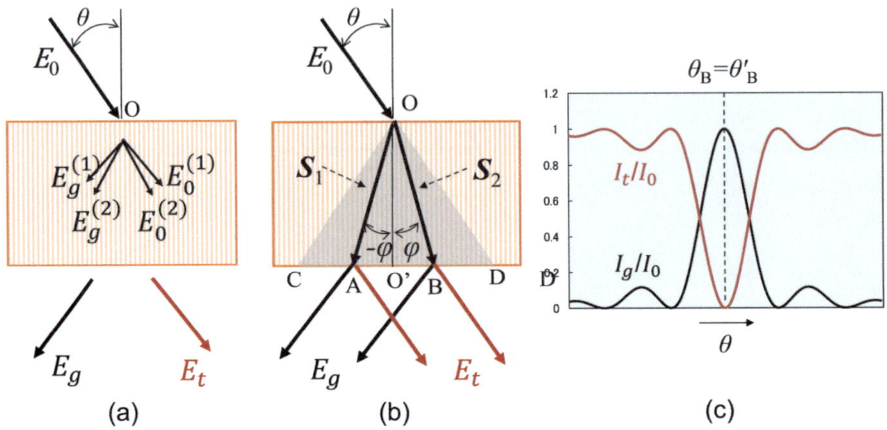

Fig. 1.6 (**a**) Wavevectors inside and outside the crystal. Numbers in superscript parentheses denote branch numbers. (**b**) Trajectory of pseudo-refracted waves in the crystal. (**c**) diffraction (I_g) and transmission (I_t) intensities as a function of the incident angle (θ)

direction of the diffracted wave. These waves become transmitted waves ($\boldsymbol{E}_t$) and diffracted waves ($\boldsymbol{E}_g$) coming out of the bottom surface. The resultant diffraction intensity (I_g) and transmitted intensity (I_t) are shown in Fig. 1.6c.

1.4.3.1 Pendellösung Fringes

Figure 1.7a shows the diffraction intensity (I_g) and transmission intensity (I_t) in the Laue case as a function of the crystal thickness. Both I_g and I_t show periodic variations, and when I_g is maximum, I_t is minimum, and vice versa. Ewald called this variation Pendellösung (hereinafter referred to as Pendellösung fringes) because the appearance of this variation corresponds to the loosely coupled pendulum motion shown in Fig. 1.7b.

In a coupled pendulum, when one pendulum A reaches its maximum swing, the other pendulum B comes to rest; conversely, when pendulum B reaches its maximum swing, pendulum A comes to rest. The kinetic energy of A and that of B will be transferred to each other, and the energy is conserved during the motion. The period of the Pendellösung fringe is $T_p = 2\pi \big/ |\Delta \boldsymbol{k}|$, where $|\Delta \boldsymbol{k}|$ is the amplitude of difference in wavevector between the upper and lower branches ($|\Delta \boldsymbol{k}| = \left|\boldsymbol{k}_g^{(2)} - \boldsymbol{k}_g^{(1)}\right|$). The value of T_p is tens of micrometers. The intensity modulations seen in the rocking curves of the diffracted and transmitted waves around θ'_B shown in Fig. 1.6c are also Pendellösung fringes.

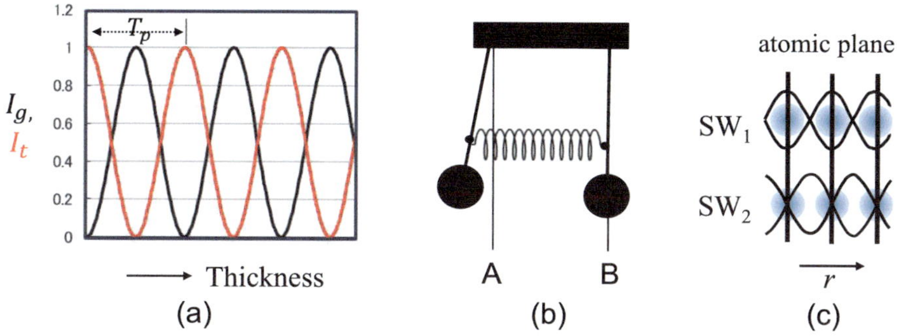

Fig. 1.7 Schematics of (**a**) Pendellösung fringe at $\theta = \theta'_B$, (**b**) a coupled pendulum, (**c**) time variation of the amplitude of a coupled pendulum, and (**d**) standing waves in a crystal

1.4.3.2 Pseudo-Refracted Waves and Angular Amplification Effect

If we express the energy flow from one tie-point as Poynting vector $\boldsymbol{S}_j$ (j denotes the branch number), it is given by

$$\boldsymbol{S}_j = \frac{\varepsilon_0 c}{2}\left(\boldsymbol{s}_0\left|\boldsymbol{E}_0^{(j)}\right|^2 + \boldsymbol{s}_g\left|\boldsymbol{E}_g^{(j)}\right|^2\right). \tag{1.3}$$

This Poynting vector $\boldsymbol{S}_j$ expresses the temporally and spatially averaged one. Figure 1.6b shows the trajectory of the Poynting vector from the entrance surface to the transmission surface. In this book, we call $\boldsymbol{S}_j$ as a pseudo-refracted wave (or a wavefield). In Fig. 1.6b, $\overline{\text{OA}}$ is the trajectory of $\boldsymbol{S}_1$, and $\overline{\text{OB}}$ is the trajectory of $\boldsymbol{S}_2$. When the pseudo-refracted waves reach the transmission surface, it is separated into a diffracted wave ($\boldsymbol{E}_g$) and a transmitted wave ($\boldsymbol{E}_t$). The direction of this pseudo-refracted wave is perpendicular to the dispersion surface. In the symmetric Laue case, both $\boldsymbol{S}_1$ and $\boldsymbol{S}_2$ are parallel to the lattice planes when $\theta = \theta'_B$. When θ deviates from θ'_B by several seconds, $\boldsymbol{S}_1$ and $\boldsymbol{S}_2$ change direction by several degrees with respect to the lattice planes due to the angular amplification effect described below. In Fig. 1.6b, the angle between the lattice plane and $\boldsymbol{S}_1$ ($\boldsymbol{S}_2$) is denoted as $-\varphi$ (φ). When $\theta = \theta'_B$, $\varphi = 0$ as described above. When the angular difference $\Delta\theta = \theta - \theta'_B$ is about several seconds, $|\varphi|$ is approximately equal to θ_B. The angle ratio between $\Delta\theta$ and $\Delta\varphi$ expressed as

$$A_\theta = \frac{\Delta\varphi}{\Delta\theta} \tag{1.4}$$

gives the angular amplification factor A_θ between the incident wave and the pseudo-refracted wave. The value of A_θ is about 10^5, and a very large angular amplification appears. The crystal acts as an extremely high magnification lens. The angle φ

changes in the range from $-\theta_B$ to $+\theta_B$. The triangle OCD shown in gray in Fig. 1.6b is called a Borrmann fan (also called a Borrmann triangle). The angle COD is $2\theta'_B$.

1.4.3.3 Borrmann Effect

The superposition of the refracted and the diffracted waves on the same branch is given by $\boldsymbol{E}^{(j)}(\boldsymbol{r}) = \boldsymbol{E}_0^{(j)}(\boldsymbol{r}) + \boldsymbol{E}_g^{(j)}(\boldsymbol{r})$. The intensity of this superposed wave $|\boldsymbol{E}^{(j)}(\boldsymbol{r})|^2$ shows a standing wave with the period of the lattice spacing corresponding to $\boldsymbol{g}$. Under a certain condition, the standing wave denoted as SW_1 corresponding to the branch 1 is always maximal at the atomic positions on the reflecting planes as shown in Fig. 1.7c. On the contrary, the standing wave SW_2 corresponding to the branch 2 is maximal between these planes. The standing wave SW_1 undergoes higher absorption than normal, which results in anomalous absorption. On the other hand, SW_2 undergoes lower absorption and propagates over long distances, which is called as anomalous transmission. The anomalous transmission was discovered experimentally by Borrmann [12], who found that anomalously strong X-ray intensities were obtained in the transmission and diffraction waves when such a diffraction condition as mentioned above was satisfied in a thick absorbing perfect crystal. Thus, this is called the Borrmann effect. Zachariasen [13] and Laue [14] theoretically showed that this is an effect of dynamical diffraction.

1.4.4 Bragg Case

In Bragg case, the diffracted wave comes out of the entrance surface as depicted in Fig. 1.8a, b. $\boldsymbol{E}_0$ is the electric field of the incident wave. $\boldsymbol{E}_g^n$ and $\boldsymbol{E}_t^n$ are the electric fields of the diffracted wave and transmitted wave after $n-1$ times reflections from the bottom surface, respectively. Figure 1.8a shows the electric field when the total reflection occurs; at this time, the pseudo-refracted wave ($\boldsymbol{S}$) runs along the entrance surface and cannot enter deep into the crystal. Figure 1.8b shows the electric fields when the total reflection does not occur. The pseudo-refracted wave $\boldsymbol{S}_1$ is excited at the incident point. When it reaches the bottom surface, a part of it becomes a transmitted wave ($\boldsymbol{E}_t^1$), and the rest of it is reflected becoming the pseudo-refracted wave $\boldsymbol{S}_2$. Gray contrast indicates the Bormann fan region.

Figure 1.8c shows rocking curves of the diffracted wave (I_g/I_0) and transmitted wave (I_t/I_0) from a thin crystal. Pendellösung fringes appear outside the total reflection region. If the length of the crystal is finite, a part of the pseudo-refracted wave that reaches the lateral surface (side surface) of the crystal is emitted to the diffracted wave direction and the rest to the transmitted wave direction. The diffraction geometry at the lateral surface is the Laue case because the diffracted wave and the transmitted wave come out of the same surface.

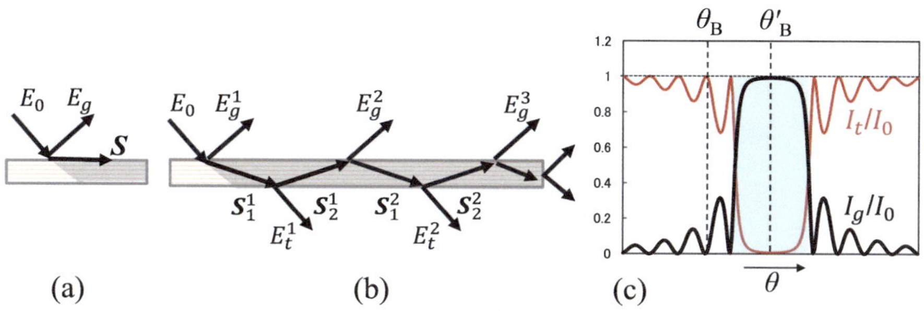

Fig. 1.8 Beam trajectories in the Bragg case in case of the total reflection (**a**), and in the other case (**b**). Diffracted intensities $I_g = |E_g|^2$ and the transmitted intensities $I_t = |E_t|^2$ from a thin crystal in the Bragg case, where $E_g = \sum_{n=1}^{n=\infty} E_g^n$ and $E_t = \sum_{n=1}^{n=\infty} E_t^n$ (**c**). See Chap. 6 in detail

In 1947, highly pure and highly perfect artificial crystals were grown and transistors were put into practical use by using them. Since then, both transistors and crystal growth techniques have advanced. Pendellösung fringes were first observed in X-ray topography from a Si single crystal [15]. This led to the use of dynamical theory of diffraction to evaluate crystal integrity. Nearly half a century after Ewald's work, research in X-ray dynamical diffraction has developed rapidly.

1.5 Plane Wave and Wave-Packet

A plane wave has a flat wavefront whose magnitude of the electric field and magnetic field are constant through any plane that is perpendicular to a fixed direction. A spherical wave is the wave generated by a point source and has a spherical wavefront. If the wave is observed in a small region sufficiently far away from a point source, the spherical wave approximately can be treated as a plane wave. In this book, unless otherwise specified, it is assumed that the incident X-ray is a monochromatic ideal plane wave. Since X-rays usually used in experiments are generated from a point source, it is not straight forward to compare calculated results for plane waves with experimental results.

X-ray photons generated in pulses from a point source are wave-packets. A wave-packet is spatially localized within Δx and has a spectral width ΔE according to the uncertainty principle. For studying the coherence of X-ray waves in experiments, it is necessary to examine the coherence of the X-ray wave-packets. The details of diffraction of wave-packet will be described in Chap. 8.

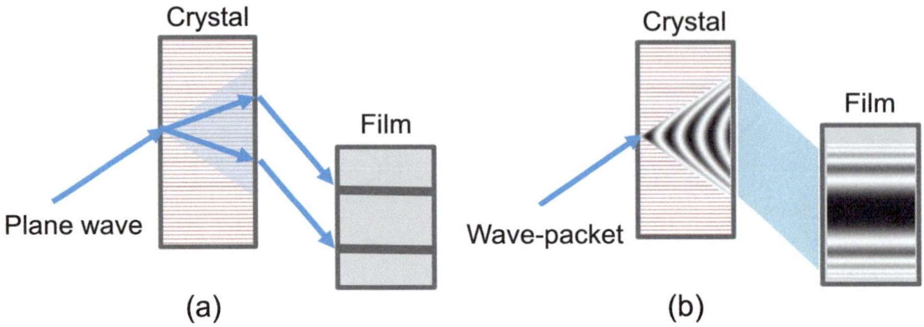

Fig. 1.9 Schematic drawings of X-ray beam geometry (left) and diffraction patterns recorded on the film (right) when the incident beam is a plane wave (**a**) and a wave-packet (**b**) in the symmetric Laue case. (**a**) is drawn using Fig. 8.4 and (**b**) is drawn using Fig. 8.8

1.5.1 Coherence

When the spatial spread Δx of a wave-packet becomes smaller, the spectral width ΔE becomes wider according to the uncertainty relation. If Δx increases, ΔE decreases, and the monochromaticity of X-rays improves. To overlap two wave-packets with each other, they must arrive at the observation point simultaneously and have certain spatial spreads to overlap. The former is called temporal coherent, and the latter transvers coherence. If these coherence conditions are satisfied, the interference fringes between these two wave-packets are observed.

Figure 1.9 shows schematic diffraction patterns recorded on the film when a plane wave (a) or a wave-packet (b) is used as the incident X-ray. In Fig. 1.9a the incident plane wave excites two pseudo-refracted waves, whose directions change greatly due to the angular amplification effect, and two diffraction lines shown in gray are recorded on the film. In the case of a wave-packet in Fig. 1.9b, interference fringes corresponding to the Pendellösung fringes in the Borrmann fan are recorded on the film.

1.6 Resonant Dynamical X-ray Diffraction

In the simplest case, suppose that only one electron is bound in an atomic system and regarded as an oscillator in one-dimension. When an external electric field of $E_i = E_0\exp(i\omega t)$ is applied, the equation of motion for the electron is written as

$$m_e\left(\frac{d^2x}{dt^2} + \gamma_d\frac{dx}{dt} + \omega_s^2 x\right) = -eE_i. \tag{1.5}$$

Here, ω is the angular frequency, m_e and e are the mass and charge of the electron. γ_d is the radiation damping factor for emitting electromagnetic waves due to the oscillation of the electron, and ω_s is the natural angular frequency of the system. By solving Eq. (1.5) for x, we have a relation

$$f(\omega) = \frac{x}{x_{\text{free}}} = \frac{\omega^2}{\omega^2 - \omega_s^2 - i\gamma_d\omega} = f_r(\omega) + if_i(\omega). \tag{1.6}$$

Here, x_{free} is the displacement of x of a free electron with $\gamma_d = \omega_s = 0$. An increase in $f(\omega)$ means that the electromagnetic wave emitted from the oscillating electrons increases.

As an example, Fig. 1.10a shows the calculated $f_r(\omega)$ and $f_i(\omega)$ by assuming $\hbar\omega_s = 10\text{keV}$ and $\gamma_d = 1\text{eV}$. The value of $f_i(\omega)$ becomes large showing a sharp peak at ω_s with the value of 10^4, which is ten thousand times larger than the value of 1 by Thomson scattering. The value of $f_r(\omega)$ is negative for $\omega < \omega_s$ and positive for $\omega > \omega_s$. The maximum value of $|f_r(\omega)|$ is 5×10^3 just below ω_s. In the resonance state at $\omega = \omega_s, f_r(\omega) = 0, f_i(\omega)$ is nonzero and the magnitude of displacement $|x|$ becomes maximum as shown in Fig. 1.10a.

When X-rays are scattered from a real atom, X-rays undergo resonant scattering accompanying core electron excitations, and the atomic scattering factor can be written as:

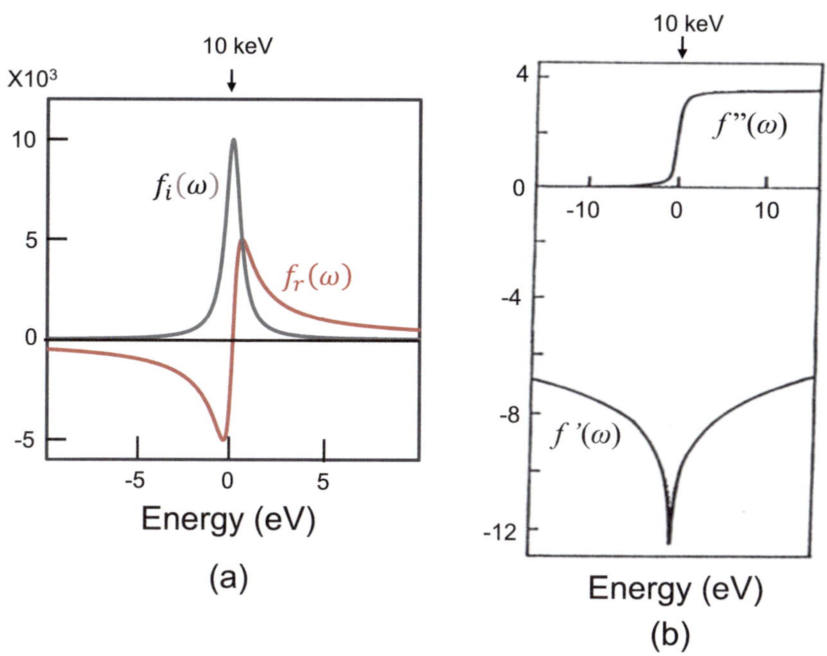

Fig. 1.10 (**a**) The real and imaginary parts of scattering factor $f_r(\omega)$ and $f_i(\omega)$ for a single mode oscillator. (**b**) Those for a real atom $f'(\omega)$ and $f''(\omega)$ near the absorption edge

$$f = f^0(s) + f'(\omega) + if''(\omega) \tag{1.7}$$

Here, $f^0(s)(> 0)$ is the normal scattering factor due to Thomson scattering, which can be obtained by using the electron density around an atom. s is the magnitude of the scattering vector $s = 4\pi \sin\theta/\lambda$, with θ being a half the scattering angle. $f'(\omega) + if''(\omega)$ is called the anomalous scattering factor due to resonant scattering.

In Fig. 1.10b are shown $f'(\omega)$ and $f''(\omega)$, which are calculated by assuming that the absorption edge energy of K electron (ω_K) is 10 keV and $\gamma_d = 1$eV. The value of $f''(\omega)$ suddenly becomes large at the absorption edge, showing a continuous spectrum that gradually decreases in the higher energy region. The maximum value is about 4. On the other hand, the value of $f'(\omega)$ is negative in the range of energy ($\hbar\omega$) shown in the figure, and its absolute value is about 10. This value is 10^{-3} times smaller than the amplitude of resonant scattering of a single oscillator at ω_s in Fig. 1.10a. The reason is that there are oscillators with different characteristic energies ($\hbar\omega_s$) continuously distributed on the higher energy side than the absorption edge. Note that the values of $|f'(\omega)|$ are comparable to the normal scattering factor $f^0(s)$ due to Thomson scattering.

The real part $f^0(s) + f'(\omega)$ of the atomic scattering factor can be positive or negative. When it is zero, the scattering factor f is pure imaginary. In dynamical theories of diffraction, it had been assumed that the real part of the atomic scattering factor is larger than the imaginary part for a long time. Since the advent of X-rays from synchrotron radiation, it is possible to set the imaginary part comparable to the real part, or even to set $f^0(s)+f'(\omega) = 0$ to obtain the diffraction only due to the imaginary part, by changing X-ray energy. To analyze the diffraction phenomenon in such a condition, especially near the absorption edge of X-rays, Kato [16] and Fukamachi & Kawamura [17] developed theories of resonant dynamical X-ray diffraction. The details of resonant dynamical X-ray diffraction are described in Chap. 7.

1.7 Dynamical Theory of Diffraction for an Elastically Bent Crystal and X-ray Mirage Diffraction

Dynamical diffraction theory is originally developed for a perfect crystal. However, Ewald-Laue's dynamical theory of diffraction can be extended to slightly strained crystals with a constant strain gradient by incorporating the concept of local reciprocal lattice. The details of dynamical diffraction for slightly strained crystals are explained in Chap. 9. Below we explain a simple concept of the dynamic diffraction from a bent crystal.

When driving on a day with strong sunlight, you may come across a mirage, which makes distant objects appear upside down. A mirage is a physical phenomenon that occurs because light is bent toward the direction of the air with a higher refractive index. In X-ray diffraction, a similar phenomenon occurs in a crystal with a constant strain gradient [18]. Figure 1.11 shows how mirage diffraction occurs in the Bragg

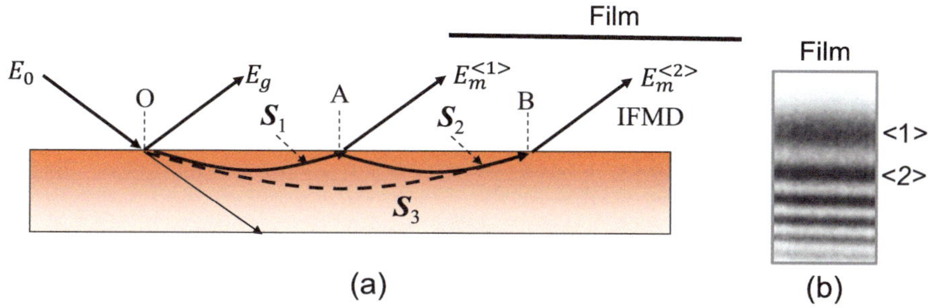

Fig. 1.11 Mirage diffraction and interference fringes of mirage diffraction waves. (**a**) Beam geometry and beam trajectories in a strained crystal, which are described using Fig. 9.21 in Chap. 9. (**b**) Interference fringes of mirage diffraction waves recorded on the film situated at the right edge of the crystal, which are taken from Fig. 9.26a (adapted from Fig. 6a in [15])

case. We assume that the lattice spacing gradually becomes larger in the depth direction. The trajectory of the pseudo-refracted wave ($\boldsymbol{S}_1$) created at the incident point is gradually bent in the upper direction due to the angular amplification effect, because the lattice spacing becomes narrower and the refractive index becomes higher when the wave propagates deeper in the crystal. This results in a north opening hyperbolic trajectory. The pseudo-refracted wave is totally reflected within the crystal and returns to the incident surface, a part of which becomes a diffracted wave ($\boldsymbol{E}_m^{\langle 1\rangle}$), and the rest is reflected as a pseudo-refracted wave ($\boldsymbol{S}_2$). In this way, the pseudo-refracted wave can propagate over long distances with almost no absorption by selecting the angle at which the anomalous transmission (Borrmann effect) occurs. If the incident wave has a wave-packet with a divergence angle of about 1 s, pseudo-refracted wave ($\boldsymbol{S}_3$) traveling in different direction from $\boldsymbol{S}_1$ can be simultaneously created at the incident point. The interference fringes between $\boldsymbol{S}_2$ and $\boldsymbol{S}_3$ appear in the diffracted wave emitted from point B. These should be called mirage interference fringes. By observing the mirage interference fringes, the strain in the crystal can be analyzed as described in Chap. 9.

References

1. V.G. Veselago, Soviet Physics. USPEKI **10**, 509–514 (1968)
2. J.B. Pendry, A.J. Holden, W.J. Stewart, I. Youngs, Phys. Rev. Lett. **76**, 4773–4776 (1996)
3. W.H. Bragg, Nature (London) **90**, 410 (1912)
4. W.H. Bragg, Proc. Cambridge Phil. Soc. **17**, 43–57 (1913)
5. W.H. Bragg, Proc. Roy. Soc. (London) A **89**, 248–277 (1913)
6. M von Laue, Sitzungsberichte der Kgl. Bayer, Akad. der Wiss. 363–373. Ann. Phys (1913) **41**, 971–989 (1912)
7. C.P. Darwin, Phil. Mag. **27**(315–333), 672–690 (1914)
8. P.P. Ewald, Physik. Z. **14**, 465–472 (1913)
9. P.P. Ewald, Ann. Phys. **54**, 519–597 (1917)
10. R.W. James, *The Optical Principles of The Diffraction of X-Rays* (G. Bell and Sons Ltd., London, 1967)

11. M. von Laue, Ergeb. Exakt. Naturwiss. **10**, 133–158 (1931)
12. G. Borrmann, Phys. Z. **42**, 157–162 (1941)
13. W.H. Zachariasen, *Theory of X-Ray Diffraction in Crystal* (John Willy Sonds Inc., New York, 1945)
14. M. von Laue, Acta Cryst. **2**, 106–113 (1949)
15. N. Kato, A.R. Lang, Acta Cryst. **12**, 787–794 (1959)
16. N. Kato, Acta Cryst. **A48**, 829–833 (1992)
17. T. Fukamachi, T. Kawamura, Acta Cryst. **A49**, 384–388 (1993)
18. A. Authier, *Dynamical Theory of X-ray Diffraction* (Oxford University Press, 2001)

Chapter 2
X-ray Diffraction from Microcrystal

X-ray is an electromagnetic wave obeying Maxwell's equation. It has two components; electric field and magnetic field. It propagates in vacuum with the velocity of light. Once one of the components is given, the other is given by Maxwell's equation. For example, the amplitude of electric field $E(=|\boldsymbol{E}|)$ and that of magnetic field H $(=|\boldsymbol{H}|)$ are related by

$$H = \left(\frac{\varepsilon_0}{\mu_0}\right)^{1/2} E \tag{2.1}$$

in vacuum. Here ε_0 and μ_0 are the dielectric constant and permeability in vacuum. The light velocity c is related to these two quantities as

$$c = \frac{1}{(\varepsilon_0 \mu_0)^{1/2}}. \tag{2.2}$$

In this chapter, we describe elastic X-ray scattering called Thomson scattering due to weakly bound electrons which are regarded to be nearly free electrons, and derive the so-called normal atomic scattering factor due to the bound electrons distributed around an atomic nucleus. We consider the single scattering of X-ray due to three-dimensional arrays of atoms in a crystal by a kinematic theory of X-ray diffraction, leading to the crystal structure factor. Then we describe the basic principle of structure determination (structure analysis) of a crystal from the structure factor.

T. Fukamachi and T. Kawamura, *Resonant Dynamical X-ray Diffraction*, Springer Series in Solid-State Sciences 208, https://doi.org/10.1007/978-981-96-8406-9_2

2.1 Thomson Scattering

The oscillating electric field of the incident X-ray accelerates a weakly bound electron which emits radiation at the same frequency as the incident X-ray, and thus the X-ray is scattered. A schematic diagram of the scattering is shown in Fig. 2.1. An electron with its mass m_e and charge $-e$ is located at the origin (O). The incident X-ray propagates along z-axis. The unit vector of the propagating direction is $\boldsymbol{s}_0$ and the electric field $\boldsymbol{E}_i$ is along x-axis with its unit vector $\boldsymbol{e}_0$. The magnitude of the incident electric field E_i is expressed as

$$E_i = E_0 \exp(i\omega t), \tag{2.3}$$

where E_0 is the amplitude of the electric field and ω is its angular frequency. The equation of motion of the accelerated electron is given by

$$m_e\left(\frac{d^2x}{dt^2} + \gamma_d \frac{dx}{dt} + \omega_s^2 x\right) = -eE_i. \tag{2.4}$$

The second term of the left side expresses the attenuation of the motion due to loss of X-ray energy. The coefficient γ_d is a radiation damping factor given as

$$\gamma_d = \frac{e^2\omega^2}{6\pi\varepsilon_0 m_e c^3} = \frac{2r_e\omega^2}{3c^2}. \tag{2.5}$$

Here r_e is the classical electron radius

$$r_e = \frac{e^2}{4\pi\varepsilon_0 m_e c^2} \tag{2.6}$$

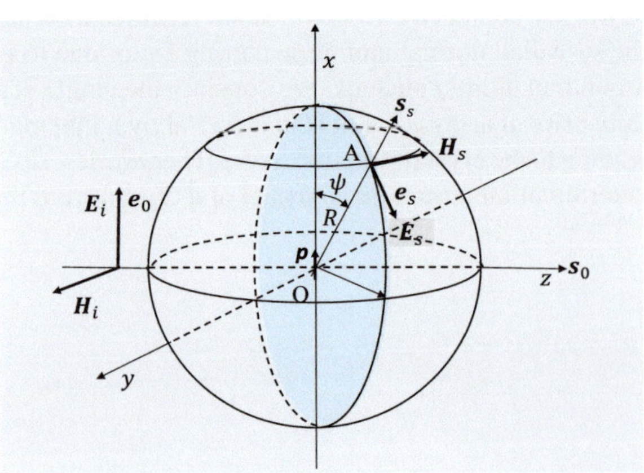

Fig. 2.1 Schematic drawing of scattered electric field $\boldsymbol{E}_s$ induced by the incident electric field $\boldsymbol{E}_i$

and $r_e = 2.8179 \times 10^{-6}$ [nm]. The third term of the left side of (2.4) expresses an oscillation around the system with its angular frequency ω_s.

When an electron is very weakly bound ($\omega \gg \omega_s$) and the damping factor is negligible, (2.4) is written as

$$m_e \frac{d^2x}{dt^2} = -eE_i. \tag{2.7}$$

If we write the diploe moment of the electron as

$$p = -ex, \tag{2.8}$$

we have

$$\ddot{p} = -e\ddot{x} = \frac{e^2}{m_e} E_0 \exp(i\omega t). \tag{2.9}$$

In Fig. 2.1, the scattered electric field $\boldsymbol{E}_s$ and the magnetic field $\boldsymbol{H}_s$ at a point A with the distance R from O are given by

$$\boldsymbol{E}_s = -\frac{1}{4\pi\varepsilon_0 c^2 R} \boldsymbol{s}_s \times \left(\boldsymbol{s}_s \times \ddot{\boldsymbol{p}}_\tau\right) \tag{2.10}$$

and

$$\boldsymbol{H}_s = \left(\frac{\varepsilon_0}{\mu_0}\right)^{1/2} \boldsymbol{s}_s \times \boldsymbol{E}_s. \tag{2.11}$$

Here, $\boldsymbol{s}_s$ is the unit vector of $\overrightarrow{\mathrm{OA}}$ direction, $\ddot{\boldsymbol{p}}_\tau$ means $\ddot{\boldsymbol{p}}$ at a time $\tau = t - R/c$ and ψ is the angle between $\boldsymbol{s}_s$ and x-axis. The relation between the electric field and the magnetic field at A is written by

$$|\boldsymbol{E}_s| = \left(\frac{\mu_0}{\varepsilon_0}\right)^{1/2} |\boldsymbol{H}_s| = \frac{\sin\psi}{4\pi\varepsilon_0 c^2 R} \left|\ddot{\boldsymbol{p}}_\tau\right| = r_e \frac{\sin\psi}{R} |\boldsymbol{E}_0|. \tag{2.12}$$

The magnitude of the electric field is

$$|\boldsymbol{E}_s|^2 = {r_e}^2 \frac{P^2}{R^2} |\boldsymbol{E}_0|^2. \tag{2.13}$$

P is the polarization factor expressed by

$$P = |\boldsymbol{e}_0 \cdot \boldsymbol{e}_s| = \sin\psi, \tag{2.14}$$

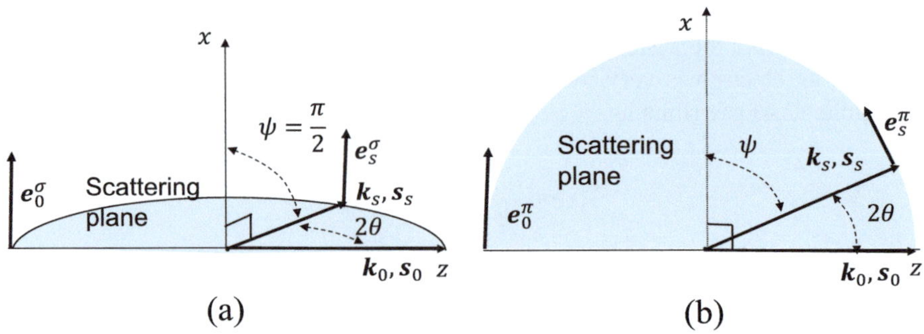

Fig. 2.2 Relation between the scattering plane and the electric fields for σ- (**a**) and π-polarized X-rays (**b**)

where $\boldsymbol{e}_s$ is the unit vector of the scattered electric field. As shown in (2.11), $\boldsymbol{E}_s$ and $\boldsymbol{H}_s$ are normal to each other. At the equator position in Fig. 2.1, $P = 1$ and the value $|\boldsymbol{E}_s|$ is the maximum. On the other hand, at the pole position, $P = 0$ and the value of $|\boldsymbol{E}_s|$ is the minimum of 0. When the electric field is normal to the scattering plane which includes both the incident wave vector $\boldsymbol{k}_0(|\boldsymbol{k}_0| = 2\pi/\lambda)$ and the scattered wave vector $\boldsymbol{k}_s$, the X-ray is σ-polarized and $P = 1$. When the electric field is parallel to the scattering plane, the X-ray is π-polarized and $P = |\cos 2\theta|$, with 2θ being the scattering angle. The relations between the scattering plane and the wave vectors for σ-polarized and π-polarized X-rays are shown in Fig. 2.2a and b, respectively.

2.2 Elastic Scattering by an Atom (Normal Scattering Factor)

When the wavelength of X-ray is the same order as the size of an atom, the electric field of the incident X-ray must be treated differently depending on the position of an electron. If the position of an electron is $\boldsymbol{r}_j$ from the center of an atom, the phase of the incident wave at $\boldsymbol{r}_j$ is different from that at the center by $\boldsymbol{k}_0 \cdot \boldsymbol{r}_j$. The electric field observed at a point A far away from the atom can be written by

$$\boldsymbol{E} = \frac{r_e}{R}\sum_j^N \boldsymbol{s}_j \times (\boldsymbol{s}_j \times \boldsymbol{E}_0)\exp\left[i\omega\left(t - \frac{R_j}{c}\right) - i\boldsymbol{k}_0 \cdot \boldsymbol{r}_j\right] \tag{2.15}$$

as the dipole moment of the electron is given by

$$\ddot{\boldsymbol{p}}_\tau = \frac{-e^2}{m_e}\boldsymbol{E}_0\exp(i\omega t) \tag{2.16}$$

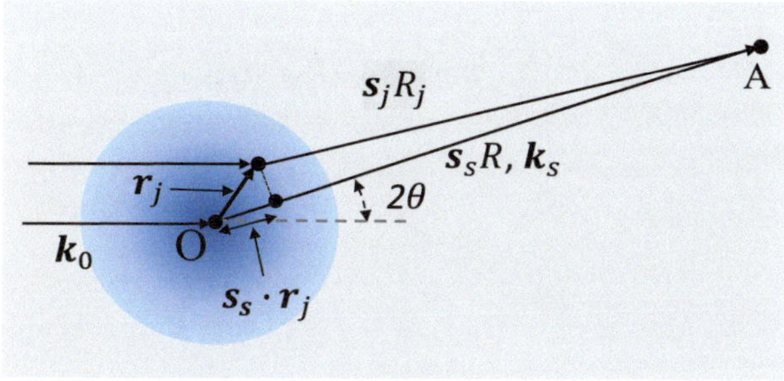

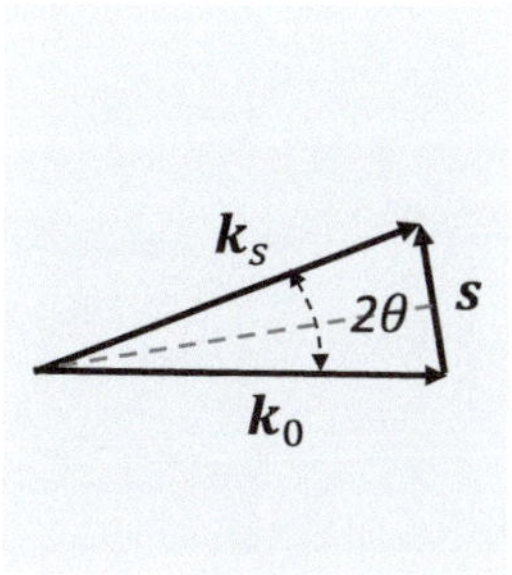

Fig. 2.3 **a** Electric field at point A scattered by the electron at $\boldsymbol{r}_j$ far away from A. **b** Vector diagram of the incident ($\boldsymbol{k}_0$), scattered ($\boldsymbol{k}_s$) and scattering ($\boldsymbol{s}$) wave vectors

according to (2.7). In (2.15), the summation is over all the electrons in an atom, R_j is the distance from the j-th electron to the point A and $\boldsymbol{s}_j$ is the unit vector in the direction from the j-th electron to the point A.

In Fig. 2.3, by taking the origin (O) at the center of an atom, a point A at the distance R from O and the unit vector $\boldsymbol{s}_s$ in the direction from O to A, we have

$$\boldsymbol{s}_s R = \boldsymbol{s}_j R_j + \boldsymbol{r}_j. \tag{2.17}$$

If we think the point A as an observation point, usually the distance R is large enough to take

$$R_j = R - \boldsymbol{s}_s \cdot \boldsymbol{r}_j. \tag{2.18}$$

We can use the approximation $\boldsymbol{s}_j \approx \boldsymbol{s}_s$, and rewrite (2.15) as

$$\boldsymbol{E} = \frac{r_e}{R}\boldsymbol{s}_j \times (\boldsymbol{s}_j \times \boldsymbol{E}_0)\exp\left[i\omega\left(t - \frac{R}{c}\right)\right]\sum_j^N \exp[i(\boldsymbol{k}_s - \boldsymbol{k}_0)\cdot \boldsymbol{r}_j]. \tag{2.19}$$

The Poynting vector $\boldsymbol{S}$ represents the directional energy flux of an electromagnetic wave. As the time average of it $\langle \boldsymbol{S} \rangle$ is given by

$$\langle \boldsymbol{S} \rangle = \frac{1}{2}\mathrm{Re}(\boldsymbol{E} \times \boldsymbol{H}^*) = \frac{\varepsilon_0 c}{2}|\boldsymbol{E}|^2 \boldsymbol{s}_s. \tag{2.20}$$

By using this, the scattering intensity of the wave at A is given by

$$I = |\langle S \rangle| = \frac{\varepsilon_0 c}{2}|\boldsymbol{E}|^2 = I_0 \frac{r_e^2}{R^2} P^2 \left| \sum_j^N \exp\left[i(\boldsymbol{k}_s - \boldsymbol{k}_0) \cdot \boldsymbol{r}_j\right] \right|^2. \tag{2.21}$$

I_0 is the intensity of the incident beam given as

$$I_0 = \frac{\varepsilon_0 c |E_0|^2}{2}. \tag{2.22}$$

If electrons are distributed around the nucleus of an atom with the density $\rho(\boldsymbol{r})$, (2.21) can be written by

$$I = I_0 \frac{r_e^2}{R^2} P^2 |S(\boldsymbol{s})|^2. \tag{2.23}$$

$S(\boldsymbol{s})$ is the Fourier transform of the density $\rho(\boldsymbol{r})$ and given by

$$S(\boldsymbol{s}) = \int \rho(\boldsymbol{r}) \exp(i\boldsymbol{s} \cdot \boldsymbol{r}) d\boldsymbol{r}. \tag{2.24}$$

$S(\boldsymbol{s})$ is called as the structure factor or structure amplitude and $\boldsymbol{s}$ is the scattering vector given by

$$\boldsymbol{s} = \boldsymbol{k}_s - \boldsymbol{k}_0, \; |\boldsymbol{s}| = 2k_0 \sin\theta = \frac{4\pi \sin\theta}{\lambda}, \tag{2.25}$$

as shown in Fig. 2.3b. Here $k_0 = |\boldsymbol{k}_0| = |\boldsymbol{k}_s|$ and 2θ is the scattering angle. As the scatterer of X-ray is electrons of an atom, $S(\boldsymbol{s})$ in (2.24) can be replaced by $f^0(\boldsymbol{s})$ defined by

$$f^0(\boldsymbol{s}) = \int_{\text{Atom}} \rho(\boldsymbol{r}) \exp(i\boldsymbol{s} \cdot \boldsymbol{r}) d\boldsymbol{r}. \tag{2.26}$$

$f^0(\boldsymbol{s})$ is called as the normal atomic scattering factor. In the forward scattering ($\theta = 0, \boldsymbol{s} = 0$), this becomes

$$f^0(0) = \int \rho(\boldsymbol{r}) d\boldsymbol{r} = Z, \tag{2.27}$$

where Z is the atomic number of the atom. According to the quantum mechanics, the electron density is given by using the wavefunction $\varphi_A(\boldsymbol{r})$ as

$$\rho(\boldsymbol{r}) = \varphi_A^*(\boldsymbol{r}) \varphi_A(\boldsymbol{r}). \tag{2.28}$$

Then the atomic scattering factor in (2.26) can be written by

$$f^0(\boldsymbol{s}) = \int \varphi_A^*(\boldsymbol{r})\varphi_A(\boldsymbol{r})\exp(i\boldsymbol{s}\cdot\boldsymbol{r})d\boldsymbol{r}. \tag{2.29}$$

Figure 2.4 shows the angular dependence of the atomic scattering factor $f^0(s)$ for an isolated atom of Si ($Z = 14$). The horizontal axis is the scattering parameter $s = \sin\theta/\lambda$. This figure also shows the scattering factor for each atomic orbital. $f^0(s)$ becomes smaller as the scattering angle increases. If we focus on 1*s* electron in the innermost shell, the scattering factor does not change much when the scattering angle becomes large. This is because the 1s electrons are localized near the atomic nucleus. On the other hand, the scattering factors of 3*s* and 3*p* electrons belonging to the outermost shell become almost zero around $s = 0.2$. This is because these electrons are widely distributed away from the atomic nucleus. When these electrons interact with neighboring atoms, the electron distribution changes from that of an isolated atom. Then, small-angle diffraction intensity measurements are important for obtaining precise information on the electron distribution of the outer electrons.

The values of the atomic scattering factor calculated by using wavefunctions are tabulated in the International Table for Crystallography [1] for almost all the neutral atoms and important ionic atoms. The atomic scattering factors are also represented by the sum of four Gaussian forms as

$$f^0(s) = \sum_{j=1}^{4} a_i\exp(-b_j s^2) + c \tag{2.30}$$

for the convenience of calculation and the coefficients a_i, b_j, c are tabulated in [1] for $0 < s < 2.0$ Å^{-1}. Here s is the scattering parameter. Note that s above is different from s in (2.25) by 4π.

Fig. 2.4 Atomic scattering factor f^0 of an Si atom in green. The atomic scattering factor for each atomic orbital is also shown. The number in parentheses is the number of electrons in the orbital. The graph is obtained by using the calculated values by Fukamachi, T. (1971). *Technical Rep. of ISSP* Series B No.12, 1–62

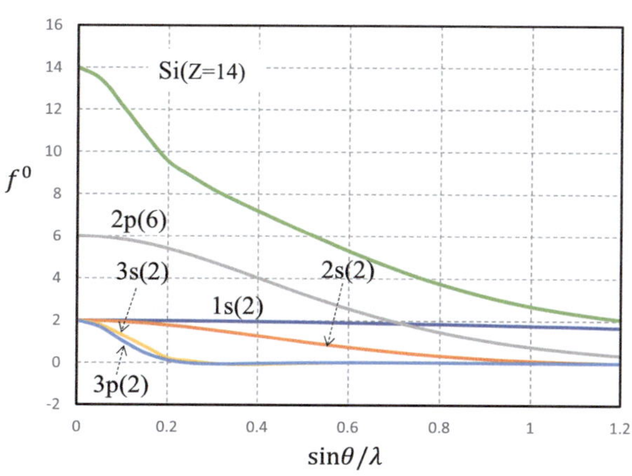

2.3 Diffraction from Microcrystal

We will consider diffraction by microcrystals, which are small enough to eliminate the effects of multiple scatterings. Then we will discuss Laue's diffraction conditions.

2.3.1 *Crystal Structure Factor*

Let's consider the structure factor $S(\boldsymbol{s})$ of a crystal. In a crystal, the electron density $\rho(\boldsymbol{r})$ is a three-dimensional periodic function. Let the fundamental vectors of the unit crystal lattice (unit cell) be $\boldsymbol{a}$, $\boldsymbol{b}$ and $\boldsymbol{c}$. Since a crystal is a three-dimensional stack of crystal lattices, $\rho(\boldsymbol{r}')$ at a certain position $\boldsymbol{r}'$ in the unit crystal lattice has a translational symmetry and is the same as that at $\boldsymbol{r}' + n_1\boldsymbol{a} + n_2\boldsymbol{b} + n_3\boldsymbol{c}$, as shown in Fig. 2.5, that is,

$$\rho(\boldsymbol{r}') = \rho(n_1\boldsymbol{a} + n_2\boldsymbol{b} + n_3\boldsymbol{c} + \boldsymbol{r}'), \tag{2.31}$$

with n_1, n_2 and n_3 being integers. Using this electron density, the crystal structure factor $S(\boldsymbol{s})$ from the whole crystal is given by

$$S(\boldsymbol{s}) = \int_{\text{crystal}} \rho(\boldsymbol{r})\exp(i\boldsymbol{s}\cdot\boldsymbol{r})d\boldsymbol{r}$$

$$= \sum_{n_1}\sum_{n_2}\sum_{n_3} \exp[i\boldsymbol{s}\cdot(n_1\boldsymbol{a} + n_2\boldsymbol{b} + n_3\boldsymbol{c})]\int_{\text{u.c.}} \rho(\boldsymbol{r}')\exp\left(i\boldsymbol{s}\cdot\boldsymbol{r}'\right)d\boldsymbol{r}'. \tag{2.32}$$

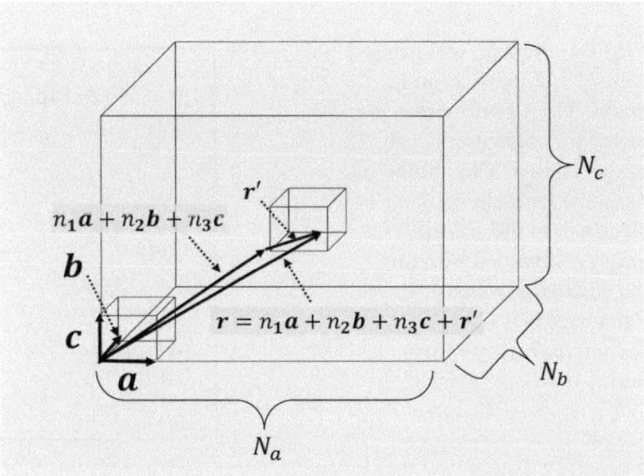

Fig. 2.5 Relationship between parallelepiped crystal and a unit cell lattice

The summations n_1, n_2 and n_3 are over the unit cell in a rectangular crystal. The integration is over a unit cell ('u.c.' stands for a unit cell) and the integral part

$$F_s = \int_{\text{u.c.}} \rho(\boldsymbol{r})\exp(i\boldsymbol{s}\cdot\boldsymbol{r})d\boldsymbol{r} \tag{2.33}$$

is called the crystal structure factor.

The electron density $\rho(\boldsymbol{r})$ of an atom in a crystal is in principle not the same as that of an isolated atom as the electrons are rearranged within the crystal. By considering that only the electrons in the outer shell are to be rearranged, it is thought to be a good approximation that $\rho(\boldsymbol{r})$ in the crystal is not much different from that of an isolated atom. We assume that the electron distribution in a crystal is a superposition of the electron densities of isolated atoms. If the electron density around an atom j at $\boldsymbol{r}_j$ is expressed as $\rho_j(\boldsymbol{r}-\boldsymbol{r}_j)$, the electron density at $\boldsymbol{r}$ is given by

$$\rho(\boldsymbol{r}) = \sum_j \rho_j(\boldsymbol{r}-\boldsymbol{r}_j). \tag{2.34}$$

Then the crystal structure factor F_s becomes

$$F_s = \int_{\text{u.c.}} \sum_j \rho_j(\boldsymbol{r}-\boldsymbol{r}_j)\exp(i\boldsymbol{s}\cdot\boldsymbol{r})d\boldsymbol{r} = \sum_j \exp(i\boldsymbol{s}\cdot\boldsymbol{r}_j)\int_{\text{Atom}} \rho(\boldsymbol{r})\exp(i\boldsymbol{s}\cdot\boldsymbol{r}')d\boldsymbol{r}'$$

$$= \sum_j f_j^0 \exp(i\boldsymbol{s}\cdot\boldsymbol{r}_j) \tag{2.35}$$

where $\boldsymbol{r}' = \boldsymbol{r}-\boldsymbol{r}_j$. F_s is now given by using the normal scattering factors f_j^0 in (2.26).

2.3.2 *Laue Function and Reciprocal Lattice*

Let us express the summation part of (2.32) as $G(\boldsymbol{s})$ and assume that the outer shape of the crystal is a rectangular parallelepiped, as shown in Fig. 2.5. Then,

$$G(\boldsymbol{s}) = \sum_{n_1}\sum_{n_2}\sum_{n_3} \exp[i\boldsymbol{s}\cdot(n_1\boldsymbol{a}+n_2\boldsymbol{b}+n_3\boldsymbol{c})]$$

$$= \sum_{n_1=0}^{N_a-1} \exp(in_1\boldsymbol{s}\cdot\boldsymbol{a}) \sum_{n_2=0}^{N_b-1} \exp(in_2\boldsymbol{s}\cdot\boldsymbol{b}) \sum_{n_3=0}^{N_c-1} \exp(in_3\boldsymbol{s}\cdot\boldsymbol{c}) \tag{2.36}$$

where the sizes of the crystal lattice are given by $N_a\boldsymbol{a}$, $N_b\boldsymbol{b}$ and $N_c\boldsymbol{c}$. Then $|G(\boldsymbol{g})|^2$ is expressed as

$$|G(s)|^2 = \frac{\sin^2\left(\frac{N_a s\cdot a}{2}\right)\sin^2\left(\frac{N_b s\cdot b}{2}\right)\sin^2\left(\frac{N_c s\cdot c}{2}\right)}{\sin^2\left(\frac{s\cdot a}{2}\right)\sin^2\left(\frac{s\cdot b}{2}\right)\sin^2\left(\frac{s\cdot c}{2}\right)}. \tag{2.37}$$

The function $G(s)$ is called the Laue function. Using the Laue function, the crystal structure factor $S(s)$ of the crystal is expressed as

$$S(s) = G(s)F_s. \tag{2.38}$$

The scattered or diffracted intensity from the crystal is given as

$$I = I_0 \frac{r_e^2 P^2}{R^2} |G(s)|^2 |F_s|^2. \tag{2.39}$$

Figure 2.6 shows the one-dimensional Laue function. It has a principal maximum at $\boldsymbol{s}\cdot\boldsymbol{a} = 2\pi n$ (n is either zero or an integer). The height of the main maximum is N_a^2 and the half-width is $2\pi/N_a$. As the crystal size of N_a increases, the main maximum peak becomes very sharp. In the same way, the three-dimensional principal maximum value is

$$|G(s)|^2 = (N_a N_b N_c)^2 = N^2. \tag{2.40}$$

Here, N is the number of unit cells in the crystal. It is noted that the diffracted intensity is proportional to the square of the absolute value of the crystal structure factor. This conclusion is a result of kinematic (single scattering) theory of diffraction.

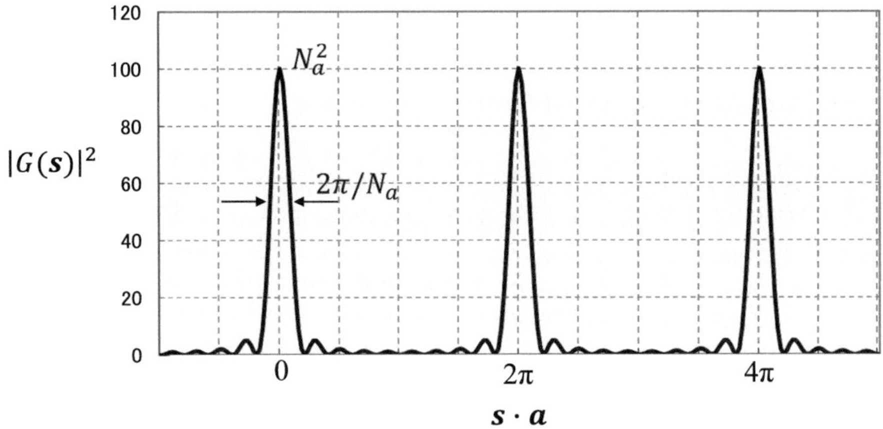

Fig. 2.6 Laue function in one dimension. $|G(s)|^2 = \sin^2(N_a \boldsymbol{s}\cdot\boldsymbol{a}/2)/\sin^2(\boldsymbol{s}\cdot\boldsymbol{a}/2)$ with $N_a = 10$ is shown by the solid line

From the Laue function in Fig. 2.6, the diffracted intensity has a large value when $\boldsymbol{s} \cdot \boldsymbol{a} = 2\pi n$. This condition is called Laue's diffraction condition. Next, consider the relationship between Laue's diffraction conditions and the reciprocal lattice vector.

Corresponding to the primitive basis vectors $\boldsymbol{a}, \boldsymbol{b}, \boldsymbol{c}$ of the direct lattice, the basis vectors $\boldsymbol{a}^*, \boldsymbol{b}^*, \boldsymbol{c}^*$ in the reciprocal lattice are defined by the equations

$$\boldsymbol{a}^* = 2\pi \frac{\boldsymbol{b} \times \boldsymbol{c}}{V_c}, \boldsymbol{b}^* = 2\pi \frac{\boldsymbol{c} \times \boldsymbol{a}}{V_c}, \boldsymbol{c}^* = 2\pi \frac{\boldsymbol{a} \times \boldsymbol{b}}{V_c}, \tag{2.41}$$

where $V_c = (\boldsymbol{a} \cdot (\boldsymbol{b} \times \boldsymbol{c}))$ is the volume of unit cell. A general reciprocal lattice vector $\boldsymbol{g}$ is expressed as

$$\boldsymbol{g} = \boldsymbol{g}_h = h\boldsymbol{a}^* + k\boldsymbol{b}^* + l\boldsymbol{c}^*. \tag{2.42}$$

In this book, a general reciprocal lattice vector is written by $\boldsymbol{g}$, and the reciprocal lattice vector of *hkl*-reflection or $\boldsymbol{h}$-reflection is written by $\boldsymbol{g}_h$. There are relations between $\boldsymbol{a}, \boldsymbol{b}, \boldsymbol{c}$ and $\boldsymbol{a}^*, \boldsymbol{b}^*, \boldsymbol{c}^*$ as follows

$$\boldsymbol{a}^* \cdot \boldsymbol{a} = \boldsymbol{b}^* \cdot \boldsymbol{b} = \boldsymbol{c}^* \cdot \boldsymbol{c} = 2\pi, \tag{2.43}$$

$$\boldsymbol{a}^* \cdot \boldsymbol{b} = \boldsymbol{a}^* \cdot \boldsymbol{c} = \boldsymbol{b}^* \cdot \boldsymbol{a} = \boldsymbol{b}^* \cdot \boldsymbol{c} = \boldsymbol{c}^* \cdot \boldsymbol{a} = \boldsymbol{c}^* \cdot \boldsymbol{b} = 0. \tag{2.44}$$

The three integers h, k, l are called Miller indices and (hkl) denotes a crystal plane orthogonal to the reciprocal vector $\boldsymbol{g}_h$. For denoting a reflection, *hkl* without any bracketing designates a reflection index. If we put $h = 1, k = l = 0, \boldsymbol{h} \equiv \boldsymbol{g}_h = \boldsymbol{a}^*$ and $\boldsymbol{h} \cdot \boldsymbol{a} = 2\pi$, a Laue's diffraction condition is satisfied. We obtain

$$\boldsymbol{s} = \boldsymbol{h}. \tag{2.45}$$

The Miller index for $\boldsymbol{h} = \boldsymbol{a}^*$ is (100) and the spacing of the lattice plane (lattice spacing) d is given by $d = |\boldsymbol{a}|$. Figure 2.7a shows the diffraction condition by the real lattice planes under a Bragg condition, while Fig. 2.7b shows the diffraction conditions in the reciprocal lattice plane. Since $|\boldsymbol{g}| = 2\pi/d$ and the magnitude of the scattering vector is $|\boldsymbol{s}| = 4\pi \sin\theta/\lambda$, the relation

$$\frac{4\pi \sin\theta}{\lambda} = \frac{2\pi}{d}$$

is obtained by using $|\boldsymbol{s}| = |\boldsymbol{h}|$. As can be seen from Fig. 2.7,

$$\lambda = 2d\sin\theta_{\mathrm{B}} = 2d\sin\theta$$

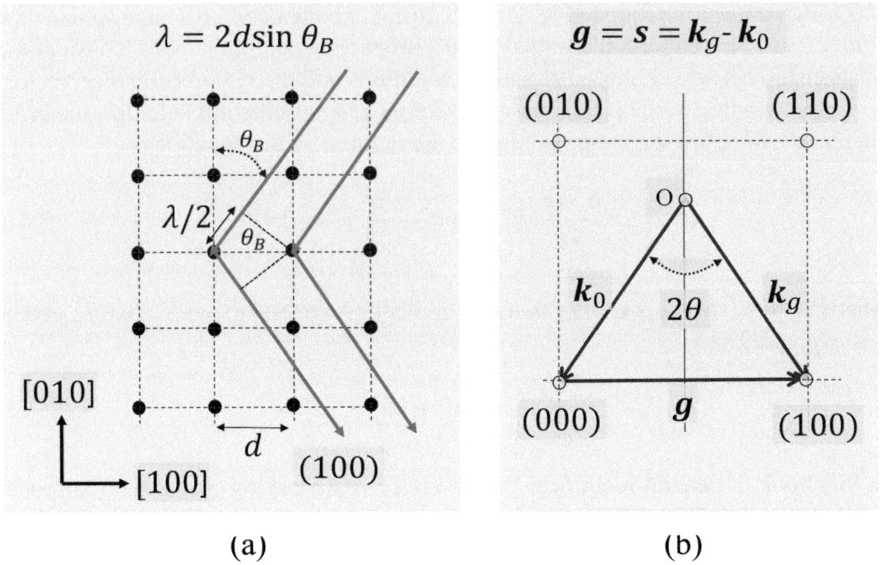

Fig. 2.7 Bragg condition in the real space (**a**) and the Laue's diffraction condition in the reciprocal space (**b**)

assuming that θ_B is a Bragg angle. Then we obtain $\theta = \theta_B$, indicating that the Laue's diffraction condition is equivalent to the Bragg's condition. For a cubic crystal with its lattice constant a, there is the following relationship between d and h, k, l as

$$d = \frac{a}{\left(h^2 + k^2 + l^2\right)^{1/2}}$$

2.4 Ewald Construction

There are an infinite number of reciprocal lattice points specified by *hkl*. When the incident beam energy and the direction are given, it is not straightforward to find the reciprocal lattice point which satisfies a *hkl* diffraction condition. The Ewald construction givens a solution to this problem. An example is shown in Fig. 2.8. This shows reciprocal points for a simple cubic lattice. A circle is drawn with the starting point O of the incident vector as the center and $|\boldsymbol{k}_0|$ as the radius (the sphere in three dimensions). The end point of $\boldsymbol{k}_0$ is taken as the point 000. When this circle intersects with a reciprocal lattice point *hkl*, the Bragg condition of *hkl* reflection is satisfied. In this figure, the reciprocal lattice point that intersects the circle is 320. This sphere is called the Ewald sphere. Usually, the incident vector of X-ray is given, that is,

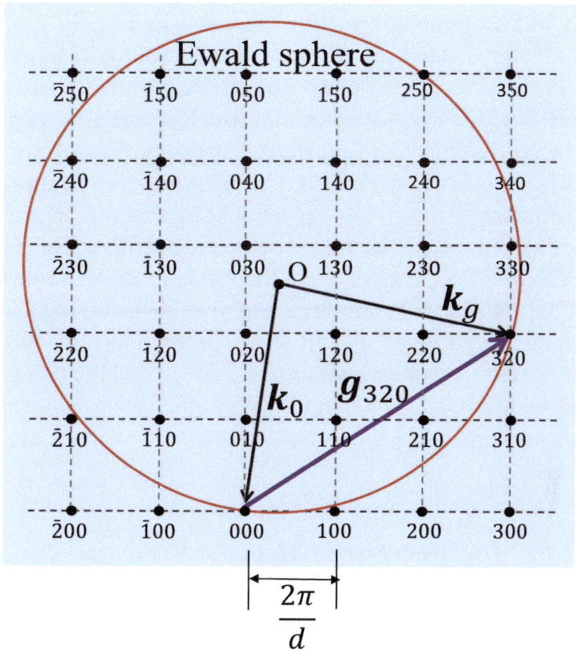

Fig. 2.8 The Ewald construction

$\boldsymbol{k}_0$ is given. In order to search for reciprocal lattice point that satisfies a diffraction condition, the azimuth of the crystal is rotated, that is, the reciprocal lattice space is rotated.

2.5 Phase Problem in Crystal Structure Analysis

As shown in (2.39), the diffraction intensity I_h is given by

$$I_h \propto |F_h|^2, \tag{2.46}$$

where

$$F_h = |F_h|\exp(i\varphi_h). \tag{2.47}$$

If the phase angle φ_h can be found in some way, the electron density $\rho(\boldsymbol{r})$ given as

$$\rho(\boldsymbol{r}) = \frac{1}{V_c}\sum_{-\infty}^{+\infty} F_h\exp\left(-i\boldsymbol{g}_h\cdot\boldsymbol{r}\right) \tag{2.48}$$

is to be obtained by using the measured I_h. In (2.48), the subscript h denotes a reflection index hkl, V_c the unit cell volume and $\boldsymbol{g}_h$ the corresponding reciprocal lattice vector. The sum goes from $-\infty$ to $+\infty$ for h, k and l. Obtaining $\rho(\boldsymbol{r})$ from the measured $|F_h|$ is called crystal structure analysis. However, as shown in (2.46), only the absolute value of the crystal structure factor $|F_h|$ is obtained from the measured diffraction intensity and the phase information φ_h shown in (2.47) is missing. Without the phase retrieval, (2.48) cannot be executed. The problem of finding this phase φ_h is called a phase problem (phase determination). There is an approach that uses X-ray resonant scattering to solve the phase problem, which will be described in detail in Chap. 3. There is another method of determining the phase using a direct method thanks to the development of computers. Once the phase problem is solved, $\rho(\boldsymbol{r})$ can be found from the measurement of $|F_h|$, and the crystal structure will be determined. In addition, as shown in (2.48), in precise structure determination, it is required to measure as many diffraction intensities I_h as possible.

2.6 Temperature Effects

Up to this point, we have assumed that the atoms in the crystal are stationary at fixed positions. In reality, however, each atom vibrates around its equilibrium position due to zero-point vibration that occurs even at the absolute zero degree (0 K) according to the uncertainty principle, and due to thermally excited elastic waves (phonons). Let us consider the influence of this lattice vibration on the crystal structure factor F_h. To simplify the problem, we deal with a crystal consisting of a single kind of atoms. By assuming that an atom is slightly displaced from the equilibrium position, the position of the atom is given by

$$\boldsymbol{r}_j(t) = \boldsymbol{r}_j + \boldsymbol{u}_j(t). \tag{2.49}$$

Here, $\boldsymbol{r}_j$ is the equilibrium position of j atom and $\boldsymbol{u}_j$ is its displacement from $\boldsymbol{r}_j$. When $\boldsymbol{g}_h \cdot \boldsymbol{u}_j$ is small, the crystal structure factor given by (2.35) can be expanded as

$$\begin{aligned} F_h &= \sum_j f_j^0 \exp\{i\boldsymbol{g}_h \cdot (\boldsymbol{r}_j + \boldsymbol{u}_j(t))\} \\ &= \sum_j f_j^0 \exp(i\boldsymbol{g}_h \cdot \boldsymbol{r}_j)\left\{1 - i\boldsymbol{g}_h \cdot \boldsymbol{u}_j(t) - \frac{1}{2}(\boldsymbol{g}_h \cdot \boldsymbol{u}_j(t))^2 + \cdots\right\}. \end{aligned} \tag{2.50}$$

As $\boldsymbol{u}_j$ represents an irregular motion independent of the direction of $\boldsymbol{g}_h$, and the time average $\langle \boldsymbol{g}_h \cdot \boldsymbol{u}_j \rangle$ must be null. Then we have an approximate relation

$$\exp\left\{-\frac{1}{2}(\langle \boldsymbol{g}_h \cdot \boldsymbol{u}_j \rangle)^2\right\} \approx 1 - \frac{1}{2}(\langle \boldsymbol{g}_h \cdot \boldsymbol{u}_j \rangle)^2,$$

and we can get

$$F_h = \sum_j f_j^0 \exp(i\boldsymbol{g}_h \cdot \boldsymbol{r}_j)\exp(-M_j). \tag{2.51}$$

The effect of thermal vibration is expressed by multiplying $\exp(-M_j)$ with f_j^0, as shown in (2.51). The term $\exp(-M_j)$ is called the Debye–Waller factor and M_j is written as

$$M_j = \frac{1}{2}\left(\langle \boldsymbol{g}_h \cdot \boldsymbol{u}_j \rangle\right)^2 = \frac{1}{2}\langle u_{jv}\rangle^2 \left(\frac{4\pi \sin\theta_B}{\lambda}\right)^2 = B_j\left(\frac{\sin\theta_B}{\lambda}\right)^2. \tag{2.52}$$

Here, u_{jv} is the component of $\boldsymbol{u}_j$ parallel to the scattering vector $\boldsymbol{s}(=\boldsymbol{g}_h)$. B_j is a parameter expressing the atomic displacement, which will be referred to as a temperature factor here (the Debye–Waller factor itself may be referred to as a temperature factor). According to (2.52),

$$B_j = 8\pi^2 \langle u_{jv}\rangle^2 \tag{2.53}$$

and for a simple cubic lattice,

$$B = \frac{6h^2T}{m_a k\theta_D^2}\left\{\phi(x) + \frac{x}{4}\right\}. \tag{2.54}$$

Here, h is the Planck constant, k the Boltzmann constant, m_a the mass of vibrating atom, θ_D Debye temperature, T the absolute temperature. $\phi(x)$ is called the Debye function and given by.

$$\phi(x) = \frac{1}{x}\int_0^x \frac{\xi}{\exp(\xi) - 1} d\xi, \tag{2.55}$$

with

$$x = \frac{\theta_D}{T}. \tag{2.56}$$

If the wavelength λ is expressed in angstroms (Å), B is expressed in square angstroms (Å^2)

$$B = 1.149 \times 10^4 \frac{T}{A_m\theta_D^2}\left\{\phi(x) + \frac{x}{4}\right\} \quad [\text{Å}^2]\,. \tag{2.57}$$

A_m is the atomic weight, and $A_m = m_a N$ with N being the Avogadro constant. As this equation shows, based on the uncertainty principle, B does not become zero due to zero-point vibration even at absolute zero (0 K).

Let's find the Debye–Waller factor for Si 220 reflection. As the atomic mass is 28.1, the Debye temperature is 645 K, and the room temperature is 293 K, $B = 0.32$ Å^2 and $\exp(-M) = 0.98$. For 333 reflection, it is 0.93. In this way, the higher the reflection index, the smaller the Debye–Waller factor. In general, hard crystals have a high Debye temperature and B tends to be small, while soft crystals have a low Debye temperature and B tends to be large. The values of the Debye temperature θ_D and Debye temperature factor B_j for many crystals are listed in [2, 3], respectively.

When the temperature factor is anisotropic, the crystal structure factor is given as

$$F_h = \sum_j f_j^0 exp\big(i\boldsymbol{g}_h \cdot \boldsymbol{r}_j\big)T_j, \tag{2.58}$$

where T_j is the correction term expressed as

$$T_j = \exp\big\{-\big(h^2B_{11j} + k^2B_{22j} + l^2B_{33j} + 2klB_{23j} + 2lhB_{31j} + 2hkB_{12j}\big)\big\}. \tag{2.59}$$

B_j has 6 values for each atom.

The vibrations of atoms in a crystal are not independent of each other but are coupled to each other. X-rays undergo diffraction for each mode of this elastic wave. As a result of combining these modes, a wide and diffuse scattering appears around a diffraction peak at which the intensity is weak. This is called thermal diffuse scattering (TDS). (TDS is an inelastic scattering by phonon).

References

1. A.J.C. Wilson, E. Prince, *International Table for Crystallography*, vol. C. (Kluwer Academic Publishers, 1999)
2. C. Kittel, *Introduction to Solid State Physics*, 8th ed. (Wiley, 2004)
3. L.-M. Peng, G. Ren, S.L. Dudarev, M.J. Whelan, Acta Cryst. **A52**, 456–470 (1996)

Chapter 3
X-ray Resonant Scattering

In this chapter, we discuss X-ray resonant scattering by atoms using classical electromagnetism (classical field theory) and quantum theory, and explain how it is useful in crystal structure analysis. Basically, we must describe atoms by using quantum theory, and cannot treat scattering by atoms based on classical field theory. However, the scattering factor due to X-ray resonant scattering, that is, the X-ray anomalous scattering factor, can also be treated within the scope of classical field theory with the help of optical theory. After explaining resonant X-ray scattering based on the classical field theory, we explain quantum field theory to describe the scattering of X-rays by atoms and clarify the process of the resonant scattering by atoms. As typical applications, we will briefly explain the polarity determination of a polar crystal, absolute structure determination of stereoisomers (chirality) and phase determination of a crystal.

3.1 X-ray Resonant Scattering Based on Classical Field Theory

Suppose that an electron is connected to the atomic nucleus by Coulomb force and is regarded as a harmonic oscillator (with one angular frequency ω_s). The solution of the equation of motion of the electron by classical mechanics given in (2.4) in Chap. 2 is

$$\boldsymbol{r} = \frac{e}{m_e(\omega^2 - \omega_s^2 - i\gamma_d\omega)}\boldsymbol{E}_0\exp(i\omega t), \tag{3.1}$$

where $\boldsymbol{r}$ is the displacement from the equilibrium position, $\boldsymbol{E}_0$ the electric field and γ_d the radiation damping factor. When $\omega \gg \omega_s$, we obtain

T. Fukamachi and T. Kawamura, *Resonant Dynamical X-ray Diffraction*, Springer Series in Solid-State Sciences 208, https://doi.org/10.1007/978-981-96-8406-9_3

$$\boldsymbol{r}_{\omega\to\infty} = \frac{e\boldsymbol{E}_0 \exp(i\omega t)}{m_e \omega^2} \tag{3.2}$$

By putting $\boldsymbol{r}_{\omega\to\infty} = \boldsymbol{r}_{\text{free}}$, $\boldsymbol{r}$ in (3.1) is rewritten as

$$\boldsymbol{r} = \boldsymbol{r}_{\text{free}} \cdot f(\omega) = \frac{e\boldsymbol{E}}{m_e \omega^2} f(\omega), \tag{3.3}$$

with $\boldsymbol{E} = e\boldsymbol{E}_0 \exp(i\omega t)$ and $f(\omega)$ in (3.3) being given by

$$f(\omega) = \frac{\omega^2}{\omega^2 - \omega_s^2 - i\gamma_d \omega} \approx \frac{\omega_s \omega}{\omega^2 - \omega_s^2 - i\gamma_d \omega} \tag{3.4}$$

As $f(\omega)$ is a complex number, we write $f(\omega) = f_r(\omega) + if_i(\omega)$ with $f_r(\omega)$ and $f_i(\omega)$ being the real and imaginary parts. By using (3.4), we can write

$$f_r(\omega) = \frac{\omega_s \omega(\omega^2 - \omega_s^2)}{(\omega^2 - \omega_s^2)^2 + \gamma_d^2 \omega^2}, \text{ and } f_i(\omega) = \frac{\gamma_d \omega_s \omega^2}{(\omega^2 - \omega_s^2)^2 + \gamma_d^2 \omega^2}. \tag{3.5}$$

Calculated values of $f_r(\omega)$ and $f_i(\omega)$ as a function of energy ($E = \hbar\omega$) is shown in Fig. 3.1. We take the energy corresponding to the eigen angular frequency as $E_s = \hbar\omega_s = 10$ keV and $\Delta E_d = \hbar\gamma_d = 1$ eV, and show the values in the energy range $E_s \pm 5$eV. The magnitudes of $f_r(\omega)$ and $f_i(\omega)$ are of the order of 10^4. $f_i(\omega)$ shows a sharp peak at $E = E_s$. $f_r(\omega)$ is zero at $E = E_s$, and shows a negative minimum at $E_s - 0.5$eV and a positive maximum at $E_s + 0.5$eV.

This scattering occurs around the resonant angular frequency $\omega \cong \omega_s$ and is called resonant scattering. The resonant scattering in X-ray region depends on photoelectric absorption of X-rays as described later. If we take $f_r(\omega)$ as the horizontal axis and $f_i(\omega)$ as the vertical axis, $f(\omega)$ in the complex plane shows a circle as shown in Fig. 3.2. This kind of circle is a common characteristic of resonance phenomena caused by a forced oscillation in the fields of mechanics, electric circuit, dispersion of dielectric materials and so on. In Fig. 3.2, if we write the maximum value of f_i as f_i^M, the radius of the circle is $f_i^M/2$, and the equation of the circle is given by

$$f_r^2(\omega) + \left(f_i(\omega) - \frac{f_i^M}{2}\right)^2 = \left(\frac{f_i^M}{2}\right)^2. \tag{3.6}$$

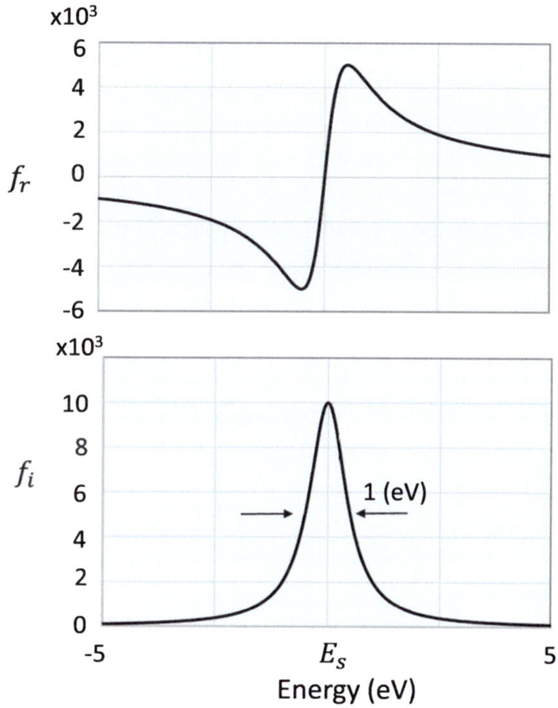

Fig. 3.1 $f_r(\omega)$ and $f_i(\omega)$ in case of single mode vibration of $E_s = \hbar\omega_s =$ 10 keV and $\hbar\gamma_d = 1$ eV

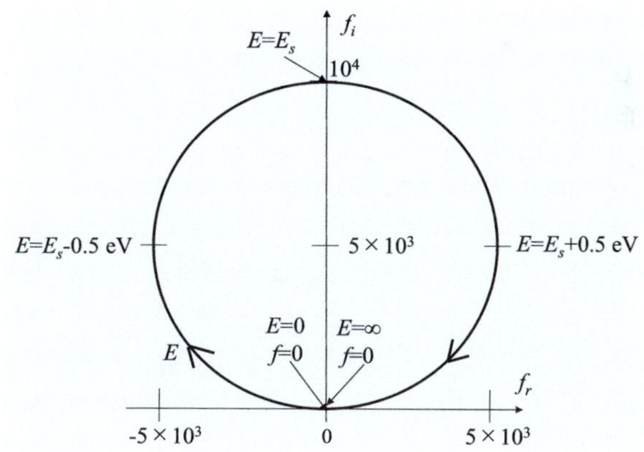

Fig. 3.2 Circle representing $f_r(\omega) + if_i(\omega)$ for a single mode oscillation in the complex plane

Such a large change in $f_i(\omega)$ as shown in Fig. 3.1 is not actually observed in actual atoms, and its magnitude $|f_i(\omega)|$ at the absorption edge is comparable to the normal atomic scattering factor, as described below. In the field of X-rays, this resonant scattering is also called "anomalous scattering." Since the scatterer is an atom, it is sometimes called "atomic resonant scattering." Now, with the help of optical theory, we will consider how large the resonant scattering is for actual atoms.

3.2 X-ray Resonant Scattering by Multiple Oscillators

The electric polarization $\boldsymbol{P}$ of an oscillating electron system is given by

$$\boldsymbol{P} = \varepsilon_0 \chi \boldsymbol{E}, \tag{3.7}$$

where χ is the electric susceptibility. Since the dipole moment for one electron is $\boldsymbol{p} = -e\boldsymbol{r}$, the electric polarization is given as

$$\boldsymbol{P} = -\frac{N_e e^2 \boldsymbol{E}}{m_e \omega^2} f(\omega), \tag{3.8}$$

when the number of electrons in the same state per unit volume is N_e. Then, χ is expressed as

$$\chi = -\frac{e^2 N_e}{\varepsilon_0 m_e \omega^2} f(\omega) = -\frac{r_e N_e}{\pi} \lambda^2 f(\omega). \tag{3.9}$$

The magnitude of χ is of the order of 10^{-4}, if the magnitude of $f(\omega)$ is estimated by the normal atomic scattering factor f^0. As the dielectric constant ε is

$$\varepsilon = \varepsilon_0(1 + \chi), \tag{3.10}$$

the relation

$$\frac{\varepsilon}{\varepsilon_0} = 1 - \frac{r_e N_e}{\pi} \lambda^2 f(\omega) \tag{3.11}$$

is obtained. The refractive index n_0 is given as

$$n_0 = \left(\frac{\varepsilon \mu}{\varepsilon_0 \mu_0}\right)^{1/2} = \left(\frac{\varepsilon}{\varepsilon_0}\right)^{1/2} \tag{3.12}$$

by assuming that the magnetic permeability of the material μ is approximately the same as that in the vacuum μ_0 for X-rays. When the electric susceptibility is small ($|\chi| \ll 1$),

$$n_0 = (1 + \chi)^{1/2} \approx 1 + \frac{\chi}{2} = 1 - \frac{r_e N_e}{2\pi} \lambda^2 f(\omega). \tag{3.13}$$

As $f(\omega)$ is complex, the refractive index n_0 is usually written as

$$n_0 = 1 - \delta - i\beta. \tag{3.14}$$

In the X-ray region, both δ and β are sufficiently smaller than 1, then they are approximately given by

$$\delta = \frac{r_e N_e}{2\pi}\lambda^2 f_r(\omega) \text{ and } \beta = \frac{r_e N_e}{2\pi}\lambda^2 f_i(\omega). \tag{3.15}$$

The refractive index n_0 is also given as

$$n_0 = \frac{k_0}{K_0} \tag{3.16}$$

by using the wave number in the vacuum (K_0) and that in a material (k_0). By assuming that the traveling direction of the X-ray is along z-axis, the two polarization components of the incident X-ray electric field in a material E_s^a (a = 1,2) is related to those in the vacuum E_0^a as

$$E_s^a = E_0^a \exp(-ik_0 z) = E_0^a \exp\{(-iK_0 z)(1-\delta)\}\exp(-\beta K_0 z). \tag{3.17}$$

The superscript 'a' on E_0^a and E_s^a represents the polarization of the X-ray. According to (3.17), $(1 - \delta)$ gives the change in the wavenumber, that is, the phase velocity, while β gives the attenuation of the amplitude. The X-ray intensity at depth z in the material P_s is given by

$$P_s = |E_s^a|^2 = P_0 \exp(-2\beta K_0 z) = P_0 \exp[-\mu_a(\omega)z] \tag{3.18}$$

using the intensity of the incident X-ray P_0. $\mu_a(\omega)$ is the linear absorption coefficient given by

$$\mu_a(\omega) = 2\beta K_0 = 2r_e N_e \lambda f_i(\omega). \tag{3.19}$$

This means that β and $f_i(\omega)$ can be obtained by measuring $\mu_a(\omega)$.

The attenuation of X-ray is caused by photoelectric effect (photo-absorption), Compton scattering and thermal diffuse scattering (TDS) due to thermal fluctuations of the atoms. Among these, photo-absorption is dominant especially near the threshold energy (called the absorption edge) characteristic to an atom in the material.

The photo-absorption coefficient is zero below the absorption edge (E_k) as shown in Fig. 3.3. It rapidly increases as X-ray energy increases above the edge and gradually decreases in the higher energy region. When $\mu_a(\omega)$ has finite values continuously for E larger than E_k, it can be considered that the oscillators with natural angular frequency ω_s are multiple oscillators that are continuously distributed above ω_k ($E_k = \hbar\omega_k$). Then, $f_i(\omega)$ in (3.5) is modified as

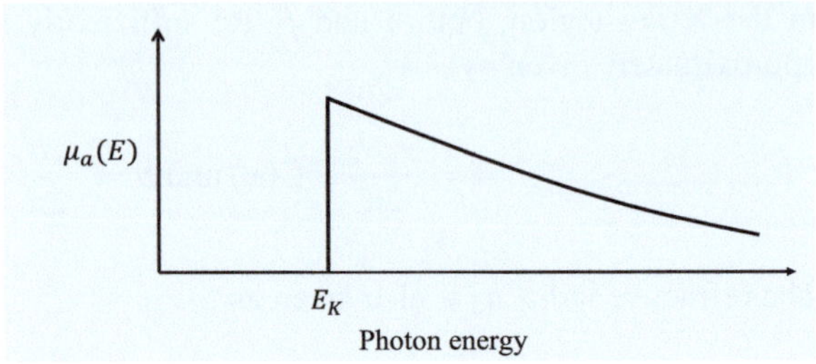

Fig. 3.3 Linear absorption coefficient by an atom as a function of X-ray energy

$$f_i(\omega) = \int_0^\infty \frac{(dg/d\omega_s)\gamma_d\omega^3}{(\omega^2 - \omega_s^2)^2 + \gamma_d^2\omega^2} d\omega_s, \tag{3.20}$$

by using the oscillator strength (density of the oscillator) for ω_s defined by

$$\frac{dg}{d\omega_s}. \tag{3.21}$$

There exists a relation

$$\int_0^\infty \frac{dg}{d\omega_s} d\omega_s = 1, \tag{3.22}$$

which is called the Thomas-Reich-Kuhn (TRK) sum rule. If $dg/d\omega_s$ in the integrand of (3.20) is more slowly varying function than the Lorentzian function which takes values only near $\omega = \omega_s$, it is possible to replace $\gamma_d^2\omega^2$ with $\gamma_d^2\omega_s^2$, and ω^2 with $\omega_s\omega$. Then, (3.20) becomes

$$f_i(\omega \approx \omega_s) = \frac{dg}{d\omega}\frac{\gamma_d\omega\omega_s}{2}\int_0^\infty \frac{2\omega d\omega}{(\omega^2 - \omega_s^2)^2 + \gamma_d^2\omega_s^2}. \tag{3.23}$$

As the definite integral gives $\pi/(\gamma_d\omega_s)$, (3.23) becomes

$$f_i(\omega) = \frac{\pi\omega}{2}\left(\frac{dg}{d\omega}\right). \tag{3.24}$$

Using β in (3.19) and $f_i(\omega)$ in (3.24),

$$\frac{dg}{d\omega} = k_\mu\mu_a(\omega) \text{ with } k_\mu = \frac{1}{2\pi^2 c r_e N_e} \tag{3.25}$$

is obtained. In a similar manner, $f_r(\omega)$ in (3.5) is modified as

$$f_r(\omega) = \int_0^\infty \frac{(dg/d\omega_s)\omega^2(\omega^2 - \omega_s^2)}{(\omega^2 - \omega_s^2)^2 + \gamma_d^2\omega^2} d\omega_s. \tag{3.26}$$

3.3 Anomalous Scattering Factor

The scattering amplitude of X-ray is proportional to $f(\omega)$ in (3.4). Without the resonant scattering, $f(\omega)$ is equal to f^0, which is equal to 1 in the forward scattering. Then we can write the amplitude as

$$f(\omega) = 1 + f'(\omega) + if''(\omega). \tag{3.27}$$

$f'(\omega)$ and $f''(\omega)$ are the increments due to resonant scattering and are related to $f_r(\omega)$ and $f_i(\omega)$ as

$$f'(\omega) = f_r(\omega) - 1 = \int_0^\infty \frac{\left(\frac{dg}{d\omega_s}\right)\left\{\omega_s^2(\omega_s^2 - \omega^2) - \gamma_d^2\omega^2\right\}}{\left(\omega^2 - \omega_s^2\right)^2 + \gamma_d^2\omega^2} d\omega_s \text{ and } f''(\omega) = f_i(\omega). \tag{3.28}$$

In deriving $f'(\omega)$ in (3.28), we use the TKR sum rule. In the limit of $\gamma_d \to 0$, we have

$$f'(\omega) = \int_0^\infty \frac{\left(\frac{dg}{d\omega_s}\right)\omega_s^2}{\omega^2 - \omega_s^2} d\omega_s \text{ and } f''(\omega) = \frac{\pi\omega}{2}\frac{dg}{d\omega}. \tag{3.29}$$

$f'(\omega)$ and $f''(\omega)$ are called the real and the imaginary parts of anomalous scattering factor. Due to the resonant scattering, the refractive index n_0 changes from an increasing function to decreasing one at ω_s as a function of X-ray energy. Because of this anomalous dispersion, the resonant scattering is also called anomalous scattering.

Anomalous scattering factor is given by using the experimental value of the absorption coefficient. According to experimental results, the absorption coefficient μ_a is expressed in the form of

$$\mu_a = \mu_k\left(\frac{\omega_k}{\omega}\right)^{p_k}, (\omega > \omega_k). \tag{3.30}$$

Here, μ_k is the linear absorption coefficient at the absorption edge, and p_k is 2.75 for 1s electrons, 2.33 for 2s electrons, and 2.5 for others. Parratt and Hempstead [1] obtained the equations to determine $f'(\omega)$ and $f''(\omega)$ by substituting (3.30) into (3.25) to obtain $(dg/d\omega)$ and by inserting thus obtained $(dg/d\omega)$ into (3.28). Using this formula, $f'(\omega)$ and $f''(\omega)$ due to photo-absorption of Ge K electrons are calculated for example. The results are shown in Fig. 3.4. The horizontal axis is $x = E/E_k$ with $E(= \hbar\omega)$ being the X-ray energy and E_k the energy of the K absorption edge of Ge (11.103 keV). $f''(\omega)$ is zero for $x<1$. It increases sharply to 3.5 at $x = 1$, and then becomes gradually small as x becomes large. On the other hand, $f'(\omega)$ is –1.2 at $x = 0$, and it gradually becomes small as x approaches 1. Near $x = 1$, it rapidly becomes small showing its minimum about -12. It rapidly increases up to 0 around $x = 1.5$, then becomes positive showing slow increase for $x>1.5$. The anomalous scattering factor for an actual atom is three orders of magnitude smaller than that for a single mode oscillator as shown in Fig. 3.1. It has a value in a wide energy range from $x = 0$ to about 4. The locus of $f'(\omega) + if''(\omega)$ shows an arc around $x = 1$, but the circle is greatly deformed for $x \neq 1$ as shown in Fig. 3.5. This indicates that oscillation modes with continuously different relaxation times are included in an actual atom.

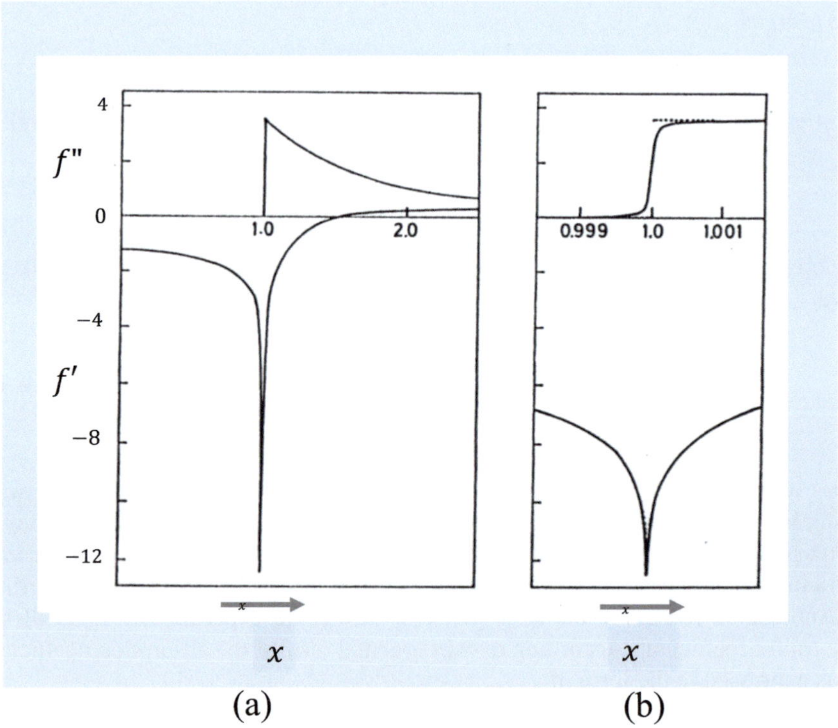

Fig. 3.4 Anomalous scattering factor for Ge K electron. The solid line shows the values obtained by taking radiative attenuation into account. The dotted line shows the values without the radiative attenuation. **a** $0 \leq x \leq 2.5$. **b** $0.9985 \leq x \leq 1.0015$. Details of calculation are also described in Sect. 3.6.1

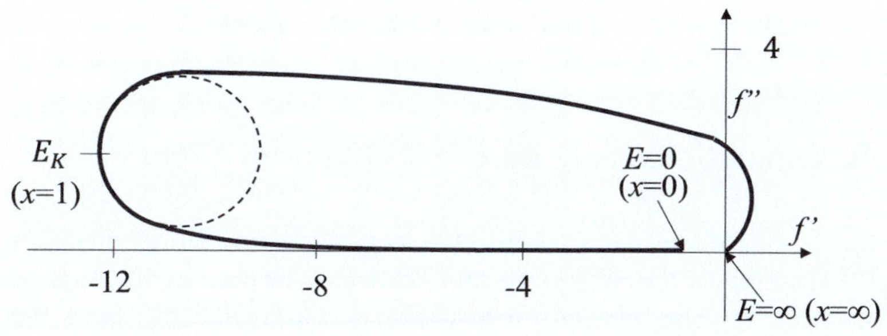

Fig. 3.5 Locus of $f'(\omega) + if''(\omega)$ for K electron in an isolated Ge atom in the complex plane

3.4 Anomalous Scattering Factor Due to Multi-electron Atom

For a multi-electron atom, the real $f'_j(\omega)$ and the imaginary $f''_j(\omega)$ parts of anomalous scattering factor are expressed by the sum of those for all the electron orbitals of the atom

$$f'(\omega) = \sum_j f'_j(\omega), f''(\omega) = \sum_j f''_j(\omega), \tag{3.31}$$

where the subscript j denotes j-th orbital electron.

3.5 Dispersion Relation (Causality)

By substituting $f''(\omega)$ in the second equation of (3.29) into the first one, we obtain

$$f'(\omega) = \frac{2}{\pi} \int_0^\infty \frac{\omega_s f''(\omega_s)}{\omega^2 - \omega_s^2} d\omega_s \tag{3.32}$$

This means that we can obtain the real part $f'(\omega)$ if we know the imaginary part $f''(\omega)$. This is a result derived from the causality principle and is a universal relation between the real and imaginary parts of a scattering factor. This relation is also called a dispersion relation. (A detailed discussion in the case of X-ray scattering can be found in the paper by Kawamura and Fukamachi [2]). By using a measured absorption coefficient, it is possible to obtain the real and imaginary parts of the anomalous scattering factor of the atom in a sample material.

3.6 Resonant Scattering Based on Quantum Theory

Since the classical theory cannot derive the electronic state of atoms, quantum theory is required to consider the electronic state. The scattering process of X-rays by atoms becomes clear by using quantum theory. Figure 3.6 graphically shows the X-ray scattering processes according to quantum theory. The electron shell states are labeled alphabetically with the letters K, L, M,..., in order of decreasing binding energy. Figure 3.6a shows elastic scattering called Thomson scattering, where the electronic state is the same before and after the scattering. (b) shows another elastic scattering called resonant scattering. An electron in the K-shell is excited to an intermediate state by absorbing X-ray with energy ($\hbar\omega_0$), and then returns to its original state by emitting the X-ray with energy ($\hbar\omega_0$). (c) shows inelastic scattering, where an electron with energy ($\hbar\omega_0 > \hbar\omega_K$) is excited to a state above vacuum level by X-ray to emit a photoelectron. Then Kα fluorescent X-ray is emitted by de-excitation of electron from the highest occupied energy level to the K-shell (this part is not included in the scattering factor). In this process, the incident X-ray energy $\hbar\omega_0$ must be larger than the absorption edge energy $E_K = \hbar\omega_K$. There are various ways to treat the electronic state of an atomic system in a radiation field using quantum theory. One way is to deal with the electronic state by quantum theory and X-ray waves by classical electromagnetism. In this semi-classical method, there is no concept of photons for X-rays. In the following, we briefly describe the quantization of X-ray radiation field and consider resonant scattering. However, we will not go into the framework of the theory of quantum optics, which takes the coherence of photons into account.

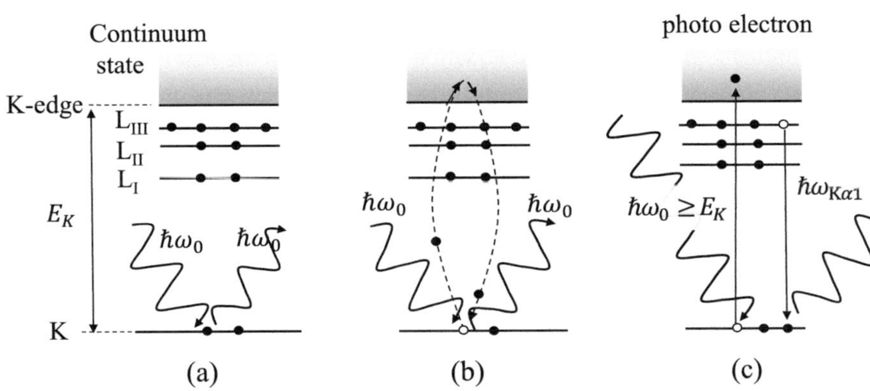

Fig. 3.6 X-ray scattering by electrons of an atom. **a** Thomson scattering of X-ray of energy $\hbar\omega_0$, **b** resonant scattering of X-ray of energy $\hbar\omega_0$ and **c** photoelectron excitation from K-shell electron by X-ray of energy $\hbar\omega_0 \geq E_K$ and accompanying fluorescent X-ray emission of energy $\hbar\omega_{K\alpha 1}$

The Hamiltonian of a one-electron system of an atom in the radiation field is written by

$$H = \frac{1}{2m_e}[\boldsymbol{P} + e\boldsymbol{A}(\boldsymbol{r}, t)]^2 + H_e + H_r \tag{3.33}$$

where $\boldsymbol{A}$ is a vector potential. If we use a Coulomb gauge, the electric field $\boldsymbol{E}$ and the magnetic field $\boldsymbol{H}$ are given by using $\boldsymbol{A}$ by

$$\boldsymbol{E} = -\frac{\partial \boldsymbol{A}}{\partial t}, \boldsymbol{H} = \mathrm{rot}\boldsymbol{A}. \tag{3.34}$$

Here, H_e is the Hamiltonian of the electron system, and H_r is that of the radiation field. In the second quantization form, $\boldsymbol{A}$ is expressed as

$$\boldsymbol{A} = \left(\frac{\hbar}{2\varepsilon_0 V}\right)^{1/2} \sum_k \sum_\alpha \frac{\boldsymbol{e}_k^\alpha}{\omega_k^{1/2}} \left\{a_{k,\alpha} e^{i\boldsymbol{k}\cdot\boldsymbol{r}} + a_{k,\alpha}^* e^{-i\boldsymbol{k}\cdot\boldsymbol{r}}\right\}. \tag{3.35}$$

The magnitude of wave vector $\boldsymbol{k}$ is $|\boldsymbol{k}| = 2\pi/\lambda$ and V is the volume. $\boldsymbol{e}_k^\alpha$ is the unit polarization vector normal to $\boldsymbol{k}$ and α denotes two polarizations as 'a' in (3.17). $\boldsymbol{e}_k^1$ and $\boldsymbol{e}_k^2$ represent the unit vector of the two polarizations and are orthogonal with each other. $a_{k,\alpha}$ and $a_{k,\alpha}^*$ are annihilation and creation operators of a photon with wave vector $\boldsymbol{k}$ and polarization α, respectively. The commutation relations between $a_{k,\alpha}$ and $a_{k,\alpha}^*$ are given as

$$[a_{k,\alpha}, a_{k',\alpha'}^*] = \delta_{kk'}\delta_{\alpha\alpha'}, \left[a_{k,\alpha}, a_{k',\alpha'}\right] = \left[a_{k,\alpha}^*, a_{k',\alpha'}^*\right] = 0. \tag{3.36}$$

Since $\boldsymbol{P}$ in (3.33) is the electron momentum and expressed as $\boldsymbol{P} = -i\hbar\nabla$, we have a relation

$$\boldsymbol{P} \cdot \boldsymbol{A}(\boldsymbol{r}, t) = -i\hbar\{\nabla\boldsymbol{A}(\boldsymbol{r}, t) + \boldsymbol{A}(\boldsymbol{r}, t)\nabla\}. \tag{3.37}$$

In the Coulomb gauge, we have

$$\nabla\boldsymbol{A}(\boldsymbol{r}, t) = \mathrm{div}\boldsymbol{A}(\boldsymbol{r}, t) = 0. \tag{3.38}$$

If we write (3.33) as

$$H = H_0 + H', H_0 = H_e + H_r \tag{3.39}$$

where H_0 is the stable state Hamiltonian and H' is the interaction between an electron and X-ray, then the first term in (3.33) becomes

$$H' = -\frac{e}{m_e}\boldsymbol{A}(\boldsymbol{r},t)\cdot\boldsymbol{P} + \frac{e^2}{2m_e}\boldsymbol{A}^2(\boldsymbol{r},t). \tag{3.40}$$

If we expand H' using $\boldsymbol{A}$, we get

$$H' = \left(\frac{\hbar e^2}{2\varepsilon_0 m_e^2 V}\right)^{1/2} \sum_{\boldsymbol{k}(0)} \sum_{\alpha=1}^{2} \frac{\boldsymbol{e}_{k(0)}^{\alpha}\cdot\boldsymbol{P}}{\omega_{k(0)}^{1/2}} [a_{k(0),\alpha} e^{i\boldsymbol{k}(0)\cdot\boldsymbol{r}} + a_{k(0),\alpha}^{*} e^{-i\boldsymbol{k}(0)\cdot\boldsymbol{r}}]$$

$$+ \frac{\hbar e^2}{4\varepsilon_0 m_e^2 V} \sum_{\boldsymbol{k}(0)} \sum_{\boldsymbol{k}(1)} \sum_{\alpha=1}^{2} \sum_{\beta=1}^{2} \frac{\boldsymbol{e}_{k(0)}^{\alpha}\cdot\boldsymbol{e}_{k(1)}^{\beta}}{(\omega_{k(0)}\omega_{k(1)})^{\frac{1}{2}}}$$

$$\left[a_{k(0),\alpha} e^{i\boldsymbol{k}(0)\cdot\boldsymbol{r}} + a_{k(0),\alpha}^{*} e^{-i\boldsymbol{k}(0)\cdot\boldsymbol{r}}\right]\cdot\left[a_{k(1),\beta} e^{i\boldsymbol{k}(1)\cdot\boldsymbol{r}} + a_{k(1),\beta}^{*} e^{-i\boldsymbol{k}(1)\cdot\boldsymbol{r}}\right] \tag{3.41}$$

Here, $\boldsymbol{k}(0)$ and $\boldsymbol{k}(1)$ are the wave vectors of the initial (0) and the final state (1), respectively.

When one photon in the initial state (0) and one photon in the final state (1) are involved, there are three possible scattering processes as shown in Fig. 3.7. The process in (a) originates from the first-order perturbation term $< 1|H'|0>$, and is derived from the second term in Eq. (3.41). The transition matrix is

$$T_{00}^{(1)} =< B; m_{k(0)} - 1, m_{k(1)} + 1|H'|A; m_{k(0)}, m_{k(1)}> \tag{3.42}$$

Here, m_k is the number of photons, A and B are the initial and final states of the atom (electron in the atom), and A is the same as B in elastic scattering. In the conventional X-ray generation mechanism, the density of X-rays is low, and we can approximate that one electron in the initial state and one electron in the final state, that is, $m_{k(0)} = 1$ and $m_{k(1)} = 0$. The transition matrix is expressed as

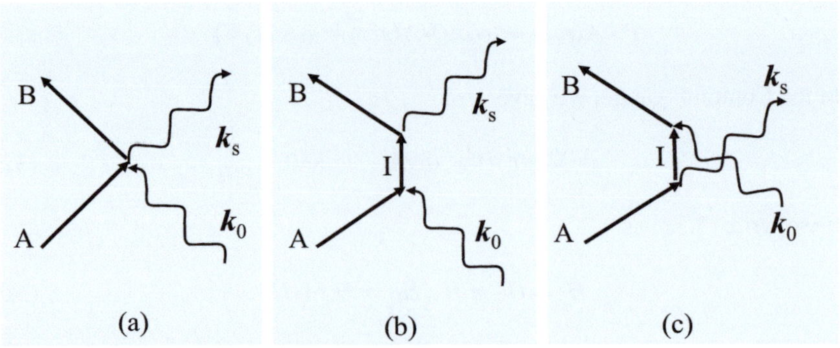

Fig. 3.7 Scattering processes involving two photons

$$T_{00}^{(1)} = <A;0,1\left|H'\right|A;1,0> = \frac{\hbar e^2}{2\varepsilon_0 m_e^2 V}\frac{\boldsymbol{e}_{k(0)}^{\alpha}\cdot\boldsymbol{e}_{k(1)}^{\alpha}}{\omega_0}<\varphi_A\left|e^{i[\boldsymbol{k}(1)-i\boldsymbol{k}(0)]\cdot\boldsymbol{r}}\right|\varphi_B> \quad (3.43)$$

Here, φ_A and φ_B are wave functions of an atom and $|\boldsymbol{k}(0)| = |\boldsymbol{k}(1)|$ because of elastic scattering. In these days, with the advent of free electron lasers (FEL), it is possible to excite multi-photons by high-intensity coherent X-rays. The approximation of (3.42) cannot be applied.

The first term in (3.41) represents the second-order perturbation term in the process of Fig. 3.7b, c, and its matrix element is given as

$$T_{00}^{(2)} = \sum_I \frac{\langle A|H'|I\rangle\langle I|H'|A\rangle}{E_A - E_I'} + \sum_{II}\frac{\langle A|H'|II\rangle\langle II|H'|A\rangle}{E_A - E_{II}'}. \quad (3.44)$$

The first term represents the process that one photon with wavevector $\boldsymbol{k}(0)$ in the initial state is absorbed by exciting an electron in an atom to the intermediate state, then one photon with $\boldsymbol{k}(1)$ is emitted when the atom returns to the ground state in the final state as shown in Fig. 3.7b. In (3.44),

$$E_A = E_0 + \hbar\omega_0,\ E'_I = E_I + i\gamma_{0I},\ \gamma_{0I} = \frac{\Gamma_{0I}}{2}, \quad (3.45)$$

with Γ_{0I} being the natural line width. The second term in (3.44) represents the process that one photon with $\boldsymbol{k}(1)$ is emitted first then one photon with $\boldsymbol{k}(0)$ is absorbed when the atom returns to the ground state as shown in Fig. 3.7c. The energy relations are

$$E_A = E_0 + \hbar\omega_0,\ E'_{II} = E_I + \hbar\omega_0 + \hbar\omega_1. \quad (3.46)$$

In the present case, $\hbar\omega_1 = \hbar\omega_0$. The interaction between a photon and an atom in which absorption and emission occur in sequence in Fig. 3.7b, c is given by the first term in (3.41). By inserting the interaction, (3.44) can be written as

$$T_{00}^{(2)} = -\frac{\hbar e^2}{2\varepsilon_0 m_e^2 V}\frac{1}{\omega_0}\sum_I\left[\frac{\langle\varphi_A|\mathbf{e}_{k(1)}^{\beta}\cdot\mathbf{P}e^{-i\mathbf{k}(1))\cdot\mathbf{r}}|\varphi_I\rangle\langle\varphi_I|\mathbf{e}_{k(0)}^{\alpha}\cdot\mathbf{P}e^{i\mathbf{k}(0))\cdot\mathbf{r}}|\varphi_A\rangle}{E_A - E_I - \hbar\omega_0 - i\gamma_{0I}}\right.$$
$$\left.+\frac{\left\langle\varphi_A\left|\mathbf{e}_{k(0)}^{\alpha}\cdot\mathbf{P}e^{-i\mathbf{k}(0))\cdot\mathbf{r}}\right|\varphi_I\right\rangle\left\langle\varphi_I\left|\mathbf{e}_{k(1)}^{\beta}\cdot\mathbf{P}e^{i\mathbf{k}(1))\cdot\mathbf{r}}\right|\varphi_A\right\rangle}{E_A - E_I + \hbar\omega_0}\right] \quad (3.47)$$

In the denominator of the second term, γ_{0I} is omitted because it is sufficiently small. By using the sum of the first order and the second order matrix elements

$$T_{00} = T_{00}^{(1)} + T_{00}^{(2)}, \quad (3.48)$$

and Fermi's golden rule, the transition probability w is given by

$$w = \frac{2\pi}{\hbar}|T_{00}|^2\rho(\omega). \tag{3.49}$$

Here, $\rho(\omega)$ is the density of photons with energy $\hbar\omega$ of the final state and given by

$$\rho(\omega) = \frac{V}{(2\pi)^3}\frac{\omega^2}{\hbar c^3}d\Omega \tag{3.50}$$

using the relation $|\boldsymbol{k}(0)| = |\boldsymbol{k}(1)| = \omega/c$. $d\Omega$ is the solid angle. Since the differential scattering cross section is

$$d\sigma = \frac{V}{c}w, \tag{3.51}$$

we have

$$\frac{d\sigma}{d\Omega} = r_e^2|-T|^2. \tag{3.52}$$

Here r_e is the classical electron radius and T is given as

$$T = \left(\mathbf{e}^{\alpha}_{k(0)} \cdot \mathbf{e}^{\alpha}_{k(1)}\right)f^0(\mathbf{s}) + \sum_I \left[\frac{\langle\varphi_A|\mathbf{e}^{\beta}_{k(1)} \cdot \mathbf{P}e^{-i\mathbf{k}(1))\cdot\mathbf{r}}|\varphi_I\rangle\langle\varphi_I|\mathbf{e}^{\alpha}_{k(0)} \cdot \mathbf{P}e^{i\mathbf{k}(0))\cdot\mathbf{r}}|\varphi_A\rangle}{E_A - E_I - \hbar\omega_0 - i\gamma_{0I}} + \frac{\left\langle\varphi_A\middle|\mathbf{e}^{\alpha}_{k(0)} \cdot \mathbf{P}e^{-i\mathbf{k}(0))\cdot\mathbf{r}}\middle|\varphi_I\right\rangle\left\langle\varphi_I\middle|\mathbf{e}^{\beta}_{k(1)} \cdot \mathbf{P}e^{i\mathbf{k}(1))\cdot\mathbf{r}}\middle|\varphi_A\right\rangle}{E_A - E_I + \hbar\omega_0}\right] \tag{3.53}$$

A negative sign in front of T in (3.52) represents that the phase difference between the incident wave and the scattered wave due to normal scattering is π. $f^0(\boldsymbol{s})$ is the normal atomic scattering factor

$$f^0(\boldsymbol{s}) = \int \varphi_A^*\varphi_A e^{-i\boldsymbol{s}\cdot\boldsymbol{r}}dV \tag{3.54}$$

with $\boldsymbol{s} = \boldsymbol{k}(0) - \boldsymbol{k}(1)$ being the scattering vector.

The second term in (3.53) corresponds to the anomalous scattering factor. We assume that the scatterer is an isolated atom. Since resonant scattering in the X-ray region occurs due to the excitation of an inner-shell electron in the atom, the radius of the scatterer is much shorter than the wavelength of the X-ray. Then we can use the approximation

$$e^{-i\boldsymbol{k}(0)\cdot\boldsymbol{r}} \approx 1, \tag{3.55}$$

which indicates that there is no angular dependence except for the polarization factor. In order to have correspondence with the classical theory and simplify the calculation, we calculate the anomalous scattering factor using forward scattering approximation. Then (3.52) is given by

$$\frac{d\sigma}{d\Omega}(\theta = 0) = r_e^2 \left[1 - \sum_I \left| \frac{<\varphi_I | \boldsymbol{e} \cdot \boldsymbol{P} | \varphi_0>}{m_e} \right|^2 \left\{ \frac{1}{\hbar\omega_{I0} - \hbar\omega - i\gamma_{0I}} + \frac{1}{\hbar\omega_{I0} + \hbar\omega} \right\} \right]^2. \tag{3.56}$$

The value of 1 of the first term in (3.56) is obtained because of $\int \varphi_A^* \varphi_A dV = 1$. The second term can be written as

$$f'(\omega) - if''(\omega) = -\sum_I \left| \frac{<\varphi_I | \boldsymbol{e} \cdot \boldsymbol{P} | \varphi_0>}{m_e} \right|^2 \left\{ \frac{1}{\hbar\omega_{I0} - \hbar\omega - i\gamma_{0I}} + \frac{1}{\hbar\omega_{I0} + \hbar\omega} \right\}. \tag{3.57}$$

In classical theory, it is customary to express a wave as $e^{i(\omega t - \boldsymbol{k} \cdot \boldsymbol{r})}$, while in quantum theory, to express it as $e^{i(\boldsymbol{k} \cdot \boldsymbol{r} - \omega t)}$. Correspondingly, the atomic scattering factor f is expressed as

$$(\text{Classical theory}) f = f^0 + f' + if'' \tag{3.58}$$

$$(\text{Quantum theory}) f = f^0 + f' - if'' \tag{3.59}$$

It should be noted that the sign of the imaginary part is opposite.

In order to have correspondence with the classical theory in later part, we put

$$g_{0I} = \frac{2}{\hbar\omega_{I0}} \left| \frac{<\varphi_I | \boldsymbol{e} \cdot \boldsymbol{P} | \varphi_0>}{m_e} \right|^2. \tag{3.60}$$

Then, from (3.57), we have

$$f'(\omega) = -\sum_I \frac{\hbar\omega_{I0} g_{0I}}{2} \left\{ \frac{\hbar\omega_{I0} + \hbar\omega}{(\hbar\omega_{I0} - \hbar\omega)^2 + \gamma_{0I}^2} + \frac{1}{\hbar\omega_{I0} + \hbar\omega} \right\} \tag{3.61}$$

$$f''(\omega) = \sum_I \frac{\hbar\omega_{I0} g_{0I}}{2} \frac{\gamma_{0I}}{(\hbar\omega_{I0} - \hbar\omega)^2 + \gamma_{0I}^2}. \tag{3.62}$$

For obtaining the oscillator strength g_{0I} in (3.61) and (3.62), by using the relation

$$\boldsymbol{P} = i\frac{m_e}{\hbar}(H\boldsymbol{r} - \boldsymbol{r}H), \tag{3.63}$$

we have the matrix element as

$$< B|\boldsymbol{P}|A> = i\frac{m_e}{\hbar} < B|H\boldsymbol{r} - \boldsymbol{r}H|A> = i\frac{m_e}{\hbar}(E_B - E_A) < B|\boldsymbol{r}|A>. \quad (3.64)$$

Then we have,

$$\frac{< \varphi_I|\boldsymbol{e}\cdot\boldsymbol{P}|\varphi_0>}{m_e} = i\omega_{I0} < \varphi_I|\boldsymbol{e}\cdot\boldsymbol{r}|\varphi_0> \quad (3.65)$$

by writing the initial state $A = \varphi_0$ and the intermediate state $B = \varphi_I$ and using $E_B - E_A = \hbar\omega_{I0}$. The oscillator strength is obtained as

$$g_{0I} = \frac{2}{\hbar}\omega_{I0}| < \varphi_I|\boldsymbol{e}\cdot\boldsymbol{r}|\varphi_0>|^2. \quad (3.66)$$

This indicates that the anomalous scattering factor can be obtained if the oscillator strength g_{0I} is obtained using the atomic wavefunction φ_0 and φ_I.

3.6.1 Selection Rule and Calculation of Anomalous Scattering Factor

The wave function in a spherically symmetric potential of an atom can be expanded as

$$\varphi_0 = R_0(r)Y_l^m(\theta,\varphi), \varphi_I = R_I(r)Y_{l'}^{m'}(\theta,\varphi) \quad (3.67)$$

where angles θ and φ represent the polar and azimuthal angle, respectively. $R_0(r)$ ($R_I(r)$) is the wavefunction in the radial direction and Y_l^m ($Y_{l'}^{m'}$) is the spherical harmonics with l (l') and m (m') being the azimuthal and the magnetic quantum numbers. If linearly polarized X-rays parallel to the z-axis are incident on such an atom, the magnitude of z-component of the oscillator strength is given by

$$z_{0I} = \int \varphi_I^*(\boldsymbol{r})z\varphi_0(\boldsymbol{r})dV$$

$$= \left(\frac{4\pi}{3}\right)^{\frac{1}{2}}\int_0^\infty R_I^*(r)R_0(r)r^2dr\int_0^\pi\int_0^{2\pi} Y_{l'}^{'m*}(\theta,\varphi)Y_l^0(\theta,\varphi)Y_l^m(\theta,\varphi)\sin\theta d\theta d\varphi. \quad (3.68)$$

We use the relation

$$z = r\cos\theta = \left(\frac{4\pi}{3}\right)^{\frac{1}{2}} rY_l^0(\theta, \varphi) \tag{3.69}$$

in deriving (3.68). The integral value of the spherical harmonic is nonzero only when $l' = l \pm 1$ and $m = m'$, which is called the selection rule regarding transition (excitation). Since the specific coordinate axes cannot be determined for an isolated atom, we need to find the oscillator strength for the x and y axes in the same way. If we calculate the average of the values for these three axes, the oscillator strength is given by

$$g_{0I} = \left(\frac{2\omega_{I0}}{3\hbar}\right)\left\{|x_{0I}|^2 + |y_{0I}|^2 + |z_{0I}|^2\right\}. \tag{3.70}$$

In this case the selection rule gives $l' = l \pm 1$. For example, if we take s-orbital ($l = 0$) as the ground state, the intermediate state must be p-orbital ($l = 1$). As can be seen from Eq. (3.70), g_{0I} can be obtained from theoretical calculations by specifying the electronic level (l) of the initial state of resonant scattering and calculating its $R_0(r)$ and $R_I(r)$. When an electron is excited or makes a transition, the final (intermediate) state must be unoccupied. As an approximation, we regard the final state as continuous one. Then we replace sum with integral as

$$\sum g_{0I} = \int \frac{dg_{0I}}{d\omega_{I0}} d\omega_{I0}, \tag{3.71}$$

to obtain

$$f'(\omega) = -\frac{1}{2}\int_{\omega_k}^{\infty}\left(\frac{dg_{0I}}{d\omega_{I0}}\right)\hbar\omega_{I0}\left\{\frac{\hbar\omega_{I0} - \hbar\omega}{(\hbar\omega_{I0} - \hbar\omega)^2 + \gamma_{0I}^2} + \frac{1}{\hbar\omega_{I0} + \hbar\omega}\right\}d\omega_{I0} \tag{3.72}$$

$$f''(\omega) = -\frac{1}{2}\int_{\omega_k}^{\infty}\left(\frac{dg_{0I}}{d\omega_{I0}}\right)\hbar\omega_{I0}\frac{\gamma_{0I}}{(\hbar\omega_{I0} - \hbar\omega)^2 + \gamma_{0I}^2}d\omega_{I0}. \tag{3.73}$$

Except for the vicinity of the absorption edge, we can put as $\gamma_{0I} \to 0$ and obtain

$$f'(\omega) = \int_{\omega_k}^{\infty}\left(\frac{dg_{0I}}{d\omega_{I0}}\right)\frac{\omega_{I0}^2}{\omega^2 - \omega_{I0}^2}d\omega_{I0}. \tag{3.74}$$

By using $\lim_{\varepsilon\to 0}\varepsilon/(x^2 + \varepsilon^2) = \pi\delta(x)$, we have

$$f''(\omega) = \left[f''(\omega)\right](\gamma_{0I} \to 0) = \frac{\pi\omega}{2}\left(\frac{dg}{d\omega}\right)\int \delta(\omega_{I0} - \omega)d\omega_{I0} = \frac{\pi\omega}{2}\left(\frac{dg}{d\omega}\right). \tag{3.75}$$

$f'(\omega)$ and $f''(\omega)$ derived above by using quantum theory agree with those derived by using classical theory.

Based on the above quantum theory, Thomson scattering was derived from the $\boldsymbol{A}^2$ term in (3.40), and resonant scattering was derived from the $\boldsymbol{P} \cdot \boldsymbol{A}$ term. The normal atomic scattering factor f^0 of Thomson scattering is obtained from the wavefunction of the ground state of an atom, and it agrees with that of classical theory. The real and the imaginary parts of anomalous scattering factor of resonant scattering $f'(\omega)$ and $f''(\omega)$ are obtained using the transition probability of an inner-shell electron in the ground state to an unoccupied state in the intermediate state. However, it is not easy to determine the wavefunction of the excited state, especially that of the absorption edge of atoms in a crystal. For this reason, μ_k thus $f''(\omega)$ at the absorption edge in (3.30) is derived as follows. By putting

$$\frac{dg}{d\omega} = k_\mu \mu_a(\omega) = k_\mu \mu_k \left(\frac{\omega_k}{\omega}\right)^{p_k} = g'_k \left(\frac{\omega_k}{\omega}\right)^{p_k}, \tag{3.76}$$

we derive g_k as

$$g_k = \int_{\omega_k}^{\infty} \left(\frac{dg}{d\omega}\right) d\omega = g'_k \int_{\omega_k}^{\infty} \left(\frac{\omega_k}{\omega}\right)^{p_k} d\omega = g'_k \frac{\omega_k}{p_k - 1}, \tag{3.77}$$

then we obtain

$$\frac{dg}{d\omega} = \frac{p_k - 1}{\omega_k} g_k \left(\frac{\omega_k}{\omega}\right)^{p_k}. \tag{3.78}$$

If we apply the TRK summation rule to K electron excitation, for example, and use the above relations, we have

$$\int_{\omega_k}^{\infty} \left(\frac{dg}{d\omega}\right) d\omega + \sum_m g_m = g_k + \sum_m g_m = 2, \tag{3.79}$$

and then

$$g_k = 2 - \sum_m g_m. \tag{3.80}$$

By using (3.80), g_k can be obtained, because g_m is calculated using wavefunctions of occupied state of the atom. The absorption coefficient at the absorption edge can be determined using (3.76).

By using g_k, the anomalous scattering factor can be obtained using the formula by Parratt and Hempstead [1]. Cromer [3] calculated the ground state wavefunction using the Hatree approximation and taking account of relativistic effects, and published a table of g_k and anomalous scattering factors for characteristic X-ray energies. The curves of $f'(\omega)$ and $f''(\omega)$ near the Ge K absorption edge in Fig. 3.4 were obtained using g_k by Cromer.

3.6.2 Determination of Anomalous Scattering Factor of Atom in a Material

Anomalous scattering factor of an isolated atom can be calculated by using the wavefunctions of the atom in quantum theory. However, for an atom in a crystal, it strongly depends on the electronic state. It is possible to determine the wavefunctions of an atom in a crystal if the crystal structure is known. But it is difficult to determine the wavefunction of a material whose structure is not known. In this section, we will discuss how to calculate the anomalous scattering factor near a certain absorption edge of an atom in a material.

Figure 3.8 shows the measured absorption coefficient by exciting K electron near the K absorption edge of a Ge crystal and converting it to f'' according to (3.19).

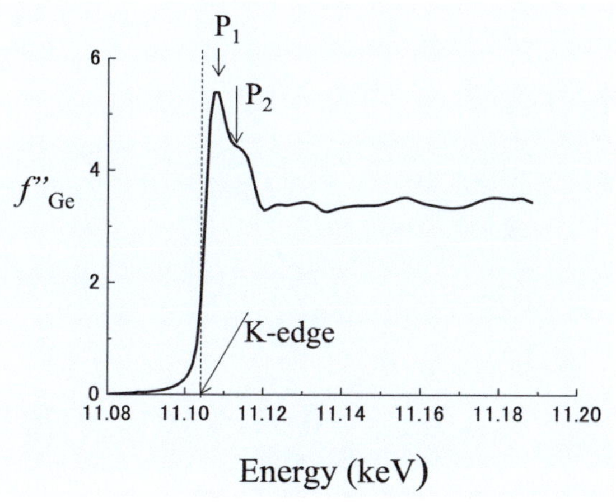

Fig. 3.8 f'' near the K absorption edge of Ge crystal showing fine structures above the edge. This is unpublished data measured by the authors. The experimental condition is the same as that described by Yoshizawa, M. et al. (2005). *Acta Cryst.* A**61**, 553–556

Unlike f'' of an isolated atom, for example in f'' in Fig. 3.8, a sharp peak denoted by P_1 appears just above the absorption edge and a shoulder structure at slightly higher energy denoted by P_2 also appears. Small fluctuations of the amplitude can be seen in further higher energy region. In this way, fine structures appear in the absorption coefficient f'' near the absorption edge, including chemical shifts. These structures are called XAFS (X-ray Absorption Fine Structures) [4] including the fine structures in higher energy region. The fine structures seen in higher energy region is called EXAFS (Extended X-ray absorption fine Structures), which are used for elucidating atomic arrangement around the absorbing atom. Once we have $f''(\omega)$, we can obtain $f'(\omega)$ using the dispersion relation in (3.32). The resultant $f'(\omega)$ is shown in Fig. 3.9 together with measured $f''(\omega)$ near the K absorption edge of crystalline Ge. In addition, the locus of $f' + if''$ obtained in Fig. 3.9 is shown in Fig. 3.10 by changing X-ray energy. Since some parts of two circular loci (shown in red and green solid lines) due to the peaks of P_1 and P_2 can be interpolated to the locus of $f' + if''$, these two loci can be attributed to resonant scatterings due to two independent oscillators.

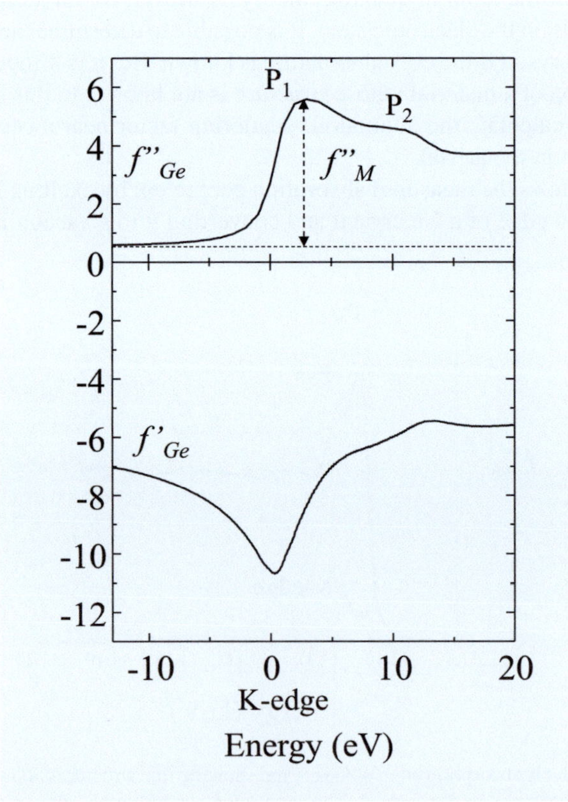

Fig. 3.9 Anomalous scattering factor near the K absorption edge of crystalline Ge. Reproduced using a part of Fig. 1 in Yoshizawa, M. et al. (2005). *Acta Cryst.* A**61**, 553–556

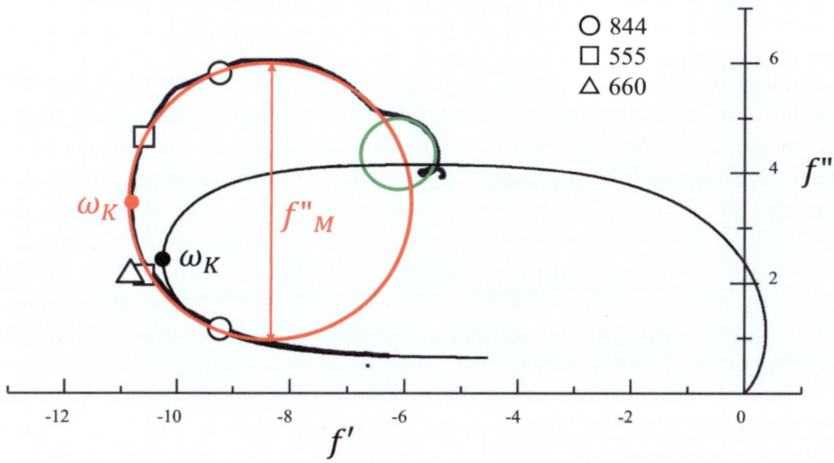

Fig. 3.10 Locus of $f' + if''$ of crystalline Ge. The thick solid line shows one obtained by using values in Fig. 3.9. The thin solid line shows one obtained using equation of Parratt & Hempstead. ○, □ and △ are measured values using the zero method by Yoshizawa et al. [5] (see Chap. 7)

According to (3.57), the anomalous scattering factor associated with P_1 peak (very near the edge) can be put as

$$f' + if'' \approx \frac{\hbar\omega_{I0}g_{0I}}{2}\left(\frac{1}{\hbar\omega - \hbar\omega_{I0} - i\gamma_{0I}}\right) + f'_b + if''_b. \tag{3.81}$$

The locus of the anomalous scattering factor in the first term is circular as shown in Fig. 3.10, which indicates that the anomalous scattering factor represented as $\Delta f' + i\Delta f''$ can be given by

$$\Delta f' + i\Delta f'' = \frac{\hbar\omega_{I0}g_{0I}}{2}\left(\frac{1}{\hbar\omega - \hbar\omega_{I0} - i\gamma_{0I}}\right) = \frac{f''_M\gamma_{0I}}{\hbar\omega - \hbar\omega_{I0} - i\gamma_{0I}}. \tag{3.82}$$

This is because the relation $\hbar\omega_{I0}g_{0I} = 2f''_M\gamma_{0I}$ holds at $\omega = \omega_{I0}$ by writing the maximum value of f'' as f''_M. Then we can obtain

$$(\Delta f')^2 + \left(\Delta f'' - \frac{f''_M}{2}\right)^2 = \left(\frac{f''_M}{2}\right)^2.$$

A large circle shown in Fig. 3.10 corresponds to one oscillator of peak P_1, and similarly a smaller circle corresponds to another oscillator of peak P_2. If the height of the peak of $f''(\omega)$ near the absorption edge is obtained, the anomalous scattering factor due to the oscillator can be obtained. These two oscillators are independent with each other.

3.7 Atomic Scattering Factor Due to Resonant Scattering

Up until now, we have considered $f'(\omega) + if''(\omega)$ in the case of forward scattering. Since the inner shell electrons are strongly bound and localized at the nucleus of the atom, the transition probability of resonant scattering shows almost no angular dependence. The atomic scattering factor including resonant scattering generally can be written as

$$f(s,\omega) = f^0(s) + f'(\omega) + if''(\omega), \tag{3.83}$$

where s is the magnitude of scattering vector. This can be rewritten in polar coordinates of a complex plane as

$$f(s,\omega) = |f(s,\omega)|\exp(i\theta) \tag{3.84}$$

with

$$\tan\theta = \frac{f''(\omega)}{f^0(s) + f'(\omega)}. \tag{3.85}$$

As $f''(\omega)$ is always positive, the phase θ is $0 \leq \theta \leq \pi$. In general, $f^0(s) + f'(\omega) > 0$ for low-order reflections. For higher-order reflections, $f^0(s) + f'(\omega)$ can become negative as shown in Fig. 3.11. When $f^0(s) + f'(\omega)$ is zero, scattering only by $f''(\omega)$ occurs and θ is equal to $\pi/2$.

In the kinematical theory of diffraction, the scattering intensity is proportional to $|f(s,\omega)|^2$ and the phase information of θ is lost in the intensity. This phase information can be obtained directly from measured intensity from a perfect crystal to which dynamical theory of diffraction is applied, which will be explained in Chap. 7.

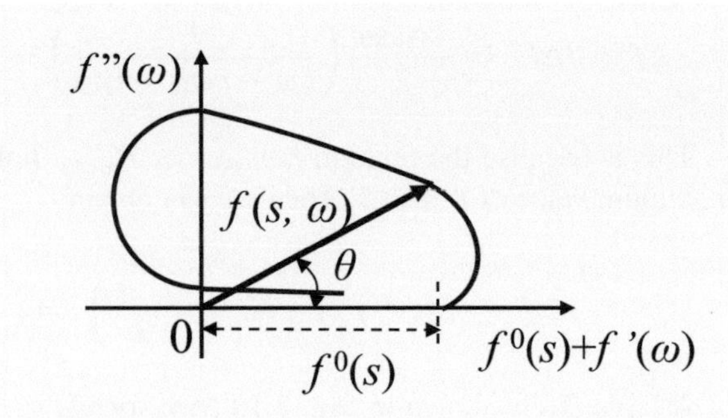

Fig. 3.11 Representation of $f(s,\omega)$ in complex plane

3.8 Some Applications of X-ray Resonant Scattering

3.8.1 Polarity Determination

The crystal structure factor F_h with only normal scattering factor is given by (2.35) in Chap. 2 as

$$F_h = \sum_j f_j^0(\boldsymbol{g}_h)\exp(i\boldsymbol{g}_h \cdot \boldsymbol{r}_j)$$

for reflection with $\boldsymbol{g}_h$. We can express F_h as

$$F_h = |F_h|\exp(i\varphi_h). \tag{3.86}$$

If we write the crystal structure factor for reflection with $-\boldsymbol{g}_h$ as $F_{\bar{h}}$, we have

$$F_{\bar{h}} = |F_h|\exp(-i\varphi_h). \tag{3.87}$$

There are relations between F_h and $F_{\bar{h}}$ given by

$$F_{\bar{h}} = F_h^*, \ |F_h| = \left|F_{\bar{h}}\right|. \tag{3.88}$$

Figure 3.12a shows the relations in (3.88) in the complex plane. The relations are called Friedel's law. The diffraction intensities in the directions of $\boldsymbol{g}_h$ and $-\boldsymbol{g}_h$ are equal.

When resonant scattering is involved, the crystal structure factor is given by

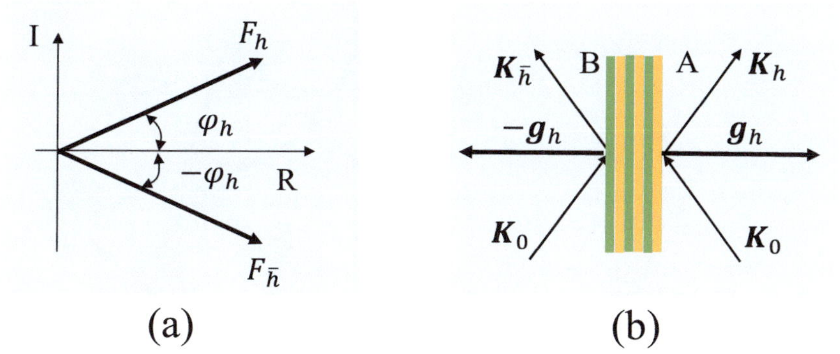

Fig. 3.12 **a** Vector diagram showing the relation between F_h and $F_{\bar{h}}$ in a complex plane. 'R' and 'I' represent the real and imaginary axes, respectively. **b** Schematic diagram of the incident and the diffracted wave vectors for a polar crystal

$$F_h = \sum_j f_j(\boldsymbol{g}_h, \omega)\exp(i\boldsymbol{g}_h \cdot \boldsymbol{r}_j) = \sum_j \left\{ f_j^0(\boldsymbol{g}_h) + f_j'(\omega) + if_j''(\omega) \right\} \exp(i\boldsymbol{g}_h \cdot \boldsymbol{r}_j), \tag{3.89}$$

as the atomic scattering factor is written by $f_j(\boldsymbol{g}_h, \omega) = f_j^0(\boldsymbol{g}_h) + f_j'(\omega) + if_j''(\omega)$. (3.88) does not always hold and the diffraction intensities in the directions of $\boldsymbol{g}_h$ and $-\boldsymbol{g}_h$ can be different depending on $\boldsymbol{g}_h$.

There is a type of crystal called a polar crystal, which does not have a center of symmetry and has different physical and chemical properties on the front and back sides. Typical examples are crystals with zinc blende structure and wurtzite structure (see Figs. 3.12b and 3.13). When X-rays are scattered only by the normal scattering factor, the diffraction intensities in the directions of $\boldsymbol{g}_h$ and $-\boldsymbol{g}_h$ show no difference for any pairs of $\boldsymbol{g}_h$ and $-\boldsymbol{g}_h$, the polarity cannot be determined. Because the difference in polarity does not appear in the diffraction intensity according to Friedel's law. If the anomalous scattering factor due to resonant scattering is added, Friedel's law can be broken, and the polarity can be determined by the X-ray diffraction intensities. Around the 1930s, Nishikawa and Matukawa [6] and Coster et al. [7] attempted to solve this interesting phenomenon and determined the polarity of ZnS. They focused on the resonant scattering of the K electron absorption edge (9.660 keV) of Zn and conducted experiments using characteristic X-rays close to the edge energy (for example, WLβ_1 (9.671 keV)).

We will explain polarity determination of GaP with a zincblende structure by resonant X-ray diffraction as an example (see Fig. 3.13). When we take coordinates of Ga (atomic number Z = 31) at (0,0,0), (0,1/2,1/2), (1/2,0,1/2), (1/2,1/2,0) in the

Fig. 3.13 Atomic arrangement of zincblende structure (GaP)

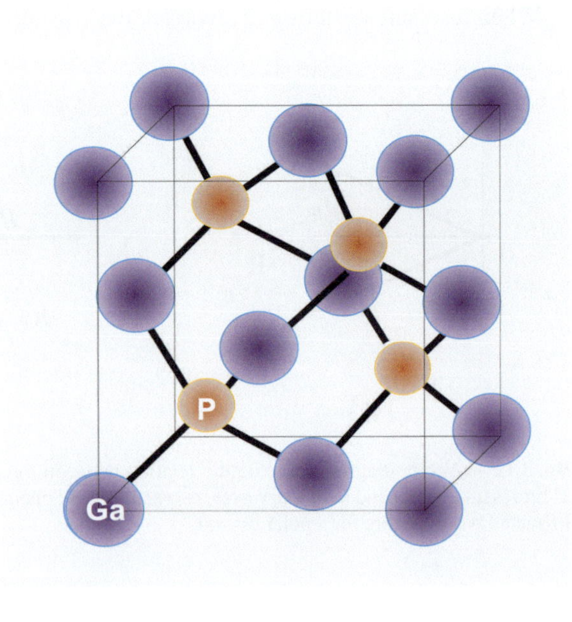

unit cell and those of P (Z = 15) at (1/4,1/4,1/4), (1/4,3/4,3/4), (3/4,1/4,3/4), (3/4,3/4,1/4), the crystal structure factor is given by

$$F = \left[1 + e^{i\pi(h+k)} + e^{i\pi(h+l)} + e^{i\pi(k+l)}\right]\left[f_{\mathrm{Ga}}(hkl) + f_{\mathrm{P}}(hkl)e^{\left(\frac{i\pi}{2}\right)(h+k+l)}\right]. \quad (3.90)$$

When $h + k + l = 4m \pm 1$ (m: integer), for example, $h = k = l = 5$,

$$F_{555} = 4\left(f_{\mathrm{Ga}}(555) - if_{\mathrm{P}}(555)\right), F_{\overline{555}} = 4\left(f_{\mathrm{Ga}}(555) + if_{\mathrm{P}}(555)\right) \quad (3.91)$$

by using $f_{\mathrm{Ga}}(555) = f_{\mathrm{Ga}}(\overline{555})$ and $f_{\mathrm{P}}(555) = f_{P}(\overline{555})$. Figure 3.14 shows the arrangement of atoms in a plane perpendicular to the (111) plane, with the [111] direction upward in the vertical direction and the $[\bar{1}01]$ direction facing left in the horizontal direction. Without resonant scattering, $f_{\mathrm{Ga}}(555)$ and $f_{\mathrm{P}}(555)$ are both real and $|F_{555}|^2 = |F_{\overline{555}}|^2$. No difference appears between 555 and $\overline{555}$ diffraction intensities. When we tune the X-ray energy near the Ga K-absorption edge, $f_{\mathrm{Ga}}(555)$ becomes complex due to resonant scattering.

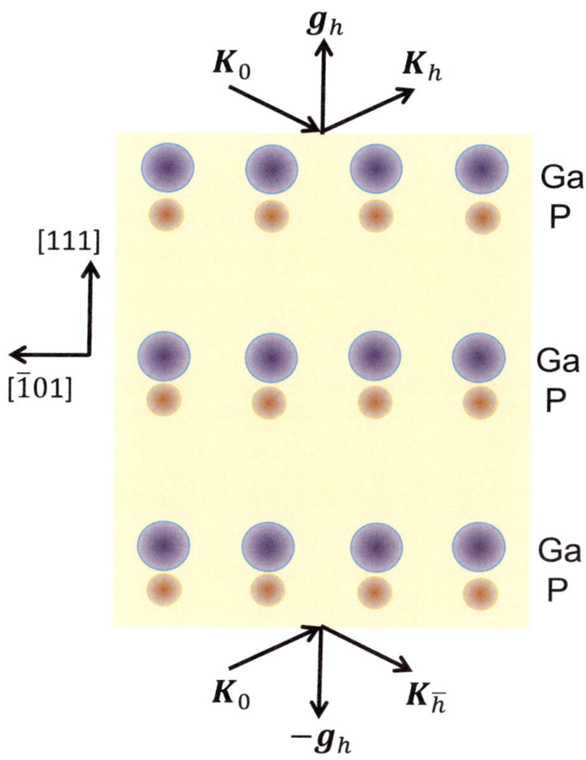

Fig. 3.14 Geometry of the incident and diffracted wave vectors for a GaP crystal. Ga atoms sit in the topmost layer in the direction of $\boldsymbol{g}_h$ and P atoms in the topmost layer in the direction of $-\boldsymbol{g}_h$

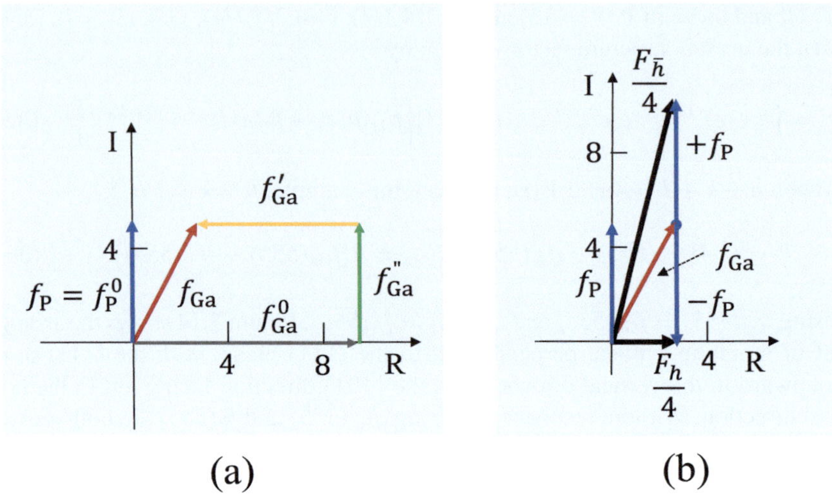

Fig. 3.15 Vector diagrams of atomic scattering factors of Ga and P **a**, and the crystal structure factors of F_{555} and $F_{\overline{555}}$ for a GaP crystal **b** in the complex plane. In (a), the values of $f_{Ga}^0(= 9.6)$ for Ga and $f_P^0(= 4.3)$ for P, and $f'_{Ga}(= -6.7)$ and $f''_{Ga}(= 4.3)$ are used. Anomalous scattering factors by P atoms are small enough to be neglected

Figure 3.15 shows the vector diagram of atomic scattering factors (a) and crystal structure factors (b) of GaP. The anomalous scattering factors are obtained for X-ray energy 10 eV higher than the Ga K absorption edge (10.368 keV) by assuming that the influence of resonant scattering of P atom (K absorption edge is 2.143 keV) should be almost negligible small. In this figure, it is clear that $|F_{555}|<|F_{\overline{555}}|$, that is,

$$|F_{555}| \neq |F_{\overline{555}}|. \tag{3.92}$$

When the anomalous scattering factor is included, the polarity of the crystal can be determined by using the fact that Friedel's law is violated.

Figure 3.16 shows the *hhh* and $\overline{hhh}$ diffraction spectra from a GaP single crystal measured by using a white X-ray diffractometer [8]. The Bragg angle is chosen so that the 555 reflection peaks move across the Ga K absorption edge (10,368 eV). Large changes are observed corresponding to large change in the anomalous scattering factor of Ga in the diffraction spectra showing $I_{555} < I_{\overline{555}}$, $I_{777} > I_{\overline{777}}$ and $I_{999} < I_{\overline{999}}$. A pair of diffractions of $\boldsymbol{g}_h$ and $-\boldsymbol{g}_h$ for which Friedel's law is violated are often called a Bijvoet pair.

Even a polar crystal, there are reflections that do not show any dependence on the polarity. For example, when $h+k+l = 4m\pm2$, the magnitude of the crystal structure factors are the same for *hkl* and $\overline{hkl}$ reflections, such as $F_{666} = F_{\overline{666}} = 4(f_{Ga} - f_P)$ and $F_{888} = F_{\overline{888}} = 4(f_{Ga} + f_P)$. The reflection intensities from the back and front planes are equal with each other. The diffraction intensities for these reflections are obtained in the experimental results shown in Fig. 3.16, where $I_{666} \approx I_{\overline{666}} < I_{888} \approx I_{\overline{888}}$. A

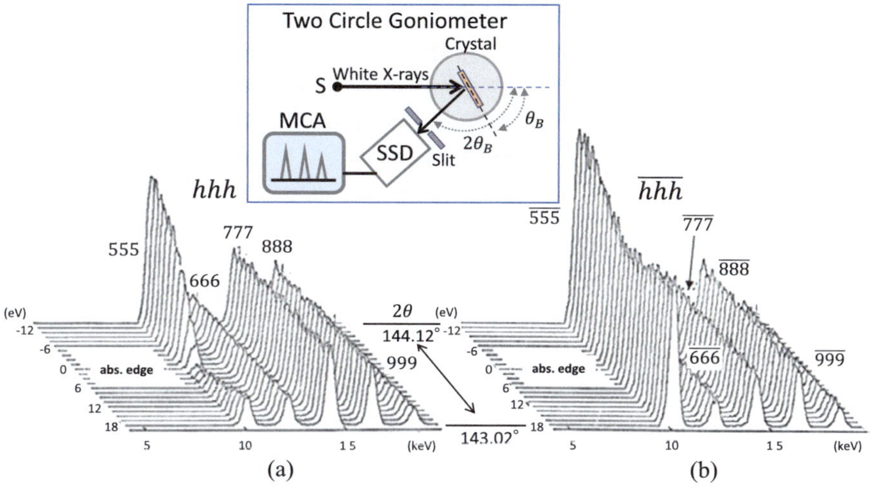

Fig. 3.16 White X-ray diffraction spectra from a GaP single crystal. **a** *hhh* diffraction and **b** $\overline{hhh}$ diffraction spectra [8]. The Bragg angle is chosen so that the 555 reflection peaks move across the Ga K absorption edge (10,368 eV) (2θ is chosen from 143.02° to 144.12°). Diffraction spectra below 9 keV have been deleted. The used diffractometer system is shown in the inset. In the measurement, white X-rays generated by bremsstrahlung radiation from the X-ray tube (S) is used. The angles θ (crystal)—2θ (detector) is set so that *hhh* reflections are obtained. SSD stands for Solid State Detector and MCA stands for Multi-channel Analyzer. Adapted from Fig. 2 in [8]

pair of diffractions whose intensities do not show the polarity dependence are called a Friedel pair. The 666-reflection and 888-reflection pairs from a GaP crystal are Friedel pairs.

3.8.2 Absolute Structure Determination

Polarity determination is to determine the arrangement of lattice planes non-symmetric in one-dimension. Determination of the absolute structure is to determine the spatial arrangement of atoms of a physically identified non-centrosymmetric crystal in three dimensions. Bijvoet et al. [9] used resonant scattering to determine the structure of stereoisomers, or chiral molecules. The property that a three-dimensional object cannot be superimposed with its mirror image is called "chirality," and molecules with this property are called "chiral molecules." The examples are molecules in which four atomic groups single-bonded to one carbon atom are different from each other. Two molecules shown in Fig. 3.17a, b are chiral with respect to the plane containing AD and the center (O). When the light passes through the isomers, the plane of polarization of light is rotated by the same angle but in opposite directions.

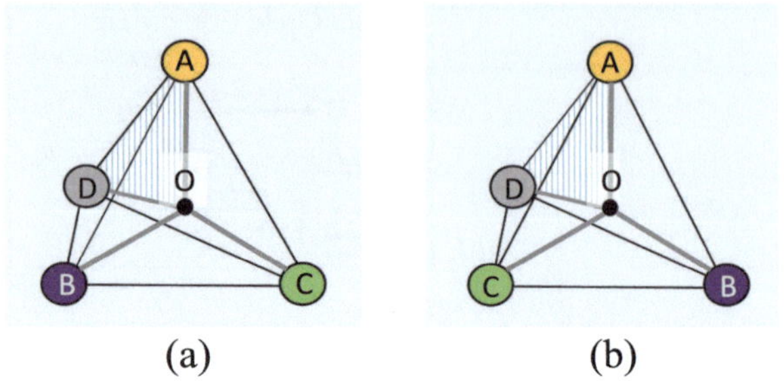

Fig. 3.17 A chiral pair of molecules. **a** and **b** are mirror images of each other with respect to the plane (with hatching) that includes AD and the center (O)

In the following, we will describe the crystal structure factors of a pair stereoisomers with reference to the explanation by Als-Nielsen and McMorrow [10]. The atoms are assumed to be heavier in the order of B, C, A and D in Fig. 3.17a, b. Figure 3.18 shows these atoms projected onto a plane perpendicular to OD. H, M and L are the projections of B, C and A in Fig. 3.17, respectively. The projected D overlaps with O. Atoms H, M and L rotate counter-clockwise in Fig. 3.18 (a), while in (b) they rotate clockwise. A molecule in (a) is called S molecule, and one in (b) is called R molecule. For simplicity, it is assumed that the scattering vector $\boldsymbol{g}$ is chosen perpendicular to OL and parallel to HM.

Figure 3.19a shows a vector diagram of the atomic scattering factors f_L, f_M and f_H of L, M and H atoms, respectively in the complex plane. Here, we ignore the anomalous scattering factor of L atom, and assume that both the real and the imaginary parts of the atomic scattering factor of H atom is larger than those of M atom. Figure 3.19b, c show the crystal structure factors F_S and F_R of S and R molecules in the complex plane. For L atoms, OL($\boldsymbol{r}_L$) and the scattering vector $\boldsymbol{g}$ are orthogonal, so $\exp(i\boldsymbol{g}\cdot\boldsymbol{r}_L) = 1$. For S molecule, there is a scattering vector $\boldsymbol{g}$ such that $\boldsymbol{g}\cdot\boldsymbol{r}_H < 0$ for H atom with OH($\boldsymbol{r}_H$) and the phase is $\exp(i\boldsymbol{g}\cdot\boldsymbol{r}_H) = \exp(-i\theta)$ with $0 < \theta < \pi$. Then, $\boldsymbol{g}\cdot\boldsymbol{r}_H > 0$ and the phase is $\exp(i\boldsymbol{g}\cdot\boldsymbol{r}_H) = \exp(i\theta)$ for H atom in R molecule. Since H and M are symmetric, the phase for M atom in S molecule is $\exp(i\theta)$, and that for R molecule is $\exp(-i\theta)$. The structure factors for S and R molecules F_S and F_R are given by,

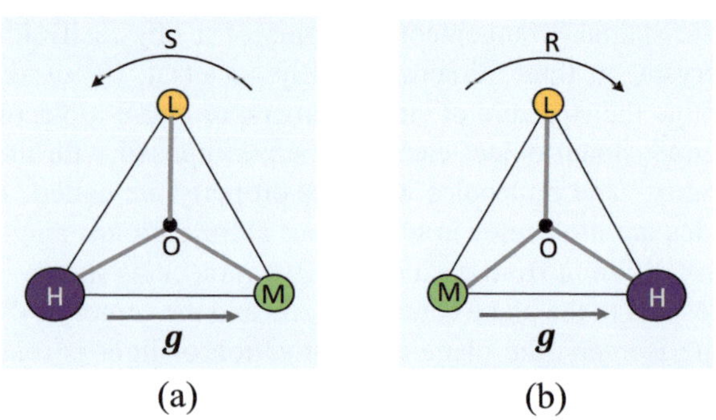

Fig. 3.18 Definition of chirality. **a** is S molecule and **b** is R molecule

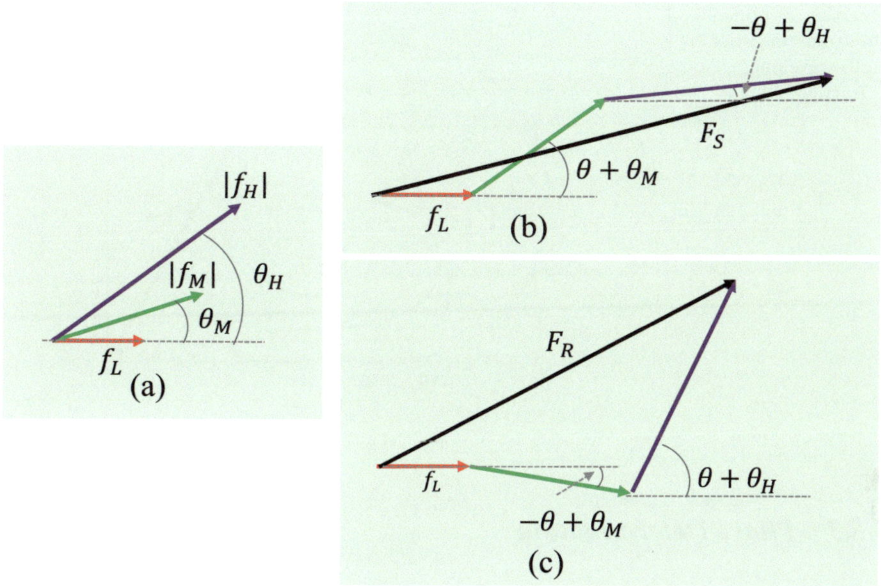

Fig. 3.19 Vector diagrams of atomic scattering factors for atoms H, M and L (**a**) and crystal structure factors for molecules of different chirality, labelled S (**b**) and R (**c**)

$$F_S = |f_L| + |f_M|\exp(i\theta_M)\exp(i\theta) + |f_H|\exp(i\theta_H)\exp(-i\theta), \quad (3.93)$$

$$F_R = |f_L| + |f_M|\exp(i\theta_M)\exp(-i\theta) + |f_H|\exp(i\theta_H)\exp(i\theta). \quad (3.94)$$

Using these equations, we obtain

$$|F_S|^2 - |F_R|^2$$

$$= 2\{|f_L||f_M|[\cos(\theta_M + \theta) - \cos(\theta_M - \theta)] + |f_L||f_H|[\cos(\theta_H - \theta) - \cos(\theta_H + \theta)]\}$$

$$-4|f_M||f_H|\sin(2\theta)\sin(\theta_M - \theta_H) \neq 0. \quad (3.95)$$

The diffraction intensities from a pair of chiral molecules (S and R) are different, and in turn, the chirality of the molecules can be determined from the measured intensities.

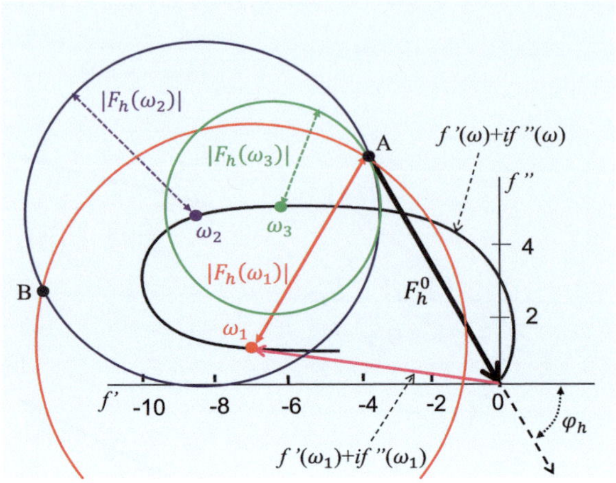

Fig. 3.20 A principle of phase determination by multi-wavelength anomalous diffraction method

3.8.3 Phase Determination

As already mentioned, the absolute value $|F_h|$ of the crystal structure factor F_h given by (3.86) can be determined from the measurement of the X-ray diffraction intensity, but its phase φ_h cannot be simply determined. Several ways to determine the phase φ_h have been proposed. One way is to utilize resonant scattering. An example of the principle is shown in Fig. 3.20. The diffraction intensities should be measured at three different X-ray energies. There must be one atom that causes resonant scattering within the unit cell. The position of that atom is taken at the origin of the unit cell. If we denote the crystal structure factor for $\boldsymbol{g}_h$ as $F_h(\omega)$ and that only due to normal scattering factors as F_h^0, we have

$$F_h(\omega) = F_h^0 + f'(\omega) + if''(\omega). \tag{3.96}$$

Suppose that $|F_h(\omega_j)|$ are obtained from measured diffraction intensities for X-ray energies at ω_1, ω_2 and ω_3. If $f'(\omega_j) + if''(\omega_j) (j = 1, 2 \text{ and } 3)$ are known in advance, we can draw a circle for each ω_j with its origin at $f'(\omega_j) + if''(\omega_j)$ and its radius $|F_h(\omega_j)|$ in the complex plane as shown in Fig. 3.20. We obtain three circles corresponding to the three energies. We have one or two intersections from two of three circles. The phase angle φ_h can be obtained from the intersection of these three circles. This method is proposed by Herzenberg and Lau [11]. This way of phase determination is called multi-wavelength anomalous diffraction (MAD) method.

Hosoya [12] proposed an approach using reflected intensities from the intersections of two Bijvoet pairs, as shown in Fig. 3.21. Two intersections of two circles A and B are obtained from a Bijvoet pair of diffraction intensities at X-ray energy of ω_1. By adding another circle of the Bijvoet pair at ω_2, the intersection point A of the three circles is obtained. Then the phase φ of F_h^0 is determined. The advantage of this approach is that it reduces measurement time and effort.

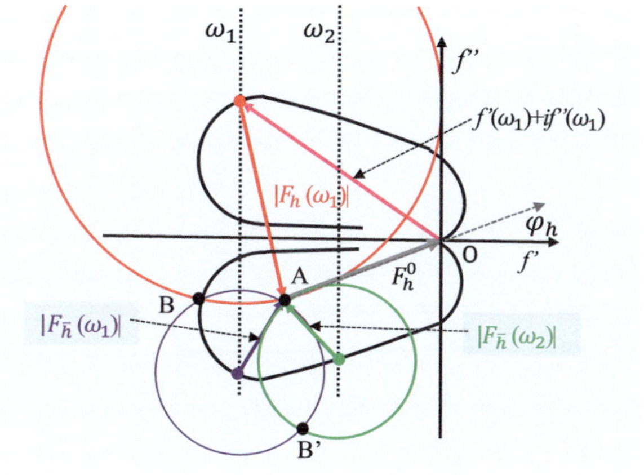

Fig. 3.21 Phase determination using diffraction intensities of Bijvoet pairs. As the propagating directions of X-rays are reversed on the front and back sides of the crystal, the sign of $f''(\omega)$ in the locus of $f''(\omega) + if''(\omega)$ is reversed for h and $\bar{h}$ reflection, in this vector diagram

By developing a four-axis X-ray diffractometer composed of white X-rays and a solid-state detector, Sakamaki et al. [13] measured the absorption coefficient near the K absorption edge of Zn in hemimorphite to investigate the anomalous scattering factor of Zn. The measured values are used to determine the phase using the MAD method and the absolute structure [14].

References

1. L. Parratt, C.F. Hempstead, Phys. Rev. **94**, 1598–1600 (1954)
2. T. Kawamura, T. Fukamachi, Jpn. J. App. Phys. **17–2**, 224–226 (1978)
3. D.T. Cromer, Acta Cryst. **18**, 17–23 (1965)
4. D.C. Koningsberger, R. Prins, *Edited, X-ray absorption: principles, applications and techniques of EXAFS* (John Wiley & Sons, SEXAFS and XANES, 1988)
5. M. Yoshizawa, S. Zhou, R. Negishi, T. Fukamachi, T. Kawamura, Acta Cryst. **A64**, 321–325 (2008)
6. S. Nishikawa, R. Matukawa, Proc. Imp. Acad. Jpn. **4**, 96 (1928)
7. D. Coster, K.S. Knol, J.A. Prins, Z. Phys. **63**, 345 (1930)
8. T. Fukamachi, S. Hosoya, Acta Cryst. **A31**, 215–220 (1975)
9. J.M. Bijvoet, A.F. Peerdeman, J.A. Van Bommed, Nature **168**, 271 (1951)
10. J. Als-Nielsen, D. McMorrow, *Elements of Modern X-ray Physics*. (John Wiley & Sons, Ltd, 2011)
11. A. Herzenberg, H.S.M. Lau, Acta Cryst. **22**, 24–28 (1967)
12. S. Hosoya, *Anomalous Scattering*. Copenhagen Munksgaard, 275–287 (1975)
13. T. Sakamaki, S. Hosoya, T. Fukamachi, J. Appl. Cryst. **13**, 433–437 (1980)
14. T. Sakamaki, S. Hosoya, T. Fukamachi, Acta Cryst. **A36**, 183–189 (1980)

Chapter 4
Optical Properties of X-rays

In this chapter, we describe the propagation of X-rays in materials using Maxwell's equations and derive the basic equation for X-ray diffraction. Using the equation, we consider the refractive index and dispersion surface of X-rays from the dispersion relation, and then explain total internal reflection, the penetration depth of X-rays, reflectance, and X-ray standing waves (XSW). We also explain an X-ray waveguide that amplifies X-rays by confining XSW in an artificially produced multilayer film after deriving the reflectance from a thin single-layer film on a substrate material.

4.1 Basic Equation of X-ray Diffraction

There is a relation between electric flux density $\boldsymbol{D}$ and electric field $\boldsymbol{E}$ in a material expressed by

$$\mathbf{D} = \varepsilon_0 \mathbf{E} + \mathbf{P}. \tag{4.1}$$

Here, ε_0 is the permittivity of a vacuum, and $\boldsymbol{P}$ is the electric polarization due to the electric field. As mentioned in Chap. 3, $\boldsymbol{P}$ can be rewritten as $\boldsymbol{P} = \varepsilon_0 \chi \boldsymbol{E}$ by using the electric susceptibility χ, the permittivity from (3.10) is written as

$$\varepsilon = \varepsilon_0(1 + \chi),$$

and $\boldsymbol{D}$ is written as

$$\boldsymbol{D} = \varepsilon \boldsymbol{E}. \tag{4.2}$$

For X-rays, χ is negative and its magnitude is small, around 10^{-4} to 10^{-6}. Between the magnetic field $\boldsymbol{H}$ and the magnetic flux density $\boldsymbol{B}$, there is a relation

T. Fukamachi and T. Kawamura, *Resonant Dynamical X-ray Diffraction*, Springer Series in Solid-State Sciences 208, https://doi.org/10.1007/978-981-96-8406-9_4

$$\boldsymbol{B} = \mu \boldsymbol{H} \cong \mu_0 \boldsymbol{H} \tag{4.3}$$

where μ is magnetic permeability. For X-rays, if we ignore a very weak magnetic effect, many materials, with some exceptions such as iron and nickel, can be considered magnetically the same as a vacuum, and $\mu \cong \mu_0$.

For oscillating electromagnetic waves such as X-rays, the effects of steady current and true charge can be ignored. Maxwell's equations for X-rays in a material are given as follows:

$$\text{rot}\boldsymbol{H} = \frac{\partial \boldsymbol{D}}{\partial t}, \tag{4.4}$$

$$\text{rot}\boldsymbol{E} = -\mu_0 \frac{\partial \boldsymbol{H}}{\partial t}, \tag{4.5}$$

$$\text{div}\, \boldsymbol{D} = 0, \tag{4.6}$$

$$\text{div}\, \boldsymbol{H} = 0. \tag{4.7}$$

Taking curl of vectors on both sides of (4.5), the right-hand side becomes

$$-\mu_0 \frac{\partial}{\partial t}\, \text{rot}\, \boldsymbol{H} = -\mu_0 \frac{\partial^2}{\partial t^2} \boldsymbol{D} = \frac{K_0^2 \boldsymbol{D}}{\varepsilon_0}$$

with the aid of (4.4). Here, K_0 is the wavenumber in a vacuum and $K_0 = \omega/c$. The left-hand side of (4.5) becomes

$$\text{rot}(\text{rot}\, \boldsymbol{E}) = \frac{-1}{\varepsilon_0}\left[\nabla^2 \boldsymbol{D} + \text{rot}(\text{rot}\chi \boldsymbol{D})\right] \tag{4.8}$$

using the relation rot(rot) = grad(div) − ∇^2. Then (4.5) is rewritten as

$$\nabla^2 \boldsymbol{D} + \text{rot}\{\text{rot}(\chi \boldsymbol{D})\} + K_0^2 \boldsymbol{D} = 0, \tag{4.9}$$

by using the relation $\boldsymbol{E} = \varepsilon_0^{-1}(1+\chi)^{-1}\boldsymbol{D} \approx \varepsilon_0^{-1}(1-\chi)\boldsymbol{D}$. (4.9) is the basic equation for X-ray diffraction called the propagation equation of an electromagnetic wave in a material.

4.1.1 Propagation in Isotropic Medium

In an isotropic medium, the dielectric constant ε does not depend on the direction, and we can write the permittivity $\chi = 1 - \varepsilon_0/\varepsilon$. The second term on the left side of

(4.9) can be modified as

$$\mathrm{rot}\{\mathrm{rot}(\chi \boldsymbol{D})\} = \left(1 - \frac{\varepsilon_0}{\varepsilon}\right)\mathrm{rot}\{\mathrm{rot}(\boldsymbol{D})\} = -\nabla^2 \boldsymbol{D} + \frac{\varepsilon_0}{\varepsilon}\nabla^2 \boldsymbol{D}.$$

Then, (4.9) is written as

$$\nabla^2 \boldsymbol{D} + \frac{\varepsilon}{\varepsilon_0} K_0^2 \boldsymbol{D} = 0. \tag{4.10}$$

The plane wave as a special solution to (4.10) is (excluding the temporal variation term) given by

$$\boldsymbol{D}(\boldsymbol{r}) = \boldsymbol{D}\exp(-i\boldsymbol{k}_0 \cdot \boldsymbol{r}), \tag{4.11}$$

with $\boldsymbol{k}_0$ being the wave vector in the medium. By inserting (4.11) into (4.10), we have

$$\left(\frac{\varepsilon}{\varepsilon_0} K_0^2 - \boldsymbol{k}_0^2\right)\boldsymbol{D} = 0. \tag{4.12}$$

Usually $\boldsymbol{D} \neq 0$, and we obtain

$$\boldsymbol{k}_0^2 = \frac{\varepsilon}{\varepsilon_0} K_0^2 = \frac{\varepsilon}{\varepsilon_0}\left(\frac{\omega}{c}\right)^2. \tag{4.13}$$

This expresses the relation between X-ray energy ($\hbar\omega$) and wavenumber ($|\boldsymbol{k}_0|$ or K_0), which is called the dispersion relation.

In general, for a vector $\boldsymbol{A}(\boldsymbol{r}) = \boldsymbol{B}\exp(-i\boldsymbol{k} \cdot \boldsymbol{r})$ ($\boldsymbol{B}$ is a constant vector), we have relations

$$\mathrm{rot}\boldsymbol{A} = -i\boldsymbol{k} \times \boldsymbol{A} \tag{4.14}$$

and

$$\mathrm{div}\boldsymbol{A} = -i\boldsymbol{k} \cdot \boldsymbol{A}. \tag{4.15}$$

The left-hand side of (4.4) is written by

$$\mathrm{rot}\,\boldsymbol{H} = -i\boldsymbol{k}_0 \times \boldsymbol{H}$$

and the right-hand side is written by

$$\frac{\partial \boldsymbol{D}}{\partial t} = i\omega\varepsilon \boldsymbol{E}.$$

Equation (4.5) can be written as

$$-i\boldsymbol{k}_0 \times \boldsymbol{H} = i\omega\varepsilon \boldsymbol{E}.$$

We can have a relation

$$\boldsymbol{E} = \frac{-1}{\varepsilon\omega}\boldsymbol{k}_0 \times \boldsymbol{H}. \tag{4.16}$$

Equation (4.16) shows that the electric field $\boldsymbol{E}$ and magnetic field $\boldsymbol{H}$ are orthogonal to each other, and that the electromagnetic wave is a transverse wave (see Fig. 4.1). By using (4.13) and $c = 1/\sqrt{\varepsilon_0\mu_0}$ in (4.16), we have

$$|\boldsymbol{E}| = \sqrt{\frac{\mu}{\varepsilon}}|\boldsymbol{H}| \tag{4.17}$$

By inserting (3.11) into (4.6), we obtain

$$\mathrm{div}\boldsymbol{D}(\boldsymbol{r}) = -i\boldsymbol{k}_0 \cdot \boldsymbol{D}(\boldsymbol{r}) = 0$$

and

$$\boldsymbol{k}_0 \cdot \boldsymbol{D}(\boldsymbol{r}) = 0. \tag{4.18}$$

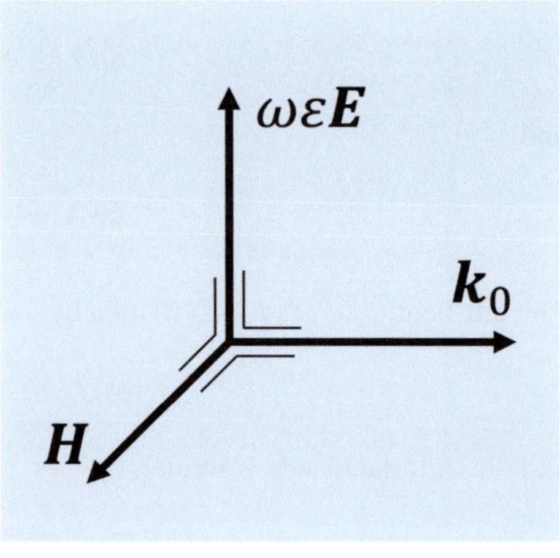

Fig. 4.1 Orthogonal relations between electric field $\boldsymbol{E}$, magnetic field $\boldsymbol{H}$ and wavevector $\boldsymbol{k}_0$

4.2 Refractive Index

The refractive index of X-rays in a material has been described in Chap. 3. If the wave number in the material is k_0, the refractive index n_0 is given by the dispersion relation (4.13) with (3.10) as

$$n_0 = \frac{k_0}{K_0} = \sqrt{\frac{\varepsilon}{\varepsilon_0}} = (1+\chi)^{\frac{1}{2}}.$$

χ is a complex number as shown in Chap. 3, however, for the moment we will ignore absorption and write the real part of χ as χ_r. Then we obtain

$$n_0 = 1 + \frac{1}{2}\chi_r = 1 - \delta = 1 - \frac{r_e\lambda^2}{2\pi}\rho = 1 - \frac{r_e\lambda^2}{2\pi}\frac{N_e}{V_c}. \quad (4.19)$$

Here, ρ is the electron density N_e/V_c, N_e the number of electrons in the unit cell, and V_C is the unit cell volume. Since χ_r is negative and its magnitude is approximately 10^{-4} to 10^{-6} for X-rays, n_0 is slightly smaller than 1. Figure 4.2 shows the geometry between the incident wave and refracted wave in the case of σ polarization (the electric field is normal to the incident plane).

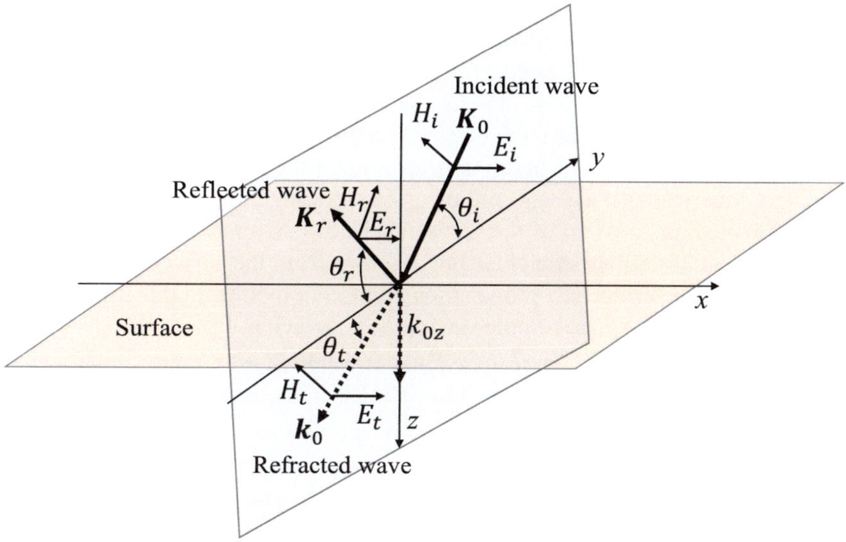

Fig. 4.2 The reflected and refracted waves for σ-polarized X-rays

4.2.1 Boundary Conditions at the Interface

The boundary conditions at the interface between two media are the same as those in electromagnetism.

(1) The tangential component of electric field $\boldsymbol{E}(\boldsymbol{r})$
(2) The tangential component of magnetic field $\boldsymbol{H}(\boldsymbol{r})$
(3) The normal component of electric flux density $\boldsymbol{D}(\boldsymbol{r})$
(4) The normal component of magnetic flux density $\boldsymbol{B}(\boldsymbol{r})$

are continuous across the interface. Two conditions out of these four are enough to determine the field of X-ray wave.

4.3 Dispersion Surface and Total Internal Reflection

We explain the total internal reflection of X-rays using a dispersion surface.

4.3.1 Dispersion Surface

Equation (4.13) describes the dispersion relation that the magnitude of the wave vector $k_0 = |\boldsymbol{k}_0|$ does not depend on the direction and is proportional to angular frequency (energy). Figure 4.3 is the cross section of a sphere of radius k_0 as well as a sphere of radius K_0, including the origin O and perpendicular to the surface of the medium. The origin O is taken at the starting point of the wave vector of the incident X-ray. The spherical surface with radius of k_0 represents the dispersion surface of X-ray in the medium and that of K_0 represents the dispersion surface in the vacuum. The intersection of the perpendicular line v drawn from the surface of the medium with the dispersion surface is a propagation point (a tie-point). In this figure, the tie-point of the dispersion surface in the vacuum is A, which is determined by the wave vector of the incident wave $\boldsymbol{K}_0 = \overrightarrow{\mathrm{AO}}$. The corresponding tie-point in the medium is the intersection point A′ of line v with the dispersion surface of radius k_0, then the incident wave vector in the medium is given by $\boldsymbol{k}_0 = \overrightarrow{\mathrm{A'O}}$. The tie-point of the reflected wave is A″, and the reflected wave vector is given by $\boldsymbol{K}_r = \overrightarrow{\mathrm{A''O}}$.

In Fig. 4.3, the incident glancing angle θ_i is represented by $\angle\mathrm{AOB} = \theta_i$ and the reflection angle θ_r by $\angle\mathrm{A''OB} = \theta_r$. Using the boundary condition (1) of 4.2.1, $K_0\cos\theta_i = K_r\cos\theta_r$ and $K_0 = K_r$, we obtain the law of reflection

$$\theta_i = \theta_r. \tag{4.20}$$

And using the boundary condition (3)

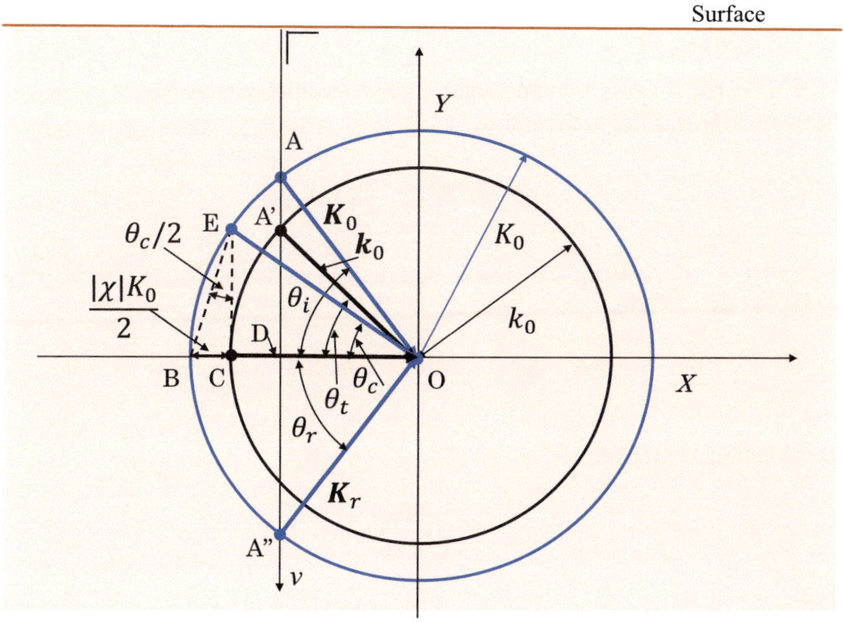

Fig. 4.3 Dispersion surface in a vacuum is shown by a blue circle and that in the medium by a black circle

$$K_0 \cos\theta_i = k_0 \cos\theta_t, \tag{4.21}$$

we obtain Snell's law. Since $K_0 > k_0$, θ_t becomes pure imaginary and there is no refracted wave inside the medium corresponding to the incident wave when the incident glancing angle θ_i is less than $\theta_c = \cos^{-1}(k_0/K_0)$. The incident wave cannot enter the medium and total internal reflection occurs. θ_c is called the critical angle.

4.3.2 *Total Internal Reflection*

Let us consider total internal reflection, which occurs as a result of using a dispersion surface with a refractive index (n_0) slightly less than one. In Fig. 4.3, we take a point C at the distance k_0 from the origin (O), i.e., $\overline{\mathrm{CO}} = k_0$ parallel to the surface. We take a point E at the intersection of the tangent at point C (this tangent is perpendicular to the surface) and the dispersion surface in the vacuum. Then the angle ∠COE is the critical angle θ_c, where the refracted angle θ_t is zero. X-rays are totally reflected when the glancing angle θ_i is less than the critical angle ($\theta_i \leq \theta_c$) as mentioned above. Since $K_0 - k_0$ ($\overline{\mathrm{BC}}$) is very small compared to K_0 on the order from 10^{-4} to 10^{-6}, θ_c is also very small. As can be seen from Fig. 4.3, $K_0^2 = k_0^2 + K_0^2 \sin^2\theta_c$, and $\sin\theta_c \approx \theta_c$, using $(1 - \theta_c^2)^{1/2} = k_0/K_0$, we obtain

$$\theta_c = (1 - n_0^2)^{1/2} = \sqrt{-\chi_r}. \tag{4.22}$$

Here, χ_r is negative, which can be determined if θ_c can be measured.

The line segment $\overline{\mathrm{BC}}$ is expressed as

$$\overline{\mathrm{BC}} = \frac{k_0\theta_c^2}{2} = \frac{k_0|\chi_r|}{2} \tag{4.23}$$

using $\angle\mathrm{BEC} = \theta_c/2$. When the glancing angle is close to the critical angle, we can write $\overline{\mathrm{BD}} = \overline{\mathrm{BC}} + \overline{\mathrm{CD}}$ as

$$\frac{K_0\theta_i^2}{2} = \frac{k_0\theta_c^2}{2} + \frac{k_0\theta_t^2}{2}. \tag{4.24}$$

If we define a parameter W as

$$W = \frac{\sin\theta_i}{\sin\theta_c}, \tag{4.25}$$

and solve (4.24) for θ_t using the relations $K_0 = k_0(1 - \chi_r/2)$ and $K_0 \approx k_0$, we obtain

$$\theta_t = \theta_c(W^2 - 1)^{1/2}. \tag{4.26}$$

When $W < 1$, θ_t is pure imaginary, which means that the incident wave cannot enter the medium. For a typical value of $|\chi_r|$ of 10^{-4}, the critical angle θ_c is 0.57° (0.01 rad).

4.3.3 Reflectance

As shown in Fig. 4.2, we consider the reflectance of σ-polarized X-rays as an example. The boundary conditions from (1) and (2) of 4.2.1 at the interface between a vacuum and a medium give

$$E_i + E_r = E_t \tag{4.27}$$

and

$$H_i\sin\theta_i - H_r\sin\theta_r = H_t\sin\theta_t. \tag{4.28}$$

From (4.17), $H = \sqrt{\varepsilon/\mu}E$ and the refractive index $n_0 = \sqrt{\varepsilon/\varepsilon_0}$, (4.28) can be written as

$$E_i\sin\theta_i - E_r\sin\theta_r = E_t n_0 \sin\theta_t. \tag{4.29}$$

In the case of X-rays, the refractive index is nearly equal to 1 for most materials. From (4.28) and (4.29), we obtain the reflection coefficient r and the transmission coefficient t as

$$r = \frac{E_r}{E_i} = \frac{\sin\theta_i - n_0 \sin\theta_t}{\sin\theta_i + n_0 \sin\theta_t} \approx \frac{W - \left(W^2 - 1\right)^{\frac{1}{2}}}{W + \left(W^2 - 1\right)^{\frac{1}{2}}} \left(= \frac{W - i\left(1 - W^2\right)^{\frac{1}{2}}}{W + i\left(1 - W^2\right)^{\frac{1}{2}}} \right), \quad (4.30)$$

$$t = \frac{E_t}{E_i} = 1 + \frac{E_r}{E_I} = \frac{2n_0 \sin\Theta_i}{\sin\theta_i + n \sin\theta_t} \approx \frac{2W}{W + \left(W^2 - 1\right)^{\frac{1}{2}}} \left(= \frac{2W}{W + i\left(1 - W^2\right)^{\frac{1}{2}}} \right). \quad (4.31)$$

The expression in the last parentheses apply when $W \leq 1$. The reflected intensity $I_r = |r|^2$ and the transmitted intensity $I_t = |t|^2$ are given by

$$I_r = \left| \frac{W - \left(W^2 - 1\right)^{\frac{1}{2}}}{W + \left(W^2 - 1\right)^{\frac{1}{2}}} \right|^2 (= 1), I_t = \frac{4W^2}{\left| W + \left(W^2 - 1\right)^{\frac{1}{2}} \right|^2} (= 4W^2). \quad (4.32)$$

The value in the parentheses in both equations apply for $W \leq 1$. The solid black lines in Fig. 4.4 show I_r (a) and I_t (b) as a function of W. I_r is 1 for $W \leq 1$ (total internal reflection) and decreases rapidly when W becomes larger than 1. On the other hand, I_t shows maximum value of 4 at $W = 1$.

For the glancing angle much larger than the critical angle $\theta_i \gg \theta_c$, by multiplying both the denominator and the numerator of (4.30) by $(\sin\theta_i + n_0 \sin\theta_t)$, the reflectance

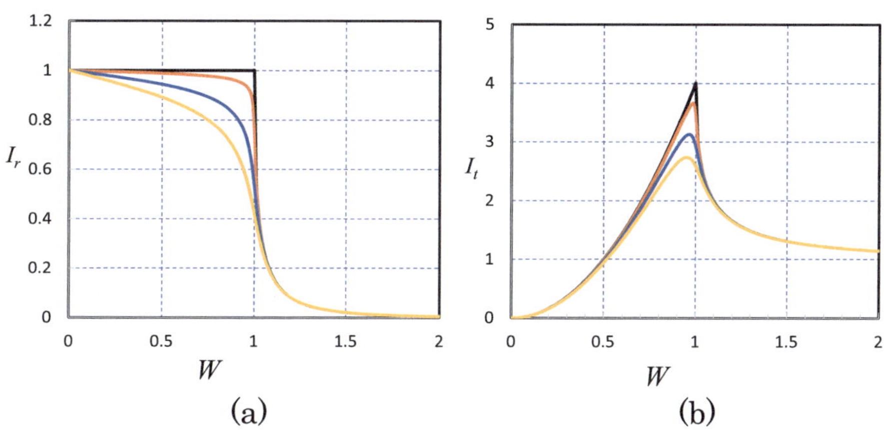

Fig. 4.4 Reflectance I_r **a** and transmittances **b**. Black, brown, blue and yellow lines show those for $k_a = 0$ (no absorption), $k_a = 0.01$, $k_a = 0.05$ and $k_a = 0.1$, respectively

r can be written as

$$r = \frac{\sin^2\theta_i - n_0^2\sin^2\theta_t}{(\sin\theta_i + n_0\sin\theta_t)^2} \approx \frac{\theta_c^2}{(2\sin\theta_i)^2} = (2W)^{-2}. \quad (4.33)$$

In the denominator of (4.33), the relation in (4.24) was used.

4.3.4 Penetration Depth (for Non-absorption Medium)

We will examine the refracted waves when the total internal reflection occurs. The surface normal component (z-direction) k_{0z} of $\boldsymbol{k}_0$ is written as

$$k_{0z} = k_0\sin\theta_t \approx k_0\theta_t. \quad (4.34)$$

From (4.26), the refracted angle θ_t for $W < 1$ is expressed as

$$\theta_t = -i\theta_c(1 - W^2)^{1/2}. \quad (4.35)$$

The sign on the right side was chosen so that the magnitude of the z-component E_{tz} of electric field decays into the medium (z-direction). E_{tz} can be written as

$$E_{tz} = E_t\exp(-ik_{0z}z) = E_t\exp\left[-k_0\theta_c\left(1 - W^2\right)^{1/2}z\right]. \quad (4.36)$$

The depth l at which $|E_{tz}/E_t|^2 = \exp(-1)$ is given by

$$l = \frac{1}{2k_0\theta_c\left(1 - W^2\right)^{\frac{1}{2}}} = \frac{\lambda}{4\pi\theta_c\left(1 - W^2\right)^{\frac{1}{2}}}.(W < 1) \quad (4.37)$$

The distance l is called the penetration depth of the X-ray under the total internal reflection condition. This l is very small, on the order of several nm. For example, when $|\chi_r|$ is 10^{-4}, λ is 0.1 nm and when $W = 0.1$, we get $l = 0.8$ nm. The X-rays penetrate only slightly into the medium and are localized near the surface. A similar effect of localization of X-rays on the surface is also caused in the Bragg case of dynamical X-ray diffraction from a perfect no-absorbing crystal, as described later. The wave whose intensity decays exponentially when it goes into the medium is called an evanescent wave. When the glancing angle is equal to the critical angle ($\theta_i = \theta_c$), $W = 1$ from (4.24). Then from (4.29), we have

$$\frac{E_r}{E_i} = 1, \frac{E_t}{E_i} = 2. \quad (4.38)$$

E_i and E_r are in phase and enhance with each other. The amplitude of the evanescent wave $|E_t|$ is twice that of the incident wave. When $\theta_i = 0(W = 0)$, $E_r = -E_i$ and $E_t = 0$, E_i and E_r are in antiphase and cancel each other out, no evanescent wave is generated.

4.4 Complex Refractive Index and Absorption

As mentioned in Chap. 3, the refractive index n_0 is generally a complex number and written as

$$n_0 = \frac{k_0}{K_0} = 1 - \delta - i\beta = \frac{1}{(1-\chi)^{\frac{1}{2}}} \approx 1 + \frac{1}{2}\chi_r + i\frac{1}{2}\chi_i. \tag{4.39}$$

Here, χ_r and χ_i are the real and imaginary parts of the electric susceptibility χ, respectively. The wave number K_0 in a vacuum is a real number, but the wave number k_0 in a medium is a complex number. δ and β are given by

$$\delta = -\frac{1}{2}\chi_r, \beta = -\frac{1}{2}\chi_i. \tag{4.40}$$

The electric field component of X-rays in a medium is

$$E = E_0 \exp\{i(\omega t - k_0 z)\} = E_0 \exp\{i(\omega t - K_0 \text{z})\}\exp(i\delta K_0 z)\exp(-\beta K_0 z). \tag{4.41}$$

In the right side, the first term of the exponential function represents the propagation in the vacuum, and the second and third terms represent phase shift and damping in the medium, respectively. Since the intensity of X-rays is proportional to $|E|^2$, the intensity of X-rays in the medium is

$$I(z) = I_0 \exp(-2\beta K_0 z) = I_0 \exp(-\mu_a z). \tag{4.42}$$

Here, $I_0 = |E_0|^2$ and μ_a is the linear absorption coefficient given by

$$\mu_a = 2\beta K_0 = -\chi_i K_0. \tag{4.43}$$

The wave vector $\boldsymbol{k}_0$ in the medium is also a complex quantity and should be written as

$$\boldsymbol{k}_0 = \boldsymbol{k}_{0r} + i\boldsymbol{k}_{0i} \tag{4.44}$$

with $\boldsymbol{k}_{0r}$ and $\boldsymbol{k}_{0i}$ being real vectors. Then, we obtain

$$\boldsymbol{k}_0^2 = \boldsymbol{k}_{0r}^2 - \boldsymbol{k}_{0i}^2 + 2i\boldsymbol{k}_{0r} \cdot \boldsymbol{k}_{0i}. \tag{4.45}$$

Using the dispersion relation (4.13), (4.45) becomes

$$\boldsymbol{k}_0^2 = K_0^2(1+\chi) = K_0^2(1+\chi_r+i\chi_i). \tag{4.46}$$

As $|\boldsymbol{k}_{0r}|^2 \gg |\boldsymbol{k}_{0i}|^2$, the real parts of (4.45) and (4.46) are

$$|\boldsymbol{k}_{0r}|^2 \approx K_0^2(1+\chi_r). \tag{4.47}$$

For the imaginary parts, $2\boldsymbol{k}_{0r}{\cdot}\boldsymbol{k}_{0i} = K_0^2\chi_{0i}$, which can be rewritten as

$$\frac{\boldsymbol{k}_{0r}}{K_0}\cdot\boldsymbol{k}_{0i} \approx \boldsymbol{s}_t\cdot\boldsymbol{k}_{0i} = \frac{K_0}{2}\chi_i. \tag{4.48}$$

Here, $\boldsymbol{s}_t$ is the unit vector along the refracted beam direction ($\boldsymbol{s}_t = \boldsymbol{k}_{0r}/|\boldsymbol{k}_{0r}|$). The left side of this equation gives a plane perpendicular to $\boldsymbol{s}_t$ in $\boldsymbol{k}_{0i}$ space. The wave vectors of waves inside and outside the medium must be the same in the direction parallel to the surface. This means that the parallel component of $\boldsymbol{k}_{0i}$ of the wave vector in the medium must be the same with that of $\boldsymbol{K}_0$ outside the medium (vacuum). Then, $\boldsymbol{k}_{0i}$ is given by $\overline{\mathrm{OB}} = k_{0iz}$ in Fig. 4.5, and is given by

$$k_{0iz} = \frac{K_0\chi_i}{2\sin\theta_t} = \frac{-\mu_a}{2\sin\theta_t}. \tag{4.49}$$

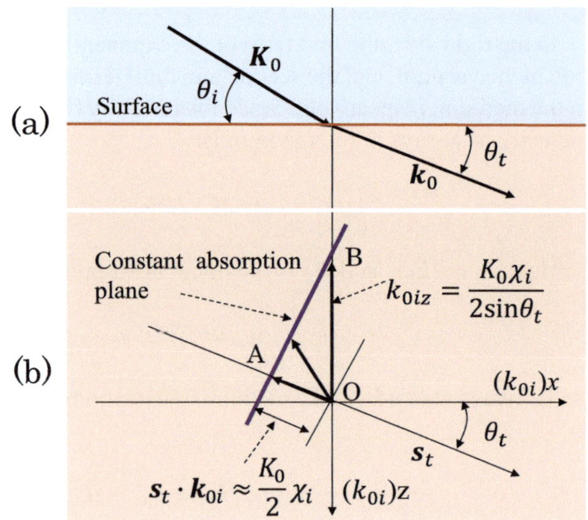

Fig. 4.5 Schematic diagram of the incident and refracted waves (**a**) and constant absorption plane in $\boldsymbol{k}_{0i}$-space (**b**)

4.4.1 Complex Dispersion Surface Under Total Internal Reflection Condition

We consider the dispersion surface and reflectance under the total internal reflection. We define the critical angle θ_c as

$$\theta_c = \sqrt{|\chi_r|}. \tag{4.50}$$

by ignoring absorption effect and using the relation $\cos\theta_c = n_0$ and (4.39). The real part of the dispersion surface is shown in Fig. 4.6a, b. When $\theta_i > \theta_c$ it is a sphere with radius k_0. When $\theta_i \leq \theta_c$, the dispersion surface deforms from a sphere as shown in Fig. 4.6b, c.

When absorption effect is ignored, θ_t becomes pure imaginary for $\theta_i \leq \theta_c$. When absorption effect is included, it becomes complex, we write $\theta_t = \theta_{tr} + i\theta_{ti}$ with θ_{tr} and θ_{ti} being the real and imaginary parts of θ_t, respectively. Since $|\theta_{tr}| \ll 1$, $|\theta_{ti}| \ll 1$, $|\chi_r| \ll 1$ and $|\chi_i| \ll 1$, we can put as

$$n_0 \sin\theta_t \approx \theta_{tr} + i\theta_{ti}. \tag{4.51}$$

Here we use the relation

$$n_0 \sin\theta_t = \left(n_0^2 - n_0^2 \cos^2\theta_t\right)^{1/2} = \left(n_0^2 - \cos^2\theta_i\right)^{1/2} \tag{4.52}$$

and $n_0 \cos\theta_t = \cos\theta_i$ obtained from Snell's law. Using $\cos^2\theta_i = 1 - \sin^2\theta_i$, (4.22), (4.39) and (4.50), we obtain

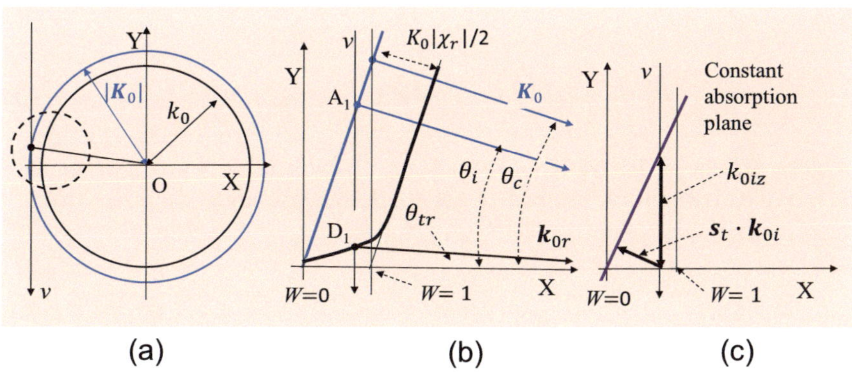

Fig. 4.6 Dispersion surfaces under the total internal reflection condition when the absorption effect is included (**a**). (**b** and **c**) are enlarged views inside the dashed circle in (**a**). (**b**) shows the real part and (**c**) shows the imaginary part

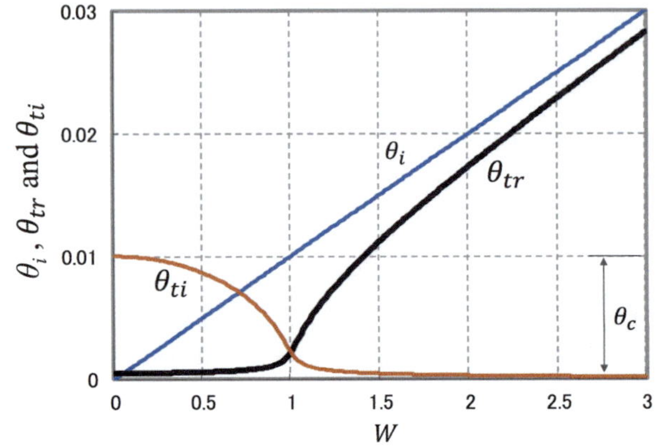

Fig. 4.7 θ_{tr} and θ_{ti} (in radian) as a function of W. The values of $\chi_r = 10^{-4}$, $k_a = 0.1$ and $= \theta_c 0.57°$ (0.01 rad) are used

$$\theta_{tr} + i\theta_{ti} = (n_0^2 - 1 + \sin^2\theta_i)^{1/2} \approx \left[W^2\theta_c^2 - |\chi_r| - i|\chi_i|\right]^{\frac{1}{2}} = \theta_c\left(W^2 - 1 + ik_a\right)^{\frac{1}{2}}. \tag{4.53}$$

Here, (4.50) is used and the parameter k_a is defined as

$$k_a = \left|\frac{\chi_i}{\chi_r}\right|. \tag{4.54}$$

θ_{tr} and θ_{ti} in (4.53) are written by

$$\theta_{tr}^2 = \frac{\theta_C^2}{2}[\{(W^2 - 1)^2 + k_a^2\}^{1/2} + (W^2 - 1)] \tag{4.55}$$

and

$$\theta_{ti}^2 = \frac{\theta_C^2}{2}[\{(W^2 - 1)^2 + k_a^2\}^{1/2} - (W^2 - 1)]. \tag{4.56}$$

Figure 4.7 shows the variations of θ_{tr} and θ_{ti} with respect to W. When $W < 1(\theta_i < \theta_c)$, θ_{tr} is significantly smaller than θ_c. θ_{ti} increases as W decreases from 1 to 0, and becomes $(\theta_c/\sqrt{2})[(1 + k_a^2)^{1/2} + 1]^{1/2}$ at $W = 0$.

4.4.2 *Penetration Depth (Including Absorption Effect)*

We consider the penetration depth in a medium when the absorption effect is included. Let x be the coordinate parallel to the surface and z be that perpendicular to the surface. The spatial dependence of the refracted wave is written by

$$\exp(-i\boldsymbol{k}_0 \cdot \boldsymbol{r}) = \exp[-i(k_0 x\cos\theta_t + k_0 z\sin\theta_t)]$$

$$= \exp(-K_0\theta_{ti}z)\exp[-iK_0(x\cos\theta_i + z\theta_{tr})]. \quad (4.57)$$

In deriving the above equation, the relations $K_0\cos\theta_i = k_0\cos\theta_t$, $n_0 = k_0/K_0$ and $n_0\sin\theta_t = \theta_{tr} + i\theta_{ti}$ are used. The sign of θ_{ti} is set to be negative so that the wave is attenuated as it propagates into the medium. A plane wave in a vacuum forms a plane with constant phase and amplitude, but the amplitude of the refracted wave in the medium given by (4.57) is constant in the z-plane (a plane perpendicular to z-axis). Such a plane wave is called a nonuniform plane wave (see Fig. 4.8). The penetration depth, at which the exponent of the damping term $|\exp(-K_0\theta_{ti}z)|^2$ is equal to -1, is given by

$$l = \frac{1}{2K_0|\theta_{ti}|} = \frac{\lambda}{4\pi|\theta_{ti}|}. \quad (4.58)$$

Figure 4.9 shows the calculated results of penetration depth (l) as a function of W. For $\theta_i < \theta_c$ the penetration depth l is significantly smaller than that predicted by the average absorption coefficient (in this case, it is several nm). The value of l increases rapidly as θ_i approaches θ_c from the low angle side. For $\theta_i \gg \theta_c$, when W increases, l converges to the depth of the wave with the mean absorption coefficient l_m given by

$$l_m = \frac{\sin\theta_i}{\mu_a}. \quad (4.59)$$

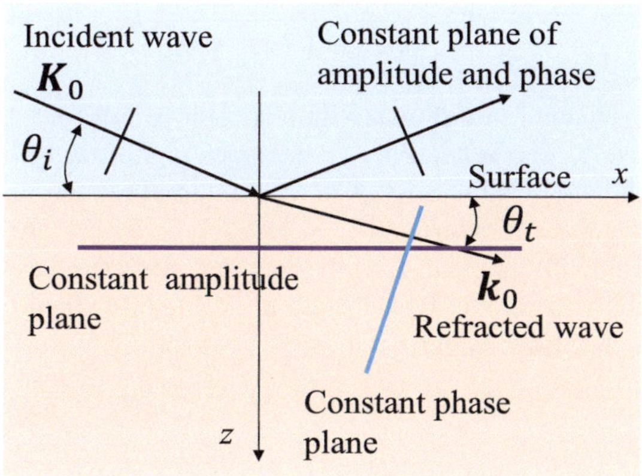

Fig. 4.8 Ununiform plane wave. The phase of the refracted wave is constant in the direction perpendicular to the propagating direction. The amplitude is constant in the direction perpendicular to the surface normal

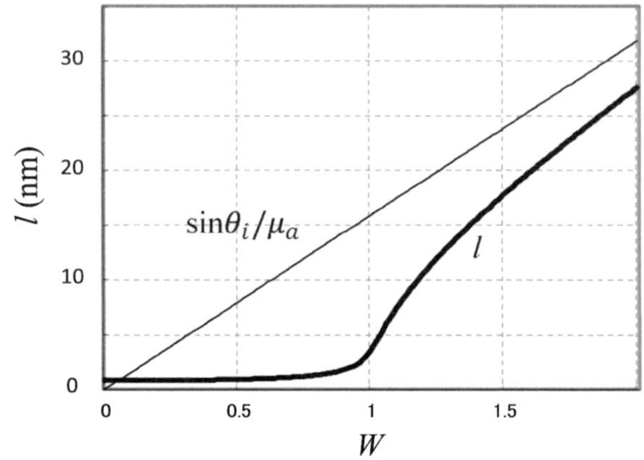

Fig. 4.9 Penetration depth with respect to W. The values of $\chi_r = 10^{-4}$, $k_a = 0.1$, $\lambda = 0.1$ nm and $\theta_c = 0.57°$ (0.01 rad) are used

4.4.3 Reflectance from Absorbing Medium

When θ_i is close to θ_c, we can set $n_0 \sin\theta_t \approx \theta_{tr} + i\theta_{ti}$ as in (4.51). By using the electric field of the σ-polarized incident wave E_i, that of the reflected wave E_r, and that inside the crystal E_t, the reflection and transmission coefficients r and t are given by

$$r = \frac{\sin\theta_i - n_0\sin\theta_t}{\sin\theta_i + n_0\sin\theta_t} = \frac{\sin\theta_i - (\theta_{tr} + i\theta_{ti})}{\sin\theta_i + (\theta_{tr} + i\theta_{ti})}, \; t = \frac{2\sin\theta_i}{\sin\theta_i + (\theta_{tr} + i\theta_{ti})}. \tag{4.60}$$

From (4.60), the reflectance I_r and the transmittance I_t at the surface are

$$I_r = |r|^2 = \frac{(\sin\theta_i - \theta_{tr})^2 + \theta_{ti}^2}{(\sin\theta_i + \theta_{tr})^2 + \theta_{ti}^2}, \; I_t = |t|^2 = \frac{4\sin^2\theta_i}{(\sin\theta_i + \theta_{tr})^2 + \theta_{ti}^2}. \tag{4.61}$$

I_r and I_t are shown with respect to W for four values of k_a in Fig. 4.4a, b, respectively. The reflectivity I_r becomes significantly large for $W < 1$, that is, $\theta_i < \theta_c$, showing characteristics of total internal reflection. I_r for $W < 1$ becomes smaller as the absorption effect ($k_a = |\chi_i/\chi_r|$) becomes larger, and the critical angle becomes gradually unclear. I_t at the medium surface takes the maximum value at $W = 1$. I_t shows a peak at $W = 1$ with the value of 4 when k_a is zero, and becomes smaller as k_a becomes larger.

4.5 X-ray Standing Waves

We consider the electric field strength $I(\boldsymbol{r})$ of the composite wave in the interference region of the incident wave and reflected wave above the surface of the medium, as shown in Fig. 4.10. The intensity $I(\boldsymbol{r})$ of the composite wave is written as

$$I(\boldsymbol{r}) = |E_i(\boldsymbol{r}) + E_r(\boldsymbol{r})|^2$$

$$= |E_i\exp(-i\boldsymbol{K}_0 \cdot \boldsymbol{r}) + E_r\exp(-i\boldsymbol{K}_r \cdot \boldsymbol{r})|^2$$

$$= |E_i|^2\left[1 + |R|^2 + 2|R|\cos(\boldsymbol{s} \cdot \boldsymbol{r} - \omega)\right], \tag{4.62}$$

where $\boldsymbol{s}$ is the scattering vector given by

$$\boldsymbol{s} = \boldsymbol{K}_r - \boldsymbol{K}_0,\ |\boldsymbol{s}| = \frac{4\pi\sin\theta_i}{\lambda}.$$

The reflection coefficient R is given as

$$R = \frac{E_r}{E_i} = |R|\exp(i\omega). \tag{4.63}$$

Since $\boldsymbol{s}$ is perpendicular to the surface, we obtain

$$\boldsymbol{s} \cdot \boldsymbol{r} = 2K_0 z\sin\theta_i = \frac{4\pi z\sin\theta_i}{\lambda} \approx \frac{4\pi W\theta_c z}{\lambda}. \tag{4.64}$$

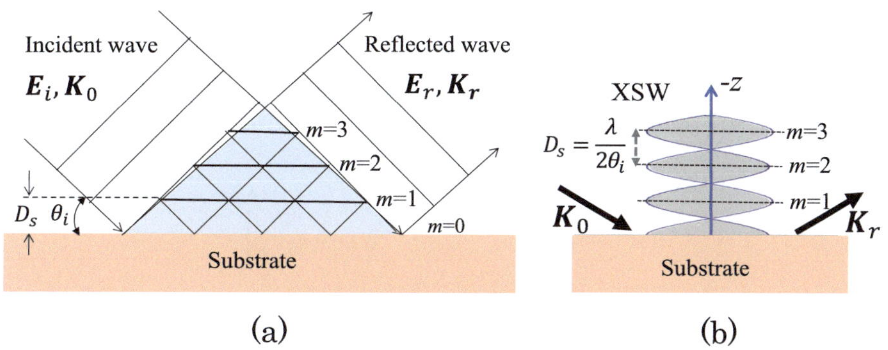

Fig. 4.10 **a** X-ray standing wave (XSW) above a substrate medium. **b** A schematic diagram of XSW in case of $W = 1$

The last equation is obtained when the glancing angle θ_i is around θ_c or less. The $\cos(\boldsymbol{s} \cdot \boldsymbol{r})$ term in (4.62) shows that a standing wave is formed with its period D_s as

$$D_s = \frac{\pi}{K_0 \sin\theta_i} = \frac{\lambda}{2\sin\theta_i} \approx \frac{\lambda}{2W\theta_c}. \tag{4.65}$$

This standing wave is called XSW (X-ray standing wave). Figure 4.10b shows the XSW for $W = 1$. In addition, Fig. 4.11 shows the phase angle ω of R in (4.63). When $W = 0$, $\omega = \pi$ and E_i and E_r are in opposite phase. When $W \geq 1$, $\omega = 0$ and E_i and E_r are in phase. Figure 4.12 shows the XSW ($I_{\mathrm{XSW}}(W)$) with respect to W at positions $m = 0$, 1, and 2 in Fig. 4.10b. Here, $I_{\mathrm{XSW}}(W)$ is the intensity of the composite wave normalized by the intensity of the incident wave given by

$$I_{\mathrm{XSW}}(W) = 1 + |R|^2 + 2|R|\cos(2\pi mW + \omega). \tag{4.66}$$

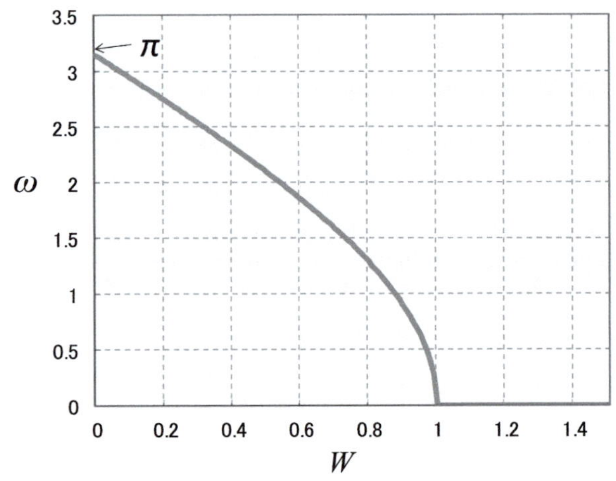

Fig. 4.11 Phase ω as a function of W

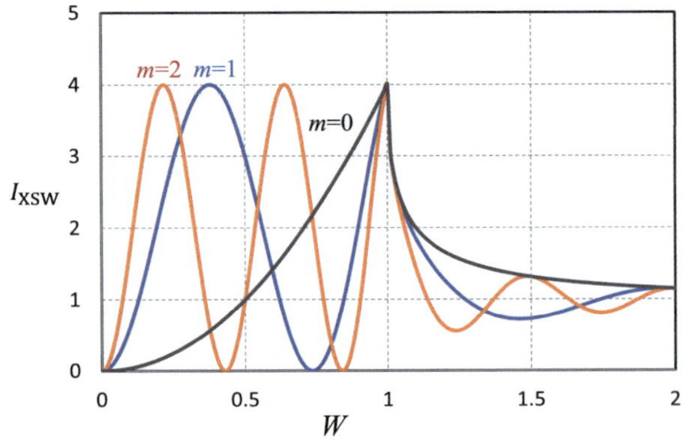

Fig. 4.12 I_{XSW} without absorption ($k_a = 0$) as a function of W at $m = 0$, 1 and 2

4.6 Reflection from Thin Film

We consider reflection from a thin film formed on a substrate with a smooth interface. A typical configuration is shown in Fig. 4.13. The refractive index of the first medium (vacuum) is n_1, that of the second medium (thin film) is n_2, and that of the third medium (substrate) is n_3.

According to Snell's law, we have

$$n_1\cos\theta_{1i} = n_2\cos\theta_{2i} = n_3\cos\theta_t. \tag{4.67}$$

The boundary conditions at the interface between the first and the second mediums are given by,

$$E_{1i} + E_{1r} = E_{2i} + E_{2r}, \tag{4.68}$$

$$n_1\sin\theta_{1i}(E_{1i} - E_{1r}) = n_2\sin\theta_{2i}(E_{2i} - E_{2r}). \tag{4.69}$$

And those at the interface between the second and the third mediums are given by

$$(E_{2i} + E_{2r})\exp\left(-\frac{i\gamma}{2}\right) = E_t\exp\left(-\frac{i\gamma}{2}\right), \tag{4.70}$$

$$n_2\sin\theta_{2i}(E_{2i} - E_{2r})\exp(-i\gamma/2) = n_3\sin\theta_{3i}E_t\exp(-i\gamma/2). \tag{4.71}$$

Here, the parameter γ is given as

$$\gamma = \frac{4\pi}{\lambda}n_2\sin\theta_{2i}d, \tag{4.72}$$

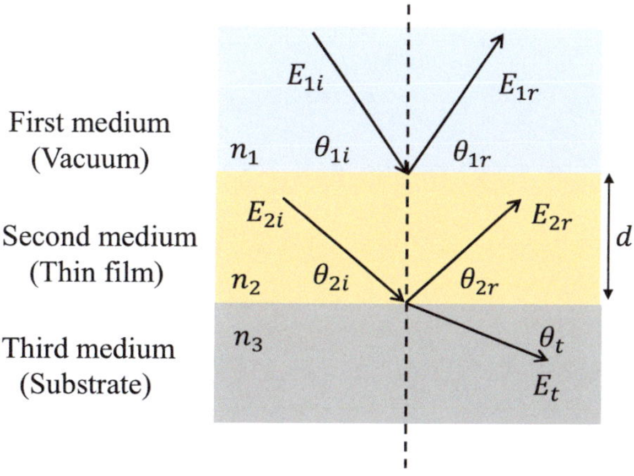

Fig. 4.13 A schematic diagram of reflections from a thin film on a substrate medium

where d is the thickness of the film. Using (4.68)–(4.71), the reflection and transmission coefficients are obtained as

$$r_s = \frac{E_{1r}}{E_{1i}} = \frac{r_{21} + r_{32}\exp(-i\gamma)}{1 + r_{32}r_{21}\exp(-i\gamma)},\ t_s = \frac{E_t}{E_{1i}} = \frac{2n_1\sin\theta_{1i}}{1 + r_{32}r_{21}\exp(-i\gamma)}. \tag{4.73}$$

Here, r_{21} and t_{12} are the reflection and transmission coefficients at the interface between the first medium and the second medium given by

$$r_{21} = \frac{n_1\sin\theta_{1i} - n_2\sin\theta_{2i}}{n_1\sin\theta_{1i} + n_2\sin\theta_{2i}},\ t_{12} = \frac{2n_1\sin\theta_{1i}}{n_1\sin\theta_{1i} + n_2\sin\theta_{2i}}. \tag{4.74}$$

Similarly, r_{32} and t_{23} are the reflection and transmission coefficients at the interface between the second medium and the third medium given by

$$r_{32} = \frac{n_2\sin\theta_{2i} - n_3\sin\theta_{3i}}{n_2\sin\theta_{2i} + n_3\sin\theta_{3i}},\ t_{23} = \frac{2n_2\sin\theta_{2i}}{n_2\sin\theta_{2i} + n_3\sin\theta_{3i}}. \tag{4.75}$$

The reflectance (R) and the transmittance (T) at $z = d$ are given by

$$R = |r_s|^2,\ T = |t_s|^2. \tag{4.76}$$

If we assume

$$r_{12} = \frac{n_2\sin\theta_{2i} - n_1\sin\theta_{1i}}{n_2\sin\theta_{2i} + n_1\sin\theta_{1i}} = -r_{21},\ t_{21} = \frac{2n_2\sin\theta_{2i}}{n_2\sin\theta_{2i} + n_1\sin\theta_{1i}} = \frac{n_2\sin\theta_{2i}}{n_1\sin\theta_{1i}}t_{12}, \tag{4.77}$$

then we have a relation between the reflection and the transmission coefficient as

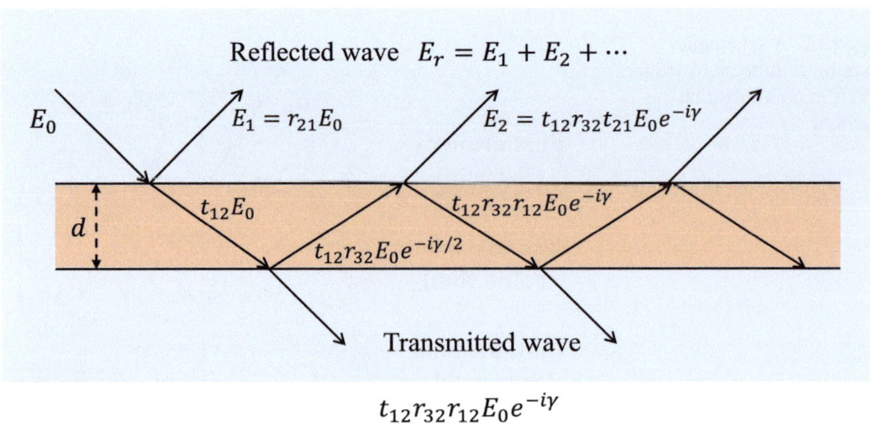

Fig. 4.14 Partial wave expansion in a thin film

$$t_{12}t_{21} - r_{21}r_{12} = 1. \tag{4.78}$$

Let's calculate the reflectance from the film by assuming that the refracted wave propagates by repeating reflections within the film as shown in Fig. 4.14. The resultant reflectance is given by

$$\begin{aligned} r_s &= r_{21} + t_{12}r_{32}t_{21}\exp(-i\gamma) + t_{12}r_{32}r_{12}r_{32}t_{21}\exp(-i2\gamma) + \cdots \\ &= r_{21} + t_{12}r_{32}t_{21}\exp(-i\gamma)[1 + r_{12}r_{32}\exp(-i\gamma) + \cdots] \\ &= r_{21} + \frac{t_{12}r_{32}t_{21}\exp(-i\gamma)}{1 - r_{12}r_{32}\exp(-i\gamma)} = \frac{r_{21} + r_{32}\exp(-i\gamma)}{1 + r_{21}r_{32}\exp(-i\gamma)}, \end{aligned} \tag{4.79}$$

which agrees with (4.73). This way of calculation is called the partial wave method. Partial wave expansion is quite useful in calculating the diffraction intensity from a thin crystal as shown in Chap. 6.

Figure 4.15 shows the calculated results of R and T from a thin film of 50 nm thick by using (4.73). For the film, the values are used of χ_{r2}=10^{-5}, $k_{a2} = 0.02$ and $\theta_{c1} = 0.0032$rad, and for the substrate, the values are used of χ_{r3}=10^{-4}, $k_{a3} = 0.1$ and $\theta_{c2} = 0.01$rad. Interference fringes reflecting the thickness of the film appear on the higher angle side than the critical angle θ_{c1} of the film. As R decreases, T increases. When the glancing angle θ_{2i} exceeds the critical angle θ_{c2} of the substrate, both R and T decreases as θ_{1i} becomes large.

4.7 Reflection from Very Thin Film

We consider reflection from a very thin film. Since we can assume that the mediums on both sides of the film are the vacuum, we have r_{21}= $-r_{32}$. When $\theta_i \gg \theta_c$, we can put $r = r_{21} = \theta_c^2/(2\sin\theta_i)^2$ from (4.33), and r_s in (4.73) is expressed as

$$r_s = \frac{r_{21}[1 - \exp(-i\gamma)]}{1 - r_{21}^2\exp(-i\gamma)} \cong \left(\frac{\theta_c}{2\sin\theta_i}\right)^2 [1 - \exp(-i\gamma)]. \tag{4.80}$$

The right side of (4.80) is derived by assuming that the denominator r_{21}^2 is much smaller than 1 and can be ignored. We further use the relation

$$[1 - \exp(-i\gamma)] = \exp\left(-i\frac{\gamma}{2}\right)\left\{\exp\left(i\frac{\gamma}{2}\right) - \exp\left(-i\frac{\gamma}{2}\right)\right\} = -i2\sin\frac{\gamma}{2}\exp\left(-i\frac{\gamma}{2}\right),$$

together with $\gamma = 2kd\ \sin\theta_i$ and $\theta_c^2 = -\chi_r = 2\delta$, then we have

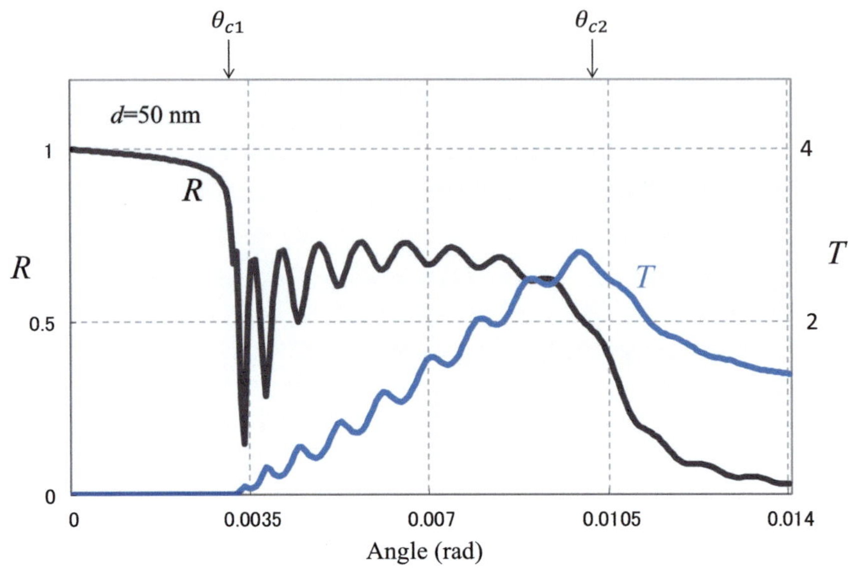

Fig. 4.15 Reflectance and transmittance from a thin film of 50 nm thick on a substrate as a function of θ_{1i}

$$r_s = -i\frac{\delta}{\sin\theta_i}dk\frac{2\sin(kd\ \sin\theta_i)}{2dk\sin\theta_i}\ \exp(-ikd\ \sin\theta_i).$$

If the film thickness d is sufficiently smaller than the wavelength of the X-ray, the exponential term can be approximated to 1, the reflectance r_s becomes

$$r_{s,slab} = -i\frac{\delta dk}{\sin\theta_i} = -i\frac{r_e\lambda\rho d}{\sin\theta_i}. \tag{4.81}$$

Here δ is given by $\delta = r_e\lambda^2\rho/\pi$ (4.81) holds true when θ_i is sufficiently larger than the critical angle and $|r_s|$ is sufficiently small. The equation is quite useful for structural analysis of laminated thin films and for some researches on dynamical theory of diffraction by Darwin (see 1.4.1 and Appendix A).

4.8 Waveguide Using X-ray Standing Wave (XSW)

Figure 4.10 shows the X-ray standing wave formed by the superposition of the incident and reflected waves on the substrate medium. As shown in (4.65), the period of this XSW D_s is given by

$$D_s = \frac{\lambda}{2\sin\theta_i}.$$

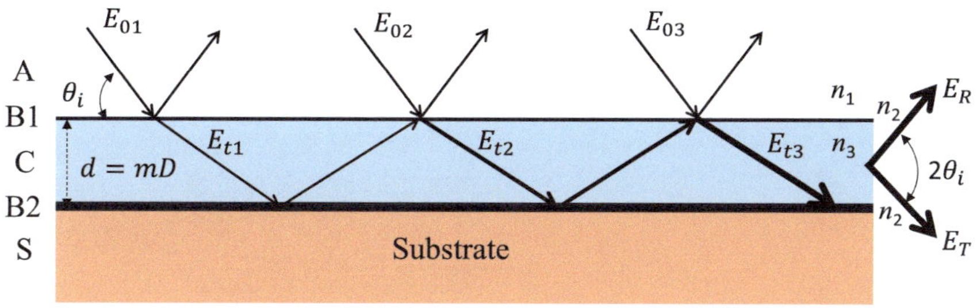

Fig. 4.16 Schematic diagram of an X-ray waveguide using XSW

If the relation

$$d = mD_s, m\lambda = 2d\sin\theta_i \tag{4.82}$$

holds between D_s and the thin film of thickness d for positive integer m, the thin film is regarded as a resonator and functions as an X-ray waveguide.

Figure 4.16 shows a schematic diagram of an example of X-ray waveguide [1]. The structure is designed to efficiently amplify refracted waves. The thin film B2 composed of heavy atoms is formed on a flat substrate S. On top of B2, a film C composed of light atoms is formed. The refractive indices of B2 and C are denoted n_2 and n_3 ($n_2 > n_3$). On top of C, another thin film B1 with the same refractive index of n_2 is formed.

The thickness of B1 is very thin, a few nanometers, so that a part of the incident X-rays can easily pass through the film even under the total internal reflection condition. With this structure, the refracted wave that has entered the film C is totally reflected by B2, and the refracted wave is efficiently reflected again by B1 to form XSW. The X-rays of this XSW are emitted from the edge of the film C separately in the reflection and the transmission directions. If X-rays of large width along the traveling direction is incident on this waveguide, the XSW density within the film C increases along the traveling direction. As a result, the emitted X-rays have a higher density than the incident beam and have good coherence. By optimizing the structure of the waveguide, the emitted beam density can be as much as 100 times higher than that of the incident beam. This emitted X-rays have a width equal to the thickness of the film C of the order of nanometers and good coherence. It is used in the development of phase contrast microscopy. The fabrication of this X-ray waveguide was made possible by the recent development of nano-sized thin film formation technology.

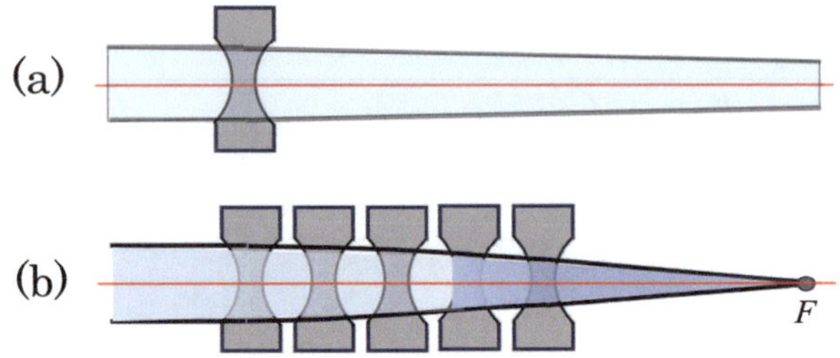

Fig. 4.17 Focus of X-ray beams by **a** a double-sided concave lens and **b** a series of double-sided concave lenses

4.9 Focusing X-rays Using Lenses and Mirrors

We explain how to obtain a high-density X-ray beam by using two methods. One uses an X-ray lens that takes advantage of the refraction effect of X-rays, and the other uses an X-ray mirror that uses total internal reflection.

4.9.1 X-ray Lens Method

The refractive index n_0 of X-rays is slightly smaller than 1, as already mentioned. A concave lens will be used to focus the X-rays. Figure 4.17a shows refractions of X-ray beams by a double-sided concave lens. The focal length is about 100 m, which is not so useful for practical use. By combining about 100 double-sided concave lenses as shown in Fig. 4.17b (this lens system is called a compound refractive lens, abbreviated as CRL), it is possible to shorten the focal length to several tens of mm. By arranging a suitable number of double-sided concave lenses vertically and horizontally, X-rays can be focused on a single point. CRL is used to focus X-rays from synchrotron radiation.

4.9.2 X-ray Mirror Method

X-rays can have a high reflectance of nearly 100% at a glancing angle (θ_i) smaller than the critical angle (θ_c) of total internal reflection, which can be used as a mirror. We explain how to curve a mirror and focus the X-rays to obtain a high-density beam. We also explain how to obtain a high-density beam by focusing the reflected beams from a curved mirror. From (4.50),

$$\theta_c = \sqrt{|\chi_r|} = \lambda\left(\frac{r_e\rho}{\pi}\right)^{1/2}.$$

By expressing the wavelength λ in angstroms (Å) and the X-ray energy E in keV, and using the relation $\lambda = 12.398/E$, we have

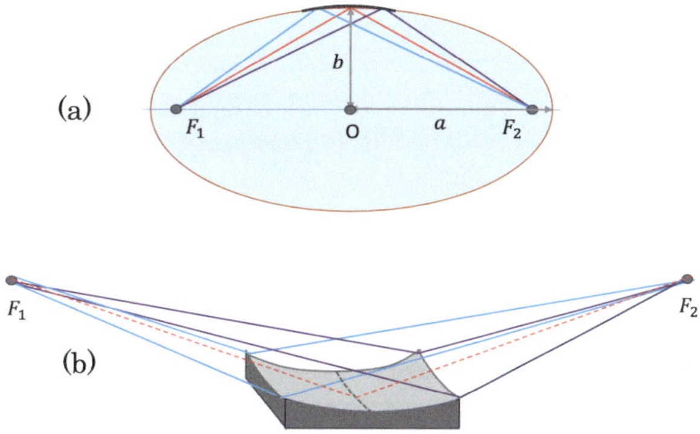

Fig. 4.18 Principle of focusing X-ray beams by curved mirrors. **a** An elliptical mirror and **b** a spheroidal mirror

$$E(\text{keV}) = \frac{12.398}{\theta_i}\left(\frac{r_e \rho}{\pi}\right)^{\frac{1}{2}}. \tag{4.83}$$

when θ_i is close to θ_c. When the glancing angle is $\theta_i = \theta_c$, E represents the maximum energy of X-rays that can be totally reflected and is called the critical energy (E_c). The reflectance of X-rays decreases significantly as the energy E becomes higher than E_c. Thus, the X-ray mirror functions as a low-pass filter and can remove high-order reflected X-rays that cannot be removed by a crystal monochromator.

The principle of the X-ray focusing mirror is shown in Fig. 4.18a. An X-ray mirror is placed on the circumference of an ellipse. The radius of the major axis of the ellipse is a, and the radius of the minor axis is b. There are two focal points F_1 and F_2 on the long axis. If a point source is placed at one of them (F_1), the beam reflected by the mirror will be focused at the other (F_2). Three beam trajectories are shown in blue, red and purple. This focusing of X-ray in the direction of the long axis is called tangential focusing. If we consider a spheroid formed by rotating this ellipse around its long axis, an X-ray that comes out of F_1 and hits any point on this spheroid passes through F_2, in other words, focuses on F_2. Thus, if we consider a circle on this spheroid centered on the major axis as well as perpendicular to the major axis, an X-ray emitted from F_1 and hitting any point on the circle will be focused on F_2. This focusing is called sagittal focusing. A mirror machined into a curve that satisfies both tangential and sagittal focusing (spheroidal mirror) can be used as a focusing mirror. An example of such a focusing mirror is shown in Fig. 4.18b. In order to shorten the focal length, it is essential to increase θ_i. For this, it is necessary to use a material with high electron density. Typically, a mirror is used that uses Si as the substrate and is coated with a high-density substance such as platinum (Pt, Z = 78) or rhodium (Ru, Z = 45).

The principles of the CRL lens and X-ray mirror described here are well described in literature [2].

References

1. A. Cedola, S. Lagomarsino, Synchrotron Rad. News. **17–3**, 30–36 (2004)
2. J. Als-Nielsen, D. McMorrow, *Elements of Modern X-ray Physics* (John Wiley & Sons, Ltd., 2011)

Chapter 5
Basic Equations of X-ray Dynamical Diffraction

To consider X-ray diffraction from a crystal, we derive a basic equation for the propagation of X-rays when the electric susceptibility is a periodic function of position, and use it to have the basic equation of diffraction from a perfect crystal called the dynamical theory of diffraction. From this basic equation, we derive the basic equation in the two-wave approximation in which one diffracted wave is excited in addition to the incident wave. We also derive the dispersion surface that represents the basic equation geometrically.

5.1 X-ray Susceptibility in a Crystal

In Chap. 4, the permittivity ε in an isotropic medium is assumed, which does not depend on position. In a crystal, atoms are arranged periodically in three-dimensional lattice, and the permittivity $\varepsilon(\boldsymbol{r})$ is a three-dimensionally periodic function of position. Using the electric polarization $\boldsymbol{P}$, the electric flux density $\boldsymbol{D}$ in the crystal is given by $\boldsymbol{D} = \varepsilon_0 \boldsymbol{E} + \boldsymbol{P} = \varepsilon(\boldsymbol{r})\boldsymbol{E}$. As the electric dipole moment $\boldsymbol{p}$ created by one free electron is given by $\boldsymbol{p} = -e\boldsymbol{r}$, the electric polarization $\boldsymbol{P}$ in a crystal can be given by $\boldsymbol{p}\rho(\boldsymbol{r})$ using the electron density distribution $\rho(\boldsymbol{r})$ at position $\boldsymbol{r}$. In (2.4) in Chap. 2, we can assume that a radiation damping factor γ_d is negligible and $\omega \gg \omega_s$, then we can write $\boldsymbol{P}$ as

$$\boldsymbol{P}(\boldsymbol{r}) = \boldsymbol{p}\rho(\boldsymbol{r}) = -e\boldsymbol{r}\rho(\boldsymbol{r}) = -\frac{e^2}{m_e\omega^2}\rho(\boldsymbol{r})\boldsymbol{E}(\boldsymbol{r}), \tag{5.1}$$

by using (2.9). Since $\boldsymbol{P}(\boldsymbol{r}) = \varepsilon_0\chi(\boldsymbol{r})\boldsymbol{E}(\boldsymbol{r})$, the electrical susceptibility $\chi(\boldsymbol{r})$ is given by

T. Fukamachi and T. Kawamura, *Resonant Dynamical X-ray Diffraction*, Springer Series in Solid-State Sciences 208, https://doi.org/10.1007/978-981-96-8406-9_5

$$\chi(\boldsymbol{r}) = -\frac{e^2}{\varepsilon_0 m_e \omega^2}\rho(\boldsymbol{r}) = -r_e\frac{\lambda^2}{\pi}\rho(\boldsymbol{r}). \tag{5.2}$$

Here r_e is the classical electron radius. Then, the permittivity is expressed as

$$\varepsilon(\boldsymbol{r}) = \varepsilon_0[1 + \chi(\boldsymbol{r})]. \tag{5.3}$$

In the case of a crystal, the electrical susceptibility has the periodicity of the crystal and can be expanded into Fourier series as

$$\chi(\boldsymbol{r}) = \sum_{g=-\infty}^{+\infty} \chi(\boldsymbol{g}) \exp(-i\boldsymbol{g}\cdot\boldsymbol{r}). \tag{5.4}$$

$\boldsymbol{g}$ is a reciprocal lattice vector, and the sum for $\boldsymbol{g}$ runs from $-\infty$ to ∞ in three-dimensional space. If we write the Fourier coefficient $\chi(\boldsymbol{g})$ as χ_g, we obtain

$$\begin{aligned}\chi_g &= \frac{1}{V_c}\int_{\text{u.c.}} \chi(\boldsymbol{r})\exp(i\boldsymbol{g}\cdot\boldsymbol{r})d\boldsymbol{r}\\ &= -r_e\frac{\lambda^2}{\pi V_c}\int_{\text{u.c.}} \rho(\boldsymbol{r})\exp(i\boldsymbol{g}\cdot\boldsymbol{r})d\boldsymbol{r}\\ &= -r_e\frac{\lambda^2}{\pi V_c}F_g.\end{aligned} \tag{5.5}$$

The 'u.c' below the integral sign indicates that the integration is performed over the unit cell. F_g is called crystal structure factor given as

$$F_g = \sum_{j=1}^{n} f_j^0(\boldsymbol{g})\exp(i\boldsymbol{g}\cdot\boldsymbol{r})T_j. \tag{5.6}$$

$f_j^0(\boldsymbol{g})$ is the normal atomic scattering factor of the j-th atom, and T_j gives the correction due to thermal vibration of the atom. The sum is taken over the atoms in the unit cell. $f^0(\boldsymbol{g})$ is given by

$$f^0(\boldsymbol{g}) = \int_{\text{atom}} \rho(\boldsymbol{r})\exp(i\boldsymbol{g}\cdot\boldsymbol{r})d\boldsymbol{r}. \tag{5.7}$$

In the forward scattering, $\boldsymbol{g} = 0$ and $f^0(0) = \int_{\text{atom}} \rho(\mathbf{r})d\mathbf{r} = Z$, then we obtain $F(0) = \sum Z_j$ meaning that $F(0)$ is the total electrons in the unit cell. Since $F(0)$ is positive, χ_0 is negative and its absolute value is 10^{-4} to 10^{-6}, which is much smaller than 1. In addition, $|\chi_0| \geq |\chi_g|$, and $|\chi_g|$ is also much smaller than 1.

5.2 Basic Equations

Since the X-ray susceptibility in a crystal can be described by (5.4) and (5.5), we use them here to derive the basic equation of dynamical theory of diffraction. According to Bloch's theorem, an X-ray in a medium with an electron density $\rho(\boldsymbol{r})$ of the lattice period is modulated by an incident plane wave as

$$\boldsymbol{D}(\boldsymbol{r}, t) = \exp(i\omega t) \sum_g \boldsymbol{D}_g \exp\left(-i\boldsymbol{k}_g \cdot \boldsymbol{r}\right), \tag{5.8}$$

$$\boldsymbol{H}(\boldsymbol{r}, t) = \exp(i\omega t) \sum_g \boldsymbol{H}_g \exp\left(-i\boldsymbol{k}_g \cdot \boldsymbol{r}\right). \tag{5.9}$$

The wave vector of the diffracted wave corresponding to the reciprocal lattice vector $\boldsymbol{g}$ is given by

$$\boldsymbol{k}_g = \boldsymbol{k}_0 + \boldsymbol{g}. \tag{5.10}$$

Taking curl of vector $\boldsymbol{H}(\boldsymbol{r}, t)$ in (5.9), we obtain

$$\mathrm{rot}\boldsymbol{H}(\boldsymbol{r}, t) = -i \exp(i\omega t) \sum_g (\boldsymbol{k}_g \times \boldsymbol{H}_g) \exp\left(-i\boldsymbol{k}_g \cdot \boldsymbol{r}\right). \tag{5.11}$$

By using (5.8), we obtain

$$\frac{\partial \boldsymbol{D}(\boldsymbol{r}, t)}{\partial t} = i\omega \exp(i\omega t) \sum_g \boldsymbol{D}_g \exp\left(-i\boldsymbol{k}_g \cdot \boldsymbol{r}\right). \tag{5.12}$$

Using the relation in (4.4), (5.11) and (5.12), we get the relation

$$-\sum_g (\boldsymbol{k}_g \times \boldsymbol{H}_g) \exp\left(-i\boldsymbol{k}_g \cdot \boldsymbol{r}\right) = \omega \sum_g \boldsymbol{D}_g \exp\left(-i\boldsymbol{k}_g \cdot \boldsymbol{r}\right). \tag{5.13}$$

Since the coefficients of the exponential function on both sides must be equal, we obtain

$$\omega \boldsymbol{D}_g = \boldsymbol{H}_g \times \boldsymbol{k}_g. \tag{5.14}$$

This equation shows that $\boldsymbol{D}_g$ and $\boldsymbol{H}_g$ are orthogonal to each other, and they are both perpendicular to $\boldsymbol{k}_g$, which means that the X-ray is a transverse wave.

5.2.1 Only One Wave in a Crystal (One-Wave Approximation)

Let us consider the case where no strong Bragg reflection occurs and only one wave corresponding to the incident wave propagates in a crystal. When the sum is only $\boldsymbol{g} = 0$ in (5.8), $\boldsymbol{D}(\boldsymbol{r})$ can be written as

$$\boldsymbol{D}(\boldsymbol{r}) = \boldsymbol{D}_0 \exp(-i\boldsymbol{k}_0 \cdot \boldsymbol{r}). \tag{5.15}$$

From (5.4), we have

$$\chi(\boldsymbol{r}) = \chi_0. \tag{5.16}$$

$\chi(\boldsymbol{r})$ does not depend on the position vector $\boldsymbol{r}$ and χ_0 is given by

$$\chi_0 = -r_e \frac{\lambda^2}{\pi} \rho. \tag{5.17}$$

Here, ρ is the average electron density. Substituting this relation into the propagation Eq. (4.9), we get

$$\nabla^2 \boldsymbol{D}(\boldsymbol{r}) + \mathrm{rot}\{\mathrm{rot}(\chi_0 \boldsymbol{D}(\boldsymbol{r}))\} + K_0^2 \boldsymbol{D}(\boldsymbol{r}) = 0. \tag{5.18}$$

By using

$$1 + \chi_0 = \frac{\varepsilon}{\varepsilon_0}, \tag{5.19}$$

Equation (5.18) becomes

$$\nabla^2 \boldsymbol{D}(\boldsymbol{r}) + \frac{\varepsilon}{\varepsilon_0} K_0^2 \boldsymbol{D}(\boldsymbol{r}) = 0. \tag{5.20}$$

Since $\boldsymbol{D}(\boldsymbol{r}) \neq 0$, as obtained in 4.1.1, we get the dispersion relation as

$$\boldsymbol{k}_0^2 = \frac{\varepsilon}{\varepsilon_0} \left(\frac{\omega}{c}\right)^2 = \frac{\varepsilon}{\varepsilon_0} K_0^2. \tag{5.21}$$

This dispersion relation is just the same as that for the propagation of X-rays in a homogeneous medium.

5.2.2 One Diffracted Wave in a Crystal (Two-Wave Approximation)

We consider the case where there are two waves, one primary wave corresponding to the incident wave and one diffracted wave in a crystal. This is called a two-wave approximation. If we denote the wave vector of the primary wave as $\boldsymbol{k}_0$ and that of a diffracted wave as $\boldsymbol{k}_g = \boldsymbol{k}_0 + \boldsymbol{g}$, then the Bloch wave can be written as the sum of the primary wave and the diffracted wave as

$$\boldsymbol{D}(\boldsymbol{r}) = \boldsymbol{D}_0 \exp(-i\boldsymbol{k}_0 \cdot \boldsymbol{r}) + \boldsymbol{D}_g \exp\left(-i\boldsymbol{k}_g \cdot \boldsymbol{r}\right). \tag{5.22}$$

The X-ray electrical susceptibility related to the excitation of these two waves is given by

$$\chi(\boldsymbol{r}) = \chi_0 + \chi_g \exp(-i\boldsymbol{g} \cdot \boldsymbol{r}) + \chi_{-g} \exp(i\boldsymbol{g} \cdot \boldsymbol{r}). \tag{5.23}$$

Using the first to third terms on the right-hand side, we can write

$$\begin{aligned} \chi(\mathbf{r})\mathbf{D}(\mathbf{r}) =& \chi_0 \mathbf{D}_0 \exp(-i\mathbf{k}_0 \cdot \mathbf{r}) + \chi_0 \mathbf{D}_g \exp\left(-i\mathbf{k}_g \cdot \mathbf{r}\right) \\ &+ \chi_g \mathbf{D}_0 \exp\left(-i\mathbf{k}_g \cdot \mathbf{r}\right) + \chi_{-g} \mathbf{D}_g \exp(-i\mathbf{k}_0 \cdot \mathbf{r}) + \cdots . \end{aligned} \tag{5.24}$$

Here, the terms whose exponent part is $(\boldsymbol{k}_g + \boldsymbol{g}) \cdot \boldsymbol{r}$ and $(\boldsymbol{k}_0 - \boldsymbol{g}) \cdot \boldsymbol{r}$ have wave vectors $\boldsymbol{k}_{2g}$ and $\boldsymbol{k}_{-g}$ are ignored because these wave vectors are far outside the Ewald sphere and is not strongly excited. By substituting (5.24) into (4.9) of the propagation equation, we get

$$\nabla^2 \boldsymbol{D} + K_0^2 \boldsymbol{D} = -\mathrm{rot}\{\mathrm{rot}(\chi(\boldsymbol{r})\boldsymbol{D})\}, \tag{5.25}$$

we find the solution for $\boldsymbol{D}$. $\chi_0 \boldsymbol{D}_0$ in the first term of (5.24) can be modified as

$$-\mathrm{rot}\{\mathrm{rot}(\chi_0 \boldsymbol{D}_0)\} = \nabla^2 \chi_0 \boldsymbol{D}_0. \tag{5.26}$$

By using the relations rot(rot) = grad(div) $- \nabla^2$ and $\mathrm{div}(\chi_0 \boldsymbol{D}_0) = 0$. The fourth term in (5.24) is modified as

$$-\mathrm{rot}\left\{\mathrm{rot}\left(\chi_{-g} \boldsymbol{D}_g\right)\right\} = \chi_{-g} \boldsymbol{k}_0^2 \boldsymbol{D}_{g[\perp \boldsymbol{k}_0]}. \tag{5.27}$$

Here, $\boldsymbol{D}_{g[\perp \boldsymbol{k}_0]}$ denotes a component vector of $\boldsymbol{D}_g$ perpendicular to $\boldsymbol{k}_0$, as shown in Fig. 5.1.

The geometry of $\boldsymbol{D}_{g[\perp \boldsymbol{k}_0]}$ will be explained as follows. As we have

$$\mathrm{rot}\boldsymbol{D} = -i\boldsymbol{k} \times \boldsymbol{D},$$

according to (4.14) in Chap. 4, we obtain

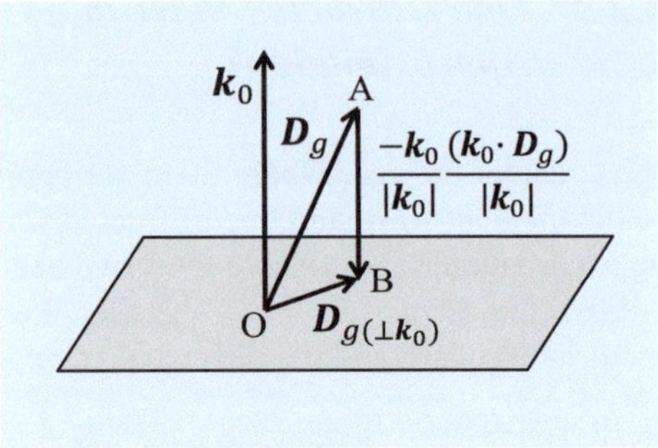

Fig. 5.1 Geometrical relations among $\boldsymbol{k}_0$, $\boldsymbol{D}_g$ and $\boldsymbol{D}_{g(\perp \boldsymbol{k}_0)}$

$$\mathrm{rot}\{\mathrm{rot}\boldsymbol{D}_g\exp(-i\boldsymbol{k}_0\cdot\boldsymbol{r})\} = -\boldsymbol{k}_0\times\boldsymbol{k}_0\times\boldsymbol{D}_g\exp(-i\boldsymbol{k}_0\cdot\boldsymbol{r}).$$

The right side is further modified by using the vector relation $\boldsymbol{A}\times(\boldsymbol{B}\times\boldsymbol{C}) = \boldsymbol{B}(\boldsymbol{A}\cdot\boldsymbol{C}) - \boldsymbol{C}(\boldsymbol{A}\cdot\boldsymbol{B})$ as

$$-\boldsymbol{k}_0\times\boldsymbol{k}_0\times\boldsymbol{D}_g = \boldsymbol{k}_0^2\boldsymbol{D}_g - \boldsymbol{k}_0(\boldsymbol{k}_0\cdot\boldsymbol{D}_g) = \boldsymbol{k}_0^2\left\{\boldsymbol{D}_g - \frac{\boldsymbol{k}_0}{|\boldsymbol{k}_0|}\frac{(\boldsymbol{k}_0\cdot\boldsymbol{D}_g)}{|\boldsymbol{k}_0|}\right\} = \boldsymbol{k}_0^2\boldsymbol{D}_{g[\perp\boldsymbol{k}_0]}.$$

In Fig. 5.1, $\overrightarrow{\mathrm{OA}} = \boldsymbol{D}_g$ and $\overrightarrow{\mathrm{AB}} = (-\boldsymbol{k}_0/|\boldsymbol{k}_0|)(\boldsymbol{k}_0\cdot\boldsymbol{D}_g/|\boldsymbol{k}_0|)$, the term in the wavy bracket is $\overrightarrow{\mathrm{OA}} + \overrightarrow{\mathrm{AB}} = \overrightarrow{\mathrm{OB}} = \boldsymbol{D}_{g[\perp\boldsymbol{k}_0]}$.

From the first and fourth terms in (5.24), we obtain

$$\nabla^2\boldsymbol{D}_0 + K_0^2\boldsymbol{D}_0 = -\mathrm{rot}\{\mathrm{rot}(\chi_0\boldsymbol{D}_0)\} - \mathrm{rot}\{\mathrm{rot}(\chi_{-g}\boldsymbol{D}_g)\} = \nabla^2\chi_0\boldsymbol{D}_0 - \boldsymbol{k}_0^2\chi_{-g}\boldsymbol{D}_{g[\perp\boldsymbol{k}_0]}$$

and from the second and third terms, we obtain

$$\nabla^2\boldsymbol{D}_g + K_0^2\boldsymbol{D}_g = -\mathrm{rot}\{\mathrm{rot}(\chi_0\boldsymbol{D}_g)\} - \mathrm{rot}\{\mathrm{rot}(\chi_g\boldsymbol{D}_0)\} = \nabla^2\chi_0\boldsymbol{D}_g - \boldsymbol{k}_g^2\chi_g\boldsymbol{D}_{0[\perp\boldsymbol{k}_g]}.$$

By using the relation $\nabla^2\boldsymbol{D} = -\boldsymbol{k}_0^2\boldsymbol{D}$, the above two equations can be written as

$$\frac{\boldsymbol{k}_0^2(1-\chi_0) - K_0^2}{\boldsymbol{k}_0^2}\boldsymbol{D}_0 = \chi_{-g}\boldsymbol{D}_{g[\perp\boldsymbol{k}_0]}, \tag{5.28}$$

$$\frac{\boldsymbol{k}_g^2(1-\chi_0) - K_0^2}{\boldsymbol{k}_g^2}\boldsymbol{D}_g = \chi_g\boldsymbol{D}_{0[\perp\boldsymbol{k}_g]}. \tag{5.29}$$

This set of equations is called the basic equations of diffraction in the two-wave approximation. As will be shown in the next section, the dispersion relation can be obtained from this set of equations, and $\boldsymbol{D}_0$ and $\boldsymbol{D}_g$ can be determined.

5.2.3 Reduced Basic Equation

We denote the average wave number in a crystal as κ_0. The refractive index n_0 is given by $n_0 = \kappa_0/K_0 = (1 + \chi_0)^{1/2}$. Since χ_0 is 10^{-4} to 10^{-6} and much smaller than 1, we can put $K_0^2 \approx \kappa_0^2(1 - \chi_0)$ and obtain

$$\frac{\boldsymbol{k}_0^2 - \kappa_0^2}{\boldsymbol{k}_0^2}\boldsymbol{D}_0 = \chi_{-g}\boldsymbol{D}_{g(\perp\boldsymbol{k}_0)}, \tag{5.30}$$

$$\frac{\boldsymbol{k}_g^2 - \kappa_0^2}{\boldsymbol{k}_g^2}\boldsymbol{D}_g = \chi_g\boldsymbol{D}_{0(\perp\boldsymbol{k}_g)} \tag{5.31}$$

from (5.28) and (5.29). This set of basic equations were derived in the two-wave approximation, assuming that there is only one diffracted beam strongly excited in the crystal. When many diffracted waves are excited, the basic equations can be written by

$$\frac{\boldsymbol{k}_g^2 - \kappa_0^2}{\boldsymbol{k}_g^2}\boldsymbol{D}_g = \sum_{g'}{}' \chi_{g-g'}\boldsymbol{D}_{g'(\perp\boldsymbol{k}_g)}, \tag{5.32}$$

where $\Sigma'_{g'}$ means the sum over $\boldsymbol{g}'$ excluding $\boldsymbol{g}' = \boldsymbol{g}$. By solving these equations under appropriate boundary conditions, the solution to the diffraction problem can be obtained. These equations are simultaneous equations of infinite order regarding $\boldsymbol{g}$. Generally, waves in a crystal consist of a superposition of an infinite number of diffracted waves, as shown in (5.8). It is difficult to analytically solve the basic equations for all these component waves. To obtain approximate solutions to these equations, we limit the number of diffracted waves, which is called a many-wave approximation. In many-wave approximation that handles a finite number of diffracted waves, such as two diffracted waves (three-wave approximation), the diffraction intensities are numerically calculated. In fact, for solving the diffraction intensities of electron, the many-wave approximation are commonly used [1] (Cowley (1975)). In the following, we explain dynamical theory of diffraction in the two-wave approximation, which is commonly used in the field of X-ray diffraction.

Let consider a plane containing $\boldsymbol{k}_0$ and $\boldsymbol{k}_g$ (scattering plane) as shown in Fig. 5.2a, b. In Fig. 5.2a, $\boldsymbol{D}_0$ is orthogonal to $\boldsymbol{k}_0$, $\boldsymbol{D}_{0\perp}$ is orthogonal to both $\boldsymbol{k}_0$ and $\boldsymbol{k}_g$, and

$$\boldsymbol{D}_{0\perp} = \boldsymbol{D}_{0(\perp\boldsymbol{k}_0)} = \boldsymbol{D}_{0(\perp\boldsymbol{k}_g)}. \tag{5.33}$$

In Fig. 5.2b, $\boldsymbol{D}_g$ is orthogonal to $\boldsymbol{k}_g$, $\boldsymbol{D}_{g\perp}$ is orthogonal to both $\boldsymbol{k}_0$ and $\boldsymbol{k}_g$, and

$$\boldsymbol{D}_{g\perp} = \boldsymbol{D}_{g(\perp\boldsymbol{k}_g)} = \boldsymbol{D}_{g(\perp\boldsymbol{k}_0)}. \tag{5.34}$$

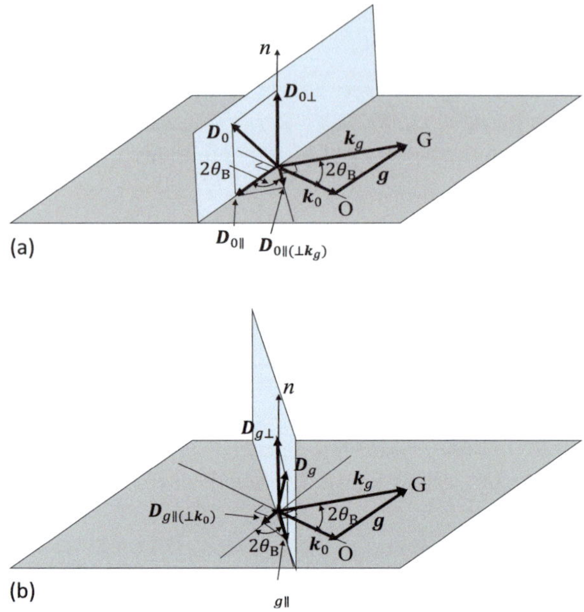

Fig. 5.2 Parallel and perpendicular components of $\boldsymbol{D}_0$ and $\boldsymbol{D}_g$ to the plane containing $\boldsymbol{k}_0$ and $\boldsymbol{k}_g$. $D_{0\parallel(\perp \boldsymbol{k}_g)} = D_{0\parallel}\cos 2\theta_{\mathrm{B}}$ is obtained from **(a)** and $D_{g\parallel(\perp \boldsymbol{k}_0)} = D_{g\parallel}\cos 2\theta_{\mathrm{B}}$ is obtained from **(b)**

In Fig. 5.2a, if we write the component of $\boldsymbol{D}_0$ parallel to the scattering plane as $D_{0\parallel}$, then $D_{0\parallel(\perp \boldsymbol{k}_g)} = D_{0\parallel}\cos 2\theta_{\mathrm{B}}$. Similarly, in Fig. 5.2b, if we write the component of $\boldsymbol{D}_g$ parallel to the scattering plane as $D_{g\parallel}$, we obtain $D_{g\parallel(\perp \boldsymbol{k}_0)} = D_{g\parallel}\cos 2\theta_{\mathrm{B}}$.

Then, for the component perpendicular to the scattering plane, we get

$$\frac{\boldsymbol{k}_0^2 - \kappa_0^2}{\boldsymbol{k}_0^2} D_{0\perp} = \chi_{-g} D_{g\perp}, \tag{5.35}$$

$$\frac{\boldsymbol{k}_g^2 - \kappa_0^2}{\boldsymbol{k}_g^2} D_{g\perp} = \chi_g D_{0\perp}. \tag{5.36}$$

For the parallel components, we get

$$\frac{\boldsymbol{k}_0^2 - \kappa_0^2}{\boldsymbol{k}_0^2} D_{0\parallel} = \chi_{-g} \cos 2\theta_{\mathrm{B}} D_{g\parallel}, \tag{5.37}$$

$$\frac{\boldsymbol{k}_g^2 - \kappa_0^2}{\boldsymbol{k}_g^2} D_{g\parallel} = \chi_g \cos 2\theta_{\mathrm{B}} D_{0\parallel}. \tag{5.38}$$

Equations (5.35)–(5.38) differ by $\cos 2\theta_{\mathrm{B}}$ on the right-hand side. If we denote those coefficients of χ_g and χ_{-g} as P, then we can write

$$\frac{\boldsymbol{k}_0^2 - \kappa_0^2}{\boldsymbol{k}_0^2} D_0 = P\chi_{-g} D_g, \tag{5.39}$$

$$\frac{\boldsymbol{k}_g^2 - \kappa_0^2}{\boldsymbol{k}_g^2} D_g = P\chi_g D_0. \tag{5.40}$$

Here, we write $D_{0\perp}$ and $D_{0\parallel}$ as D_0, and $D_{g\perp}$ and $D_{g\parallel}$ as D_g. $P = 1$ for the parallel component, and $P = \cos 2\theta_{\mathrm{B}}$ for the perpendicular component.

We introduce two parameters ξ_0 and ξ_g given by

$$|\boldsymbol{k}_0| - \kappa_0 = \xi_0, \; |\boldsymbol{k}_g| - \kappa_0 = \xi_g. \tag{5.41}$$

ξ_0 and ξ_g are called resonance errors [2] and represent the differences of $|\boldsymbol{k}_0|$ and $|\boldsymbol{k}_g|$ from κ_0. Using ξ_0 and ξ_g, the numerator of (5.39) is written by $\boldsymbol{k}_0^2 - \kappa_0^2 = (\kappa_0 + \xi_0)^2 - \kappa_0^2 = 2\xi_0\kappa_0$ and that of (5.40) is $\boldsymbol{k}_g^2 - \kappa_0^2 = \left(\kappa_0 + \xi_g\right)^2 - \kappa_0^2 = 2\xi_g\kappa_0$. Since ξ_0 and ξ_g are small compared with κ_0 by 10^{-4} to 10^{-6}, the denominators can be set as $\boldsymbol{k}_0^2 = (\kappa_0 + \xi_0)^2 \approx \kappa_0^2$ and $\boldsymbol{k}_g^2 = \left(\kappa_0 + \xi_g\right)^2 \approx \kappa_0^2$. (5.39) and (5.40) are then written as

$$\frac{2\xi_0}{\kappa_0} D_0 - P\chi_{-g} D_g = 0, \tag{5.42}$$

$$P\chi_g D_0 - \frac{2\xi_g}{\kappa_0} D_g = 0. \tag{5.43}$$

In this set of equations, the following equations must be satisfied to obtain a nontrivial solution for D_0 and D_g as

$$\det \begin{vmatrix} \frac{2\xi_0}{\kappa_0} & -P\chi_{-g} \\ P\chi_g & -\frac{2\xi_g}{\kappa_0} \end{vmatrix} = 0. \tag{5.44}$$

This gives an equation as

$$\xi_0 \xi_g = \left(\frac{P\kappa_0}{2}\right)^2 \chi_g \chi_{-g}, \tag{5.45}$$

which is called the dispersion equation in the two-wave approximation. (5.45) gives a hyperbola called dispersion surface, which means that the position of the tie-point must be on this dispersion surface. We will explain how ξ_0 and ξ_g are expressed on the dispersion surface in the next section. If $\chi_g \chi_{-g}$ is known and ξ_0 and ξ_g are given, D_0 and D_g can be solved from (5.42) and (5.43), then all the waves in the crystal can be described.

When $P = 1$, $D_{0\perp}$ and $D_{g\perp}$ are both perpendicular to the scattering plane, which is the σ-polarization component. On the other hand, when $P = \cos 2\theta_B$, $D_{0\parallel}$ and $D_{g\parallel}$ are on the scattering plane, which are the π-polarization component.

5.3 Dispersion Surface

A set of points that satisfy the dispersion equation in two-wave approximation of (5.45) form curved surfaces which are called the dispersion surfaces. A conceptual diagram is shown in Fig. 5.3. Let G be a point $\boldsymbol{g}$ away from the origin O of the reciprocal lattice. In this figure, OG is assumed to be parallel to the crystal surface for simplicity. The dispersion surfaces in vacuum are represented as two spheres T_0' and T_g' of radius K_0 and of centers O and G, respectively. The dispersion surfaces in the crystal are similarly represented as two spheres (Laue spheres) of radius $\kappa_0(= n_0 K_0)$ and of centers O and G. Note that near the intersection of the two spheres, the dispersion surfaces deform from the spheres according to (5.45). It should be noted that

$$\frac{K_0 - \kappa_0}{K_0} = 1 - n_0 = \frac{|\chi_0|}{2} \tag{5.46}$$

is a very small, on the order from 10^{-4} to 10^{-6}, so the difference between K_0 and κ_0 is greatly exaggerated in Fig. 5.3b. This figure shows a cross section of the dispersion surfaces on the plane containing $\boldsymbol{K}_0$ and $\boldsymbol{K}_g$. The intersection L_0 of two circles T_0 and T_g is called Lorentz point, and the intersection L_a of two circles T_0' and T_g' is called Laue point [2]. The starting point A on T_0' is given by the incident beam conditions and is called the tie-point (propagation point).

Once A is determined, the line v normal to the crystal surface passing through A is given as described in the dispersion surface in Chap. 4, and the tie-points A_1 and A_2 within the crystal are determined. A_1 and A_2 are points that satisfy (5.45), that is, points on the dispersion surface in the crystal. The tie-point closer to the O and G is set to A_1 in this book. In the crystal, the waves of wave vector $\boldsymbol{k}_0^{(1)}$ and $\boldsymbol{k}_g^{(1)}$ are excited from the tie-point A_1, and those of wave vector $\boldsymbol{k}_0^{(2)}$ and $\boldsymbol{k}_g^{(2)}$ from A_2.

When the reciprocal lattice vector is perpendicular to the crystal surface (referred to as the symmetric Bragg case as described later), the geometric relation between the incident wave ($\boldsymbol{K}_0$), the diffracted wave ($\boldsymbol{K}_g$) and the transmitted wave ($\boldsymbol{K}_t$) is shown in Fig. 5.4a, and the dispersion surfaces in Fig. 5.4b, c. One of the tie-point A_2 and the waves excited from A_2 are shown in Fig. 5.4b. Figure 5.4c can be obtained by rotating Fig. 5.3 by 90° clockwise, flipped horizontally, and expanded around the Lorentz point (L_o). A circle T_0 with radius κ_0 centered on O and a circle T_g with radius κ_0 centered on G are both approximated as straight lines. As can be seen from Fig. 5.4c, ξ_0 is the distance from A_2 to B parallel to $\overline{L_oO}$, and ξ_g is the distance from A_2 to D parallel to $\overline{L_oG}$. Here, $\overline{CF}$ is parallel to T_0, $\overline{EF}$ is parallel to T_g. If we use an X'-Y' orthogonal coordinate system with the X' axis parallel to the lattice plane, ξ_0

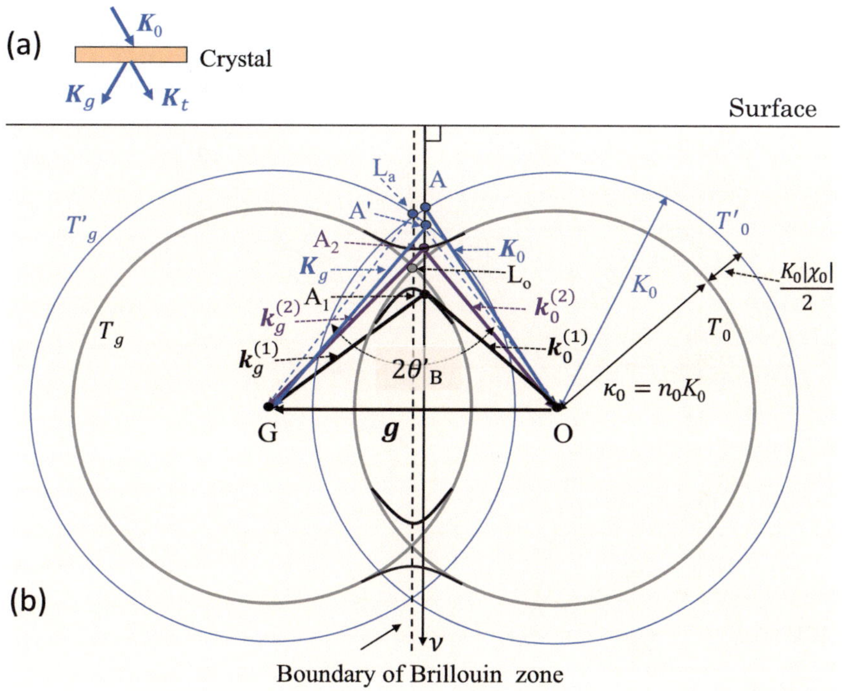

Fig. 5.3 Dispersion surfaces in the two-wave approximation. (**a**) Diffraction geometry in the symmetric Laue case. (**b**) Dispersion surfaces

and ξ_g are expressed as

$$\xi_0 = Y_0' \sin\theta_B' + X_0' \cos\theta_B', \tag{5.47}$$

$$\xi_g = X_0' \cos\theta_B' - Y_0' \sin\theta_B'. \tag{5.48}$$

Substituting these equations into (5.45), we get the dispersion equation

$$X_0'^2 \cos^2\theta_B' - Y_0'^2 \sin^2\theta_B' = \left(\frac{P\kappa_0}{2}\right)^2 \chi_g \chi_{-g}. \tag{5.49}$$

Since the crystal structure factor F_g is generally a complex number, χ_g is also a complex number, as is clear from (5.6). However, when we consider only the normal scattering, the atomic scattering factor is a real positive number, so that $|F_g| = |F_{-g}|$ and $\chi_g = \chi_{-g}^*$. Then we get

$$\chi_g \chi_{-g} = |\chi_g|^2 \tag{5.50}$$

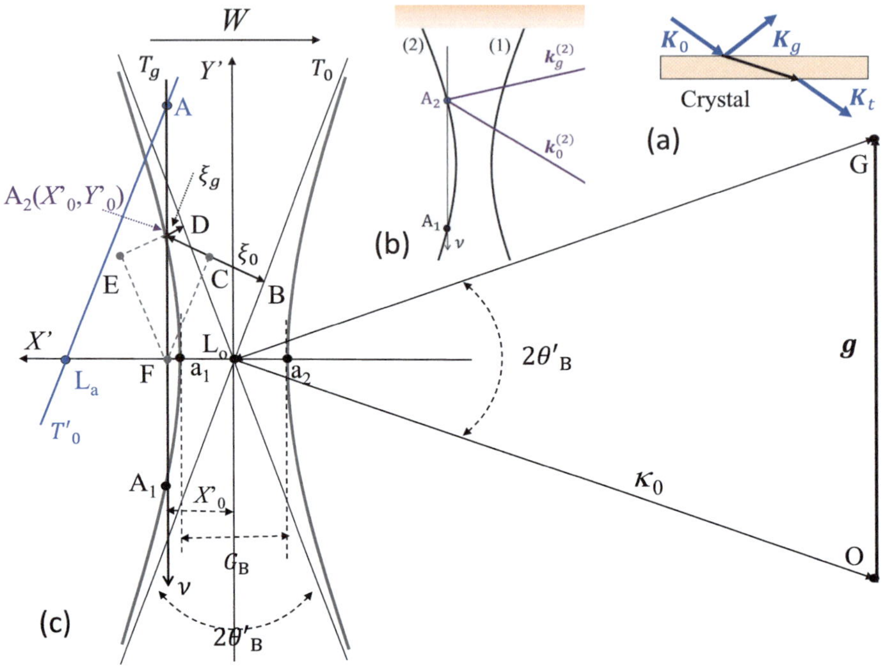

Fig. 5.4 Diffraction geometry **a** and dispersion surfaces **b, c** $\overline{A_2B} = \xi_0$, $\overline{A_2D} = \xi_g$, $\overline{BC} = \overline{DE} = X'_0\cos\theta'_B$, $\overline{CA_2} = \overline{A_2E} = Y'_0\sin\theta'_B$

and $\chi_g\chi_{-g}$ is a real positive number.

Equation (5.49) represents a hyperbola with T_0 and T_g as its asymptotes, and the distance G_B between the vertices of its two branches is

$$G_B = \overline{a_1a_2} = \frac{P\kappa_0\left(\chi_g\chi_{-g}\right)^{1/2}}{\cos\theta'_B} = \frac{2\pi}{\Lambda_0}. \tag{5.51}$$

G_B is called the Bragg gap [2] and Λ_0 is the interval of Pendellösung fringes, which will be described later. Using $\overline{a_1a_2} = G_B$, we define the parameter W to satisfy the relation

$$X' = \frac{-G_B}{2}W = -\frac{\pi}{\Lambda_0}W. \tag{5.52}$$

The reason why the coefficient of W is given a minus sign is for consistency with the incident glancing angle. This W is called a deviation parameter. At $W = 0$, X' coincides with L_o, and at $W = \pm 1$, X' is at the vertex of the hyperbola. Substituting (5.52) into (5.49), we obtain

$$Y' = \pm\frac{G_{\mathrm{B}}\cos\theta'_{\mathrm{B}}}{2\sin\theta'_{\mathrm{B}}}\left(W^2 - 1\right)^{1/2} = \pm\frac{G_{\mathrm{B}}}{2\tan\theta'_{\mathrm{B}}}\left(W^2 - 1\right)^{1/2}. \tag{5.53}$$

Then, ξ_0 and ξ_g are given by

$$\xi_0 = \frac{G_{\mathrm{B}}\cos\theta'_{\mathrm{B}}}{2}\left\{-W \pm (W^2 - 1)^{1/2}\right\}, \tag{5.54}$$

$$\xi_g = \frac{G_{\mathrm{B}}\cos\theta}{2}\left\{-W \mp (W^2 - 1)^{1/2}\right\}. \tag{5.55}$$

The value of $(W^2 - 1)^{1/2}$ in (5.53)–(5.55) becomes pure imaginary for $-1 < W < 1$. The dispersion surface in this case will be discussed in Chap. 6, and in Chap. 7. Where we will discuss in detail the resonance dynamical scattering that incorporates the effect of photo-absorption and the complex refractive index.

The dispersion surface near $\mathrm{L_o}$ point is a rotating hyperboloid obtained by rotating the hyperbola around the line connecting O and G, and all points on the surface can be tie-points. Once a tie-point is given according to the incident conditions $\boldsymbol{K}_0$, $\boldsymbol{k}_0$ and $\boldsymbol{k}_g$ are determined, then ξ_0 and ξ_g are determined. From (5.42) and (5.43), the ratio of D_0 and D_g is obtained as

$$D_0 = \frac{\kappa_0}{2\xi_0}\chi_{-g}D_g = \frac{2\xi_g}{\kappa_0}\frac{1}{\chi_g}D_g. \tag{5.56}$$

5.3.1 General Case

To consider the dispersion surface in more general case, we explain the case when the reciprocal lattice vector is at an arbitrary angle β with respect to the surface normal as shown in Fig. 5.5. X-axis is parallel to the crystal surface, and Y-axis is normal to it. The angle between Y-axis and Y'-axis is β. β is positive in the clockwise direction. The angle between Y and $\overline{\mathrm{L_oO}}$ is denoted by θ_1, and the angle between Y and $\overline{\mathrm{L_oG}}$ by θ_2. For these angles, counterclockwise is considered positive. When a plane wave with the wavevector $\boldsymbol{K}_0$ enters the crystal, the intersections of the dispersion surfaces with the line segment ν passing through the starting point A of $\boldsymbol{K}_0$ perpendicular to the crystal surface become tie-points $\mathrm{A_1}$ and $\mathrm{A_2}$. If we write the coordinates of the intersections as (X_0, Y_0), X_0 and Y_0 can be calculated by coordinate transformation from X–Y to X'-Y' as

$$X' = X_0\cos\beta + Y_0\sin\beta, \tag{5.57}$$

$$Y' = -X_0\sin\beta + Y_0\cos\beta. \tag{5.58}$$

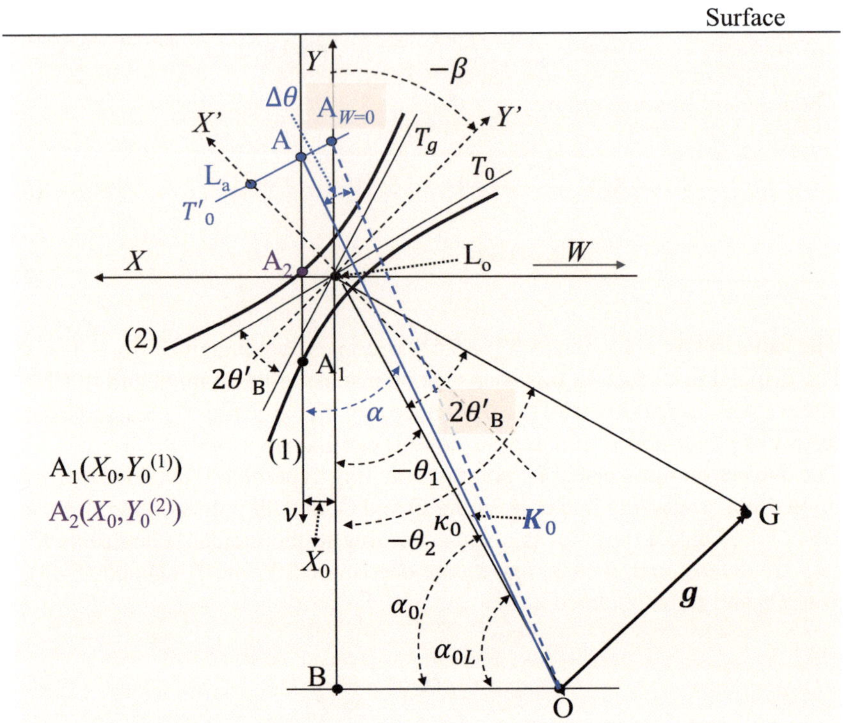

Fig. 5.5 Dispersion surface rotated by β clockwise from that in Fig. 5.4. The incident glancing angle in vacuum is $\angle\mathrm{AOB} = \alpha_0$. $\angle\mathrm{L_0OB} = \alpha_{0L}$

Substituting these into the dispersion (5.49), we get

$$X_0^2 \cos(\theta'_B + \beta)\cos(\theta'_B - \beta) - Y_0^2 \sin(\theta'_B + \beta)\sin(\theta'_B - \beta) + X_0 Y_0 \sin 2\beta = \left(\frac{P\kappa_0}{2}\right)^2 \chi_g \chi_{-g}. \tag{5.59}$$

As can be seen from Fig. 5.5

$$\theta'_B + \beta = \frac{\pi}{2} - \theta_1,\ \beta - \theta'_B = \frac{\pi}{2} - \theta_2, \tag{5.60}$$

(5.59) is rewritten as

$$X_0^2 \sin\theta_1 \sin\theta_2 + Y_0^2 \cos\theta_1 \cos\theta_2 + X_0 Y_0 \sin 2\beta = \left(\frac{P\kappa_0}{2}\right)^2 \chi_g \chi_{-g}. \tag{5.61}$$

By solving (5.61) for Y_0, we get

$$Y_0 = \frac{1}{2\cos\theta_1\cos\theta_2}$$
$$\left[-X_0\sin 2\beta \pm \left(X_0^2\sin^2 2\beta - 4X_0^2\sin\theta_1\sin\theta_2\cos\theta_1\cos\theta_2 + \cos\theta_1\cos\theta_2(P\kappa_0)^2\chi_g\chi_{-g}\right)^{1/2}\right]. \tag{5.62}$$

By using

$$\theta_1 + \theta_2 = \pi - 2\beta, \theta_2 - \theta_1 = 2\theta'_B \tag{5.63}$$

derived from (5.60) and (5.62) can be simplified as

$$Y_0 = \frac{1}{2\cos\theta_1\cos\theta_2}\left[-X_0\sin 2\beta \pm \left(X_0^2\sin^2 2\theta'_B + \cos\theta_1\cos\theta_2(P\kappa_0)^2\chi_g\chi_{-g}\right)^{1/2}\right]. \tag{5.64}$$

If we put the deviation parameter W as

$$W = -X_0\frac{\sin 2\theta'_B}{|\cos\theta_1\cos\theta_2|^{\frac{1}{2}}}\frac{1}{P\kappa_0(\chi_g\chi_{-g})^{\frac{1}{2}}}, \tag{5.65}$$

this is an extension of W defined in (5.52). By using (5.65), (5.64) can be rewritten as follows.

For $\cos\theta_2 > 0$

$$Y_0 = \frac{P\kappa_0(\chi_g\chi_{-g})^{1/2}}{2|\cos\theta_1\cos\theta_2|^{1/2}}\left[\frac{\sin 2\beta}{\sin 2\theta'_B}W \pm \left(W^2 + 1\right)^{1/2}\right]. \tag{5.66}$$

For $\cos\theta_2 < 0$

$$Y_0 = \frac{P\kappa_0(\chi_g\chi_{-g})^{1/2}}{2|\cos\theta_1\cos\theta_2|^{1/2}}\left[\frac{\sin 2\beta}{\sin 2\theta'_B}W \pm \left(W^2 - 1\right)^{1/2}\right]. \tag{5.67}$$

We consider the difference in diffraction conditions due to the diffracted wave direction θ_2 using Fig. 5.6a, b. Since the wave $\boldsymbol{k}_0$ (the primary wave) in the crystal connected to the incident wave always moves downward with respect to the crystal surface, the absolute value of θ_1 is always smaller than $\pi/2$. The angle θ_2 corresponding to the propagating direction of the diffracted wave $\boldsymbol{k}_g$ (the secondary wave) takes a value from $-\pi$ to π.

5.3.2 *In Case of* $\cos\theta_2 > 0$ *(Laue Case)*

When $\cos\theta_2 > 0$, the propagating direction of the diffracted wave is downward as shown in Fig. 5.6a. Such diffraction conditions are called the Laue case or Laue

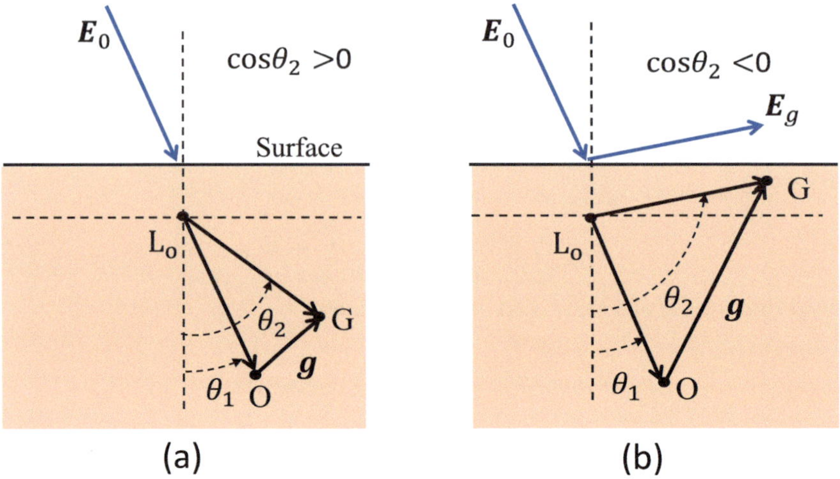

Fig. 5.6 Laue case **(a)** and Bragg case **(b)**

geometry. Substituting *X'* and *Y'* in (5.57) and (5.56) into ξ_0 and ξ_g in (5.47) and (5.48), we get

$$\xi_0 = X_0 \sin\theta_1 + Y_0 \cos\theta_1, \tag{5.68}$$

$$\xi_g = X_0 \sin\theta_2 + Y_0 \cos\theta_2 \tag{5.69}$$

by using the relations

$$\theta'_{\mathrm{B}} + \beta = -\theta_1 + \frac{\pi}{2}, \theta'_{\mathrm{B}} - \beta = \theta_2 - \frac{\pi}{2}.$$

As a result, ξ_0 and ξ_g in the Laue case are given by using W as

$$\xi_0^{(j)} = \frac{G_B \cos\theta'_{\mathrm{B}}}{2}\left(\frac{\cos\theta_1}{\cos\theta_2}\right)^{1/2}\left[W + (-1)^j\left(W^2+1\right)^{1/2}\right](j = 1, 2), \tag{5.70}$$

$$\xi_g^{(j)} = \frac{G_B \cos\theta'_{\mathrm{B}}}{2}\left(\frac{\cos\theta_2}{\cos\theta_1}\right)^{1/2}\left[-W + (-1)^j\left(W^2+1\right)^{1/2}\right](j = 1, 2). \tag{5.71}$$

ξ_0 and ξ_g have two values represented by $\xi_0^{(j)}$ and $\xi_g^{(j)}$ $(j = 1, 2)$, which correspond to two solutions of Y_0 given by (5.66).

Figure 5.7 shows the dispersion surface near the Laue point L_a. The reciprocal lattice vector $\boldsymbol{g}$ is parallel to the crystal surface, and the diffraction condition is called the symmetric Laue case. Since ξ_0 is about 10^{-4} to 10^{-6} times of κ_0, near the Lorentz point L_o, T_0 is the tangent of the circle of center O and radius κ_0 at L_o, which can be approximated as an asymptotic line of the hyperbolic dispersion surface in the crystal. Similarly, T_g is the tangent of circle at L_o, which can be approximated as another asymptotic line of the dispersion surface. The line segment T_0' (T_g') is the tangent at L_a to the circle with center O (G) and radius K_0, which can be approximated as the dispersion surface in vacuum. As can be seen from Fig. 5.7, the dispersion surface is divided into two, and correspondingly, ξ_0 and ξ_g are divided into two branches. The two values of (5.70) and (5.71) correspond to these two branches. Corresponding to tie-point A in vacuum, two tie-points A_1 and A_2 exist within the crystal. The branch inside T_0 and T_g (closer to points O and G) is labeled 1, and that outside is labeled 2.

From Fig. 5.7, for the incident wave and diffracted wave in vacuum, $\boldsymbol{K}_0 + \boldsymbol{g}$ satisfies

$$\boldsymbol{K}_0 + \boldsymbol{g} = \boldsymbol{K}_g + \Delta \boldsymbol{K}' \tag{5.72}$$

and

$$\boldsymbol{K}_0 + \boldsymbol{g} \neq \boldsymbol{K}_g,$$

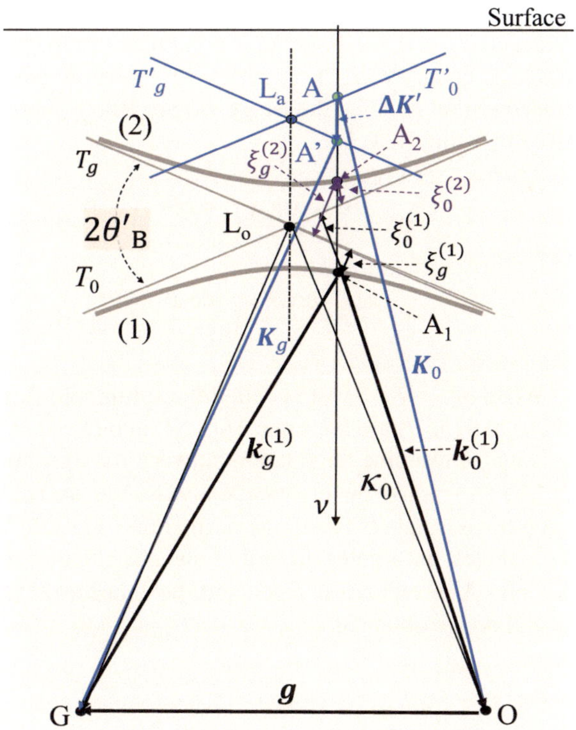

Fig. 5.7 Dispersion surface in the two-wave approximation. $\overrightarrow{AO} = \boldsymbol{K}_0$ is perpendicular to T_0'. $\overrightarrow{A'G} = \boldsymbol{K}_g$ is perpendicular to T_g'. $\overrightarrow{AA'} = \Delta \boldsymbol{K}'$

unless the tie-point A coincides with the Laue point L_a. In contrast, the wave vectors inside the crystal always satisfy the relation

$$\boldsymbol{k}_g^{(j)} = \boldsymbol{k}_0^{(j)} + \boldsymbol{g}. \tag{5.73}$$

$\boldsymbol{k}_0^{(j)}$ and $\boldsymbol{k}_g^{(j)}$ are written by using $\xi_0^{(j)}$ and $\xi_g^{(j)}$ at points A_1 ($j = 1$) and A_2 ($j = 2$) as

$$\boldsymbol{k}_0^{(1)} = \boldsymbol{s}_0^{(1)}\left(\kappa_0 - \xi_0^{(1)}\right), \boldsymbol{k}_g^{(1)} = \boldsymbol{s}_g^{(1)}\left(\kappa_0 - \xi_g^{(1)}\right), \tag{5.74}$$

$$\boldsymbol{k}_0^{(2)} = \boldsymbol{s}_0^{(2)}\left(\kappa_0 + \xi_0^{(2)}\right), \boldsymbol{k}_g^{(2)} = \boldsymbol{s}_g^{(2)}\left(\kappa_0 + \xi_g^{(2)}\right). \tag{5.75}$$

Here, $\boldsymbol{s}_0^{(j)}$ and $\boldsymbol{s}_g^{(j)}$ are unit vectors in the directions of $\boldsymbol{k}_0^{(j)}$ and $\boldsymbol{k}_g^{(j)}$.

5.3.3 *In Case of* $\cos\theta_2 < 0$ *(Bragg Case)*

When $\cos\theta_2 < 0$, as shown in Fig. 5.6b, the propagating direction of the diffracted wave (secondary wave) is upward with respect to the crystal surface. Such a diffraction condition is called the Bragg case or Bragg geometry. In this case, ξ_0 and ξ_g are given by using W as

$$\xi_0 = \frac{G_B\cos\theta'_B}{2}\left(\frac{\cos\theta_1}{|\cos\theta_2|}\right)^{1/2}\left[-W \pm \left(W^2 - 1\right)^{1/2}\right], \tag{5.76}$$

$$\xi_g = \frac{G_B\cos\theta'_B}{2}\left(\frac{|\cos\theta_2|}{\cos\theta_1}\right)^{1/2}\left[-W \mp \left(W^2 - 1\right)^{1/2}\right]. \tag{5.77}$$

When $\theta_2 = \pi - \theta_1$, it is called the symmetric Bragg case; $|\cos\theta_2| = \cos\theta_1$, and (5.76) and (5.77) are the same as (5.54) and (5.55), respectively.

We now examine the relation between the direction of the incident wave $\boldsymbol{K}_0$ and W. The direction of the incident X-rays is along $\overrightarrow{\mathrm{AO}}$ as shown by vector $\boldsymbol{K}_0$ in Fig. 5.5. Let α be the angle between the normal line and $\overline{\mathrm{AO}}$. When the incident direction of $\boldsymbol{K}_0$ changes, $\boldsymbol{K}_0$ rotates around O and the angle α changes. Once $\boldsymbol{K}_0$ is given, the tie-point A in vacuum is given and the coordinates (X_0, $Y_0^{(j)}$) of the tie-point in the crystal A_j are obtained.

5.3.4 *Deviation Parameter*

If the coordinates of L_o with respect to the origin O are written as (X_L, Y_L), then

$$X_L = \kappa_0 \sin\theta_1,\ Y_L = \kappa_0 \cos\theta_1. \tag{5.78}$$

The X coordinate of point A from L_o written as X_0 is given by

$$X_0 = K_0 \sin\alpha - X_L = K_0 \sin\alpha - \kappa_0 \sin\theta_1. \tag{5.79}$$

By inserting X_0 into (5.65), we get

$$W = -(K_0 \sin\alpha - \kappa_0 \sin\theta_1) \frac{\sin 2\theta'_{\mathrm{B}}}{|\cos\theta_1 \cos\theta_2|^{\frac{1}{2}}} \frac{1}{P\kappa_0 (\chi_g \chi_{-g})^{\frac{1}{2}}}. \tag{5.80}$$

This gives the relation between W and the incident angle α. Since $X_0 = 0$ at $W = 0$, we can write

$$K_0 \sin\alpha_L = \kappa_0 \sin\theta_1 \tag{5.81}$$

by using the incident angle α_L for $W = 0$. This means that when the wave is incident at the glancing angle $\alpha = \alpha_L$, it is refracted by the crystal with its refractive index of n_0, and satisfies the Bragg condition. If the complementary angles of α and α_L, i.e., $\pi/2 - \alpha$ and $\pi/2 - \alpha_L$ are denoted as α_0 and α_{0L}, respectively, W is rewritten as

$$W = -K_0 (\cos\alpha_0 - \cos\alpha_{0L}) \frac{\sin 2\theta'_{\mathrm{B}}}{|\cos\theta_1 \cos\theta_2|^{\frac{1}{2}}} \frac{1}{P\kappa_0 (\chi_g \chi_{-g})^{\frac{1}{2}}}. \tag{5.82}$$

Since $\sin\alpha_{0L} \approx \cos\theta_1$ and $\alpha_0 \approx \alpha_{0L}$, we get

$$\cos\alpha_0 - \cos\alpha_{0L} = -2 \sin\frac{\alpha_0 + \alpha_{0L}}{2} \sin\frac{\alpha_0 - \alpha_{0L}}{2} = -\Delta\alpha \cos\theta_1,$$

by putting $\Delta\alpha$ as

$$\Delta\alpha = \alpha_0 - \alpha_{0L}. \tag{5.83}$$

Using these relations, we get

$$W \approx \left(\frac{\cos\theta_1}{|\cos\theta_2|} \right)^{\frac{1}{2}} \frac{\sin 2\theta'_{\mathrm{B}}}{n_0 P |\chi_g|} \Delta\alpha. \tag{5.84}$$

In this book, we adopt the deviation parameter W according to Miyake [3, 4]. W is essentially similar to Zachariasen's [5] and Pinker's [6] y, James's [7, 8] p and Batterman and Cole's [9] and Authier's [2] η.

When resonance scattering is considered, $(\chi_g \chi_{-g})^{1/2}$ in the denominator of (5.82) is complex, and the deviation parameter is also complex. Since $\Delta\alpha$ is real, W is also required to be real [10, 11]. The treatment in such a case will be explained in Chap. 7.

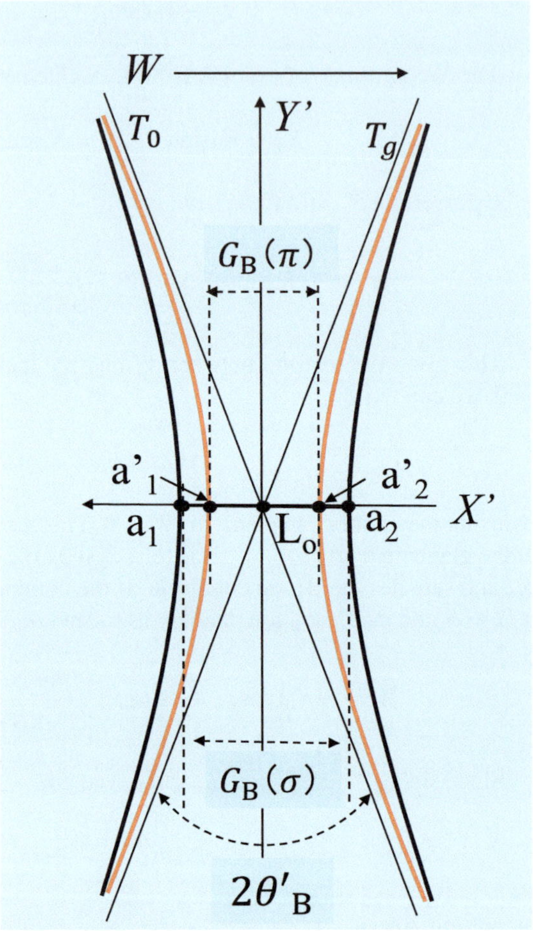

Fig. 5.8 Polarization dependence of dispersion surfaces. The black line shows the dispersion surface for σ-polarization and the brown line that for π-polarization

5.4 Effect of Polarization on Dispersion Surface

Equation (5.49) gives a dispersion surface, and the effect of polarization appears as a factor P on the right side, which is related to the Bragg gap as

$$G_B = \frac{P\kappa_0\left(\chi_g\chi_{-g}\right)^{1/2}}{\cos\theta'_{\rm B}}.$$

Figure 5.8 shows the dispersion surfaces for σ-polarized X-ray ($P(\sigma) = 1$) and π-polarized X-ray ($P(\pi) = |\cos 2\theta'_{\rm B}|$). The Bragg gap for σ-polarization is $\overline{{\rm a}_1{\rm a}_2}$ and that for π-polarization is $\overline{{\rm a}'_1{\rm a}'_2}$. The Bragg gap for π-polarization is smaller than that for σ-polarization.

References

1. J.M. Cowley, *Diffraction Physics* (North Holland, 1975)
2. A. Authier, *Dynamical Theory of X-ray Diffraction* (Oxford University Press, 2001)
3. S. Miyake, Acta Cryst. **A24**, 182–171 (1968)
4. S. Miyake, *X-ray diffraction.* (Asakura (In Japanese), 1969)
5. W.H. Zachariasen, *Theory of X-Ray Diffraction in Crystal* (John Willy Sonds Inc., New York, 1945)
6. Z. Pinsker, *Dynamical Scattering of X-rays in Crystals* (Springer, Berlin, 1977)
7. R.W. James, *The Optical Principles of The Diffraction of X-Rays* (G. Bell and Sons Ltd., London, 1967)
8. R.W. James, Solid State Physics. **15**, 53–220 (1963)
9. B.W. Battermann, H. Cole, Rev. Mod. Phys. **36**, 681–717 (1964)
10. N. Kato, Acta Cryst. **A48**, 829–833 (1992)
11. T. Fukamachi, T. Kawamura, Acta Cryst. **A49**, 384–388 (1993)

Chapter 6
Dynamical Theory of Diffraction for Non-absorbing Crystal

We discuss X-ray diffraction from a perfect crystal when X-rays are not absorbed. In practice, X-rays are always absorbed in a crystal and a dynamical theory of diffraction that takes account of the absorption effect is essential, which will be described in Chap. 7. However, it is still quite important to understand the dynamical theory of diffraction without absorption, because it gives basic ideas of the dynamical theory of diffraction and phenomena peculiar to a perfect crystal. First, we calculate the diffraction and transmission intensities in the Laue case, then consider the wave field in the crystal (Pendellösung fringes and standing waves in a crystal) and the energy flow within the crystal represented by the Poynting vector. Since the energy flow behaves as if it were a refracted wave, we call it a pseudo-refracted wave to describe its trajectory. Then, we explain diffraction and transmission intensities, Pendellösung fringes, extinction effects and so on in the Bragg case.

6.1 Laue Case ($\cos\theta_2 > 0$)

Figure 6.1 shows the diffraction geometries in the Laue case for a plane-parallel crystal of thickness H, in the real space (a) and the reciprocal space (b). Figure 6.2 shows the dispersion surface in the symmetric Laue case. This corresponds to the dispersion surface where $\beta = 0$ and $\theta_2 = -\theta_1 = \theta'_{\mathrm{B}}$ in Fig. 5.6. The hyperbola closer to the origin is labeled branch (1), and the other one is labeled branch (2). The intersection of dispersion surface of branch (1) and normal line v is defined as tie-point A_1, and that of branch (2) is defined as A_2. Two waves, $\boldsymbol{k}_0^{(1)}$ and $\boldsymbol{k}_0^{(2)}$ (referred to as primary waves or O-waves) are excited in the crystal at these tie-points in the direction of the incident X-ray. In the diffracted wave direction, two waves, $\boldsymbol{k}_g^{(1)}$ and $\boldsymbol{k}_g^{(2)}$ (referred to as secondary waves or G-waves) are excited. Among these four waves, we have relations

T. Fukamachi and T. Kawamura, *Resonant Dynamical X-ray Diffraction*, Springer Series in Solid-State Sciences 208, https://doi.org/10.1007/978-981-96-8406-9_6

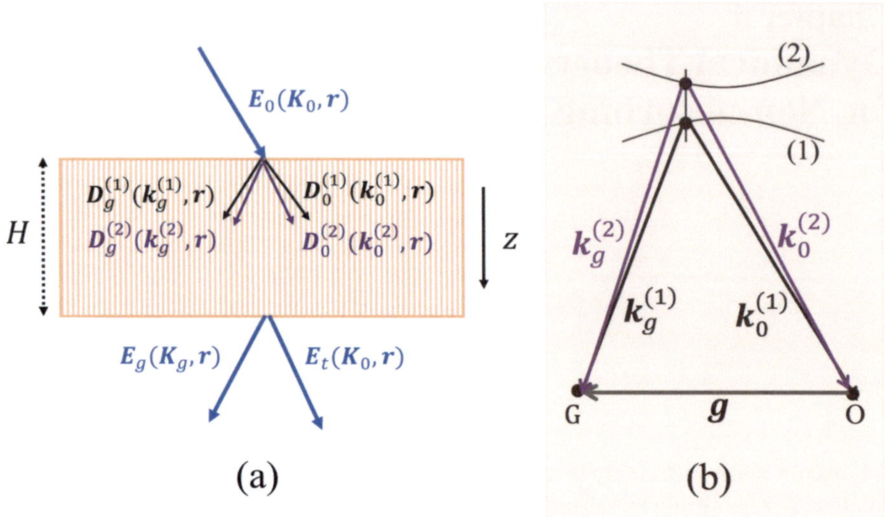

Fig. 6.1 Diffraction geometry in the Laue case in real space **a** and in reciprocal space **b**

$$k_g^{(1)} = k_0^{(1)} + g, \quad k_g^{(2)} = k_0^{(2)} + g. \tag{6.1}$$

As shown in Fig. 6.1a, when a plane wave of $E_0^a \exp(-i\boldsymbol{K}_0 \cdot \boldsymbol{r})$ is incident on the crystal from the vacuum, two primary waves, $D_0^{(1)} \exp\left(-i\boldsymbol{k}_0^{(1)} \cdot \boldsymbol{r}\right)$ and $D_0^{(2)} \exp\left(-i\boldsymbol{k}_0^{(2)} \cdot \boldsymbol{r}\right)$ are created inside the crystal by the incident wave. Two secondary waves in the diffraction direction $D_g^{(1)} \exp\left(-i\boldsymbol{k}_g^{(1)} \cdot \boldsymbol{r}\right)$ and $D_g^{(2)} \exp\left(-i\boldsymbol{k}_g^{(2)} \cdot \boldsymbol{r}\right)$ are also created inside the crystal. The two primary waves create transmitted wave $E_t^a \exp(-i\boldsymbol{K}_0 \cdot \boldsymbol{r})$ below the bottom surface, and the two secondary waves create diffracted wave $E_g^a \exp\left(-i\boldsymbol{K}_g \cdot \boldsymbol{r}\right)$ below the surface. Here, E_0^a, E_g^a and E_t^a are the electric field of the incident, diffracted, and transmitted X-rays, respectively, and the superscript 'a' denotes the polarization.

The boundary conditions at the entrance surface ($z = 0$) are

$$D_0^{(1)} + D_0^{(2)} = E_0^a, \tag{6.2}$$

$$D_g^{(1)} + D_g^{(2)} = 0. \tag{6.3}$$

The boundary conditions at the bottom surface ($z = H$) are

$$D_0^{(1)} \exp\left(-ik_{0z}^{(1)} H\right) + D_0^{(2)} \exp\left(-ik_{0z}^{(2)} H\right) = E_t^a \exp(-iK_{0z} H), \tag{6.4}$$

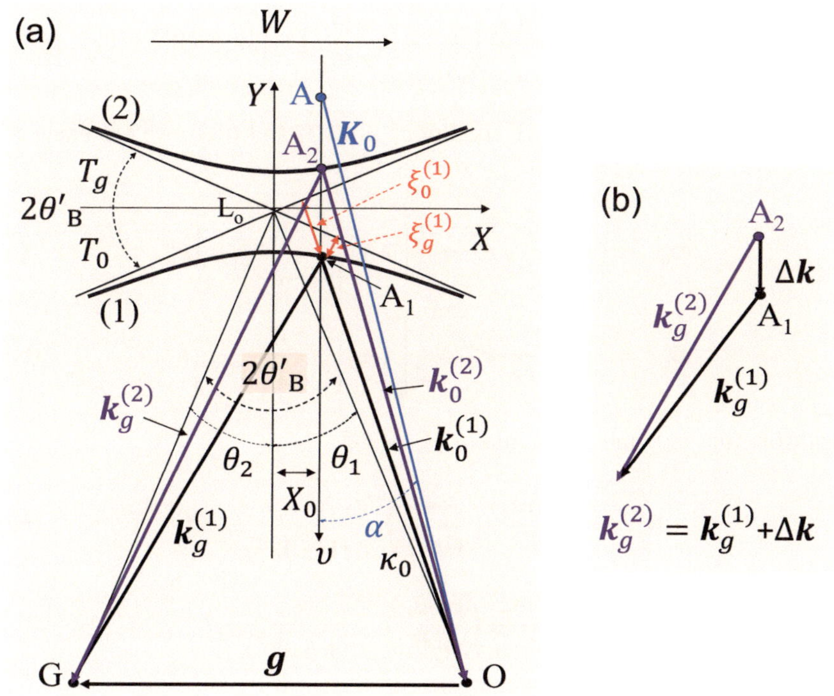

Fig. 6.2 Dispersion surface in the symmetric Laue case (**a**). Difference in the wave vectors between branch (1) and branch (2)

$$D_g^{(1)} \exp\left(-ik_{gz}^{(1)} H\right) + D_g^{(2)} \exp\left(-ik_{gz}^{(2)} H\right) = E_g^a \exp(-iK_{gz} H), \tag{6.5}$$

where k_{0z} and k_{gz} are the downward perpendicular components of $\boldsymbol{k}_0$ and $\boldsymbol{k}_g$ to the crystal surface.

Using (5.56) in Chap. 5, we get

$$D_g^{(1)} = \frac{2\xi_0^{(1)}}{\kappa_0} \frac{1}{P\chi_{-g}} D_0^{(1)} = r^{(1)} D_0^{(1)}, \quad D_g^{(2)} = \frac{2\xi_0^{(2)}}{\kappa_0} \frac{1}{P\chi_{-g}} D_0^{(2)} \\ = r^{(2)} D_0^{(2)}. \tag{6.6}$$

$\xi_0^{(1)}$ and $\xi_0^{(2)}$ are given by (5.70) and (5.71) in Chap. 5. $r^{(1)}$ and $r^{(2)}$ in (6.6) are expressed as a function of the deviation parameter W as

$$r^{(1)} = \frac{D_g^{(1)}}{D_0^{(1)}} = \frac{2}{\kappa_0 P \chi_{-g}} \xi_0^{(1)} = \frac{\left(\chi_g \chi_{-g}\right)^{1/2}}{\chi_{-g}} \left(\frac{\cos\theta_1}{\cos\theta_2}\right)^{1/2} \left[W - \left(W^2 + 1\right)^{1/2}\right], \tag{6.7}$$

$$r^{(2)} = \frac{D_g^{(2)}}{D_0^{(2)}} = \frac{2}{\kappa_0 P \chi_{-g}} \xi_0^{(2)} = \frac{\left(\chi_g \chi_{-g}\right)^{1/2}}{\chi_{-g}} \left(\frac{\cos\theta_1}{\cos\theta_2}\right)^{1/2} \left[W + \left(W^2+1\right)^{1/2}\right]. \quad (6.8)$$

Using (6.6) and the boundary conditions in (6.2) and (6.3), we eliminate $D_g^{(j)}$. Making use of the relation in (6.7) and (6.8), we get

$$D_0^{(1)} = \frac{-r^{(2)}}{r^{(1)} - r^{(2)}} E_0^a = \frac{1}{2}\left[1 + \frac{W}{\left(1+W^2\right)^{1/2}}\right] E_0^a, \quad (6.9)$$

$$D_0^{(2)} = \frac{r^{(1)}}{r^{(1)} - r^{(2)}} E_0^a = \frac{1}{2}\left[1 - \frac{W}{\left(1+W^2\right)^{1/2}}\right] E_0^a. \quad (6.10)$$

Substituting (6.9) and (6.10) into (6.6), we get

$$D_g^{(1)} = \frac{\exp(i\alpha_g)}{2} \left(\frac{\cos\theta_1}{\cos\theta_2}\right)^{1/2} \frac{-1}{\left(1+W^2\right)^{1/2}} E_0^a, \quad (6.11)$$

$$D_g^{(2)} = \frac{\exp(i\alpha_g)}{2} \left(\frac{\cos\theta_1}{\cos\theta_2}\right)^{1/2} \frac{1}{\left(1+W^2\right)^{1/2}} E_0^a, \quad (6.12)$$

indicating $D_g^{(1)} \neq D_g^{(2)}$.

For a non-absorbing crystal, $\left(\chi_g \chi_{-g}\right)^{1/2} = \left|\chi_g\right|$, the phase $\exp(i\alpha_g)$ of χ_g is given by

$$\chi_g = \left|\chi_g\right| \exp(i\alpha_g). \quad (6.13)$$

By using (6.13), we have $\chi_{-g} = \chi_g^* = \left|\chi_g\right| \exp(-i\alpha_g)$, and

$$\frac{\left|\chi_g\right|}{\chi_{-g}} = \frac{\left|\chi_g\right|}{\left|\chi_g\right| \exp(-i\alpha_g)} = \exp(i\alpha_g). \quad (6.14)$$

6.1.1 Diffraction and Transmission Intensities

We describe the diffraction and transmission intensities. The electric field of diffracted wave E_g^a is related to $D_g^{(j)}$ under the boundary condition of (6.5). By substituting (6.11) and (6.12) into this $D_g^{(j)}$, we get,

$$\frac{E_g^a}{E_0^a} = \frac{\exp(iK_{gz}H)\exp(i\alpha_g)}{2}\left(\frac{\cos\theta_1}{\cos\theta_2}\right)^{1/2}\frac{\exp\left(-ik_{gz}^{(2)}H\right)}{\left(1+W^2\right)^{1/2}}$$
$$\left[1-\exp\left\{i\left(k_{gz}^{(2)}-k_{gz}^{(1)}\right)H\right\}\right]. \tag{6.15}$$

By using the relation

$$k_{gz}^{(2)} - k_{gz}^{(1)} = Y_0^{(2)} - Y_0^{(1)} = 2s\left(W^2+1\right)^{\frac{1}{2}} \tag{6.16}$$

with the parameter s defined by

$$s = \frac{K_0 P\left(\chi_g \chi_{-g}\right)^{1/2}}{2|\cos\theta_1\cos\theta_2|^{1/2}} = \frac{G_B\cos\theta_B'}{2|\cos\theta_1\cos\theta_2|^{1/2}}, \tag{6.17}$$

we have the absolute value of (6.15) as

$$\left|\frac{E_g^a}{E_0^a}\right| = \frac{1}{2}\left(\frac{\cos\theta_1}{\cos\theta_2}\right)^{1/2}\frac{\left|1-\exp\left[2i\left(W^2+1\right)^{1/2}sH\right]\right|}{\left(1+W^2\right)^{1/2}}. \tag{6.18}$$

The square of (6.18) gives the intensity ratio. For comparing the measured intensity ratio, it is necessary to take the finite size of the incident X-ray beam into account. In an experiment, the total energy of the diffracted X-ray beam is measured when a finite-sized X-ray beam is incident on the crystal, as shown in Fig. 6.3. When the area of the crystal surface is larger than that irradiated by the X-ray flux, the incident and the diffracted beam intensities are proportional to their cross-sectional areas S_0 and S_g. The cross-sectional areas for the incident and diffracted beams are given by using the area S irradiated by the incident X-ray beam as

$$S_0 = S\cos\theta_1, S_g = S\cos\theta_2. \tag{6.19}$$

If we denote the energy flux of the incident X-ray per unit time as P_0, and that of the diffracted X-ray as P_g, we get

$$\frac{P_g}{P_0} = \left|\frac{E_g^a}{E_0^a}\right|^2\frac{S_g}{S_0} = \left|\frac{E_g^a}{E_0^a}\right|^2\frac{\cos\theta_2}{\cos\theta_1} = \frac{\sin^2[sH(1+W^2)^{\frac{1}{2}}]}{1+W^2}. \tag{6.20}$$

As for the transmitted intensity, the electric field of transmitted wave E_t^a is related to $D_0^{(j)}$ by the boundary condition of (6.4). Substituting $D_0^{(j)}$ into (6.9) and (6.10), we obtain the energy flux of the transmitted X-ray P_t as

$$\frac{P_t}{P_0} = \left|\frac{E_t^a}{E_0^a}\right|^2 = \frac{W^2+\cos^2[sH(1+W^2)^{1/2}]}{1+W^2} = 1 - \frac{\sin^2[sH(1+W^2)^{\frac{1}{2}}]}{1+W^2}. \tag{6.21}$$

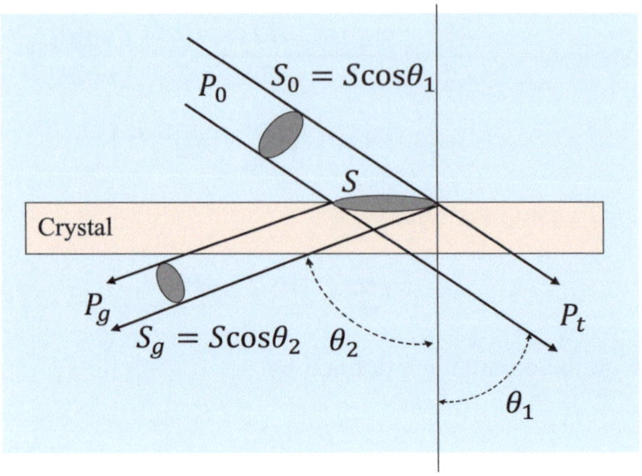

Fig. 6.3 Cross-sections of the incident and the diffracted beam fluxes

From (6.20) and (6.21), we have a relation

$$P_0 = P_g + P_t, \tag{6.22}$$

meaning that, unlike the kinematical theory of diffraction, the energy flux is conserved in the dynamical theory of diffraction.

Figure 6.4 shows the changes in P_g/P_0 and P_t/P_0 with respect to the deviation parameter W for $sH = \pi/4$, $\pi/2$ and π in the symmetric Laue case. Regardless of the value of sH, P_g/P_0 decreases while repeating oscillations as $|W|$ increases. On the other hand, P_t/P_0 gradually increases up to 1 while repeating oscillations as $|W|$ increases. The two curves P_g/P_0 and P_t/P_0 are symmetrical with respect to $W = 0$, and the increase and decrease of their oscillatory parts are reversed. The period of the oscillations becomes shorter as sH becomes large. Further, the value of P_g/P_0 at $W = 0$ is 1 when $sH = \pi/2$, but is zero when $sH = \pi$. Generally, the value of P_g/P_0 at $W = 0$ is 1 for $sH = m\pi + \pi/2$ and zero for $sH = m\pi$, with m being an integer. These variations are shown in Fig. 6.5 with sH on the horizontal axis and $P_g(W = 0)/P_0$ and $P_t(W = 0)/P_0$ on the vertical axis. This periodic variation in diffraction intensity and transmission intensity is Pendellösung fringes, which will be described later.

When s is constant, sH increases with increasing crystal thickness H. For a sufficiently thick crystal, the period of the oscillation is sufficiently short, and the oscillating part can be averaged to get

$$\frac{\overline{P}_g}{P_0} = \frac{1}{2\left(1 + W^2\right)}. \tag{6.23}$$

Figure 6.6 shows the average diffracted intensity $\overline{P}_g/P_0$ as well as the averaged transmitted intensity $\overline{P_t}/P_0$. $\overline{P}_g/P_0$ shows a peak with the maximum value of 0.5 at $W = 0$, while $\overline{P_t}/P_0$ shows the minimum of 0.5 at $W = 0$. When $W = 0$, that is, the

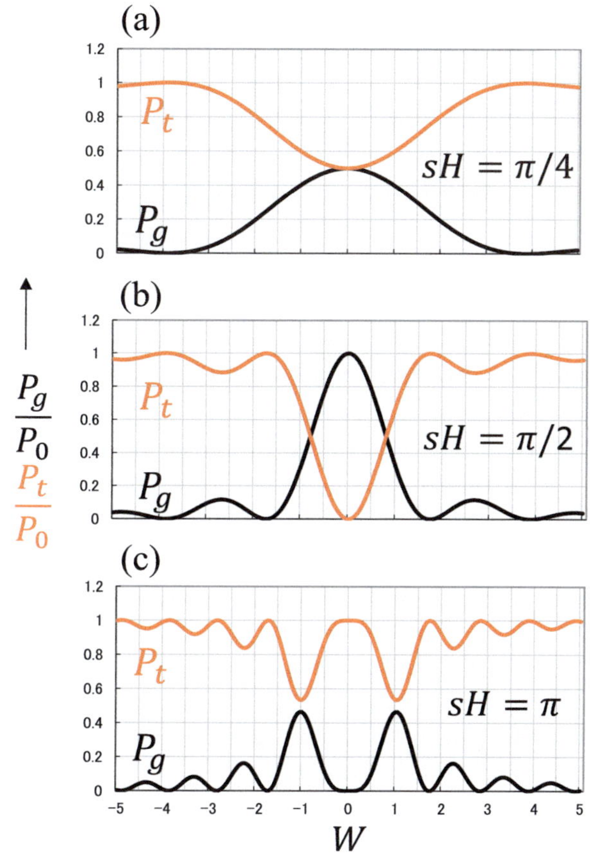

Fig. 6.4 Rocking curves in the symmetric Laue case. **a** $sH = \pi/4$, **b** $sH = \pi/2$ and **c** $sH = \pi$

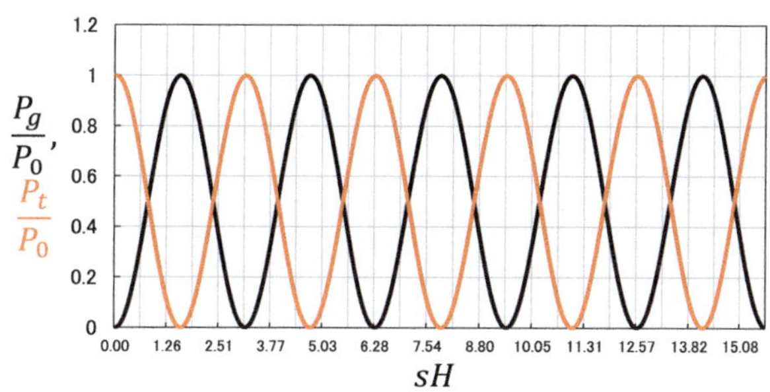

Fig. 6.5 P_g/P_0 (black line) and P_t/P_0 (orange line) at $W = 0$

Bragg condition in a crystal is satisfied exactly, the average diffracted intensity and the transmitted intensity are equal, and the energy fluxes are shared equally between the diffracted beam and the transmitted beam. When a wave is incident at a condition away from the Bragg condition, the intensity of the diffracted wave becomes smaller and only the transmitted wave has the intensity of up to 1.

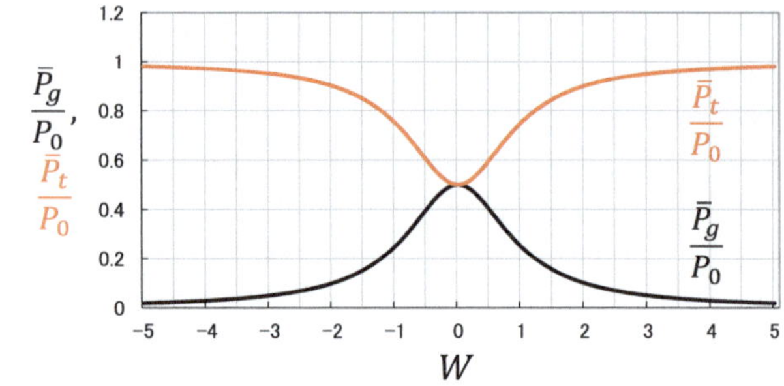

Fig. 6.6 The diffracted and transmitted rocking curves from a very thick plane parallel crystal

6.1.2 *Wavefield in a Crystal*

In a crystal, four plane waves are excited from two tie-points. We consider two kinds of interferences formed by two of these waves. One is the interference between the primary wave and secondary wave from one tie-point (standing wave), and the other is that of two waves of different branches corresponding to two tie-points (Pendellösung fringes).

6.1.2.1 Pendellösung Fringes

The intensity of the sum of the two branch waves corresponding to the secondary waves $\boldsymbol{k}_g^{(1)}$ and $\boldsymbol{k}_g^{(2)}$ is given by

$$\begin{aligned} I_g(\boldsymbol{r}) &= \left| D_g^{(1)}\exp\left(-i\boldsymbol{k}_g^{(1)}\cdot\boldsymbol{r}\right) + D_g^{(2)}\exp\left(-i\boldsymbol{k}_g^{(2)}\cdot\boldsymbol{r}\right)\right|^2 \\ &= \left|E_0^a\right|^2\frac{\cos\theta_1}{\cos\theta_2}\frac{1}{W^2+1}\sin^2\left[s\left(W^2+1\right)^{\frac{1}{2}}z\right] \end{aligned} \tag{6.24}$$

at position $\boldsymbol{r}$ inside a crystal. In the same way, that corresponding to the primary waves is

$$\begin{aligned} I_0(\boldsymbol{r}) &= \left| D_0^{(1)}\exp\left(-i\boldsymbol{k}_0^{(1)}\cdot\boldsymbol{r}\right) + D_0^{(2)}\exp\left(-i\boldsymbol{k}_0^{(2)}\cdot\boldsymbol{r}\right)\right|^2 \\ &= \left|E_0^a\right|^2\left[\cos^2\left\{s\left(W^2+1\right)^{1/2}z\right\} + \frac{W^2}{W^2+1}\sin^2\left\{s\left(W^2+1\right)^{1/2}z\right\}\right]. \end{aligned} \tag{6.25}$$

Since $I_g(z) \propto \sin^2(sz)$ and $I_0(z) \propto \cos^2(sz)$ for $W = 0$, $I_0(z)$ shows the maximum and $I_g(z)$ shows the minimum at the crystal surface ($z = 0$) as depicted in Fig. 6.7. As the depth z increases from the surface, $I_0(z)$ and $I_g(z)$ shows oscillations. The depth where $I_0(z)$ shows the first minimum and $I_g(z)$ shows the maximum is called the extinction distance, which is given by

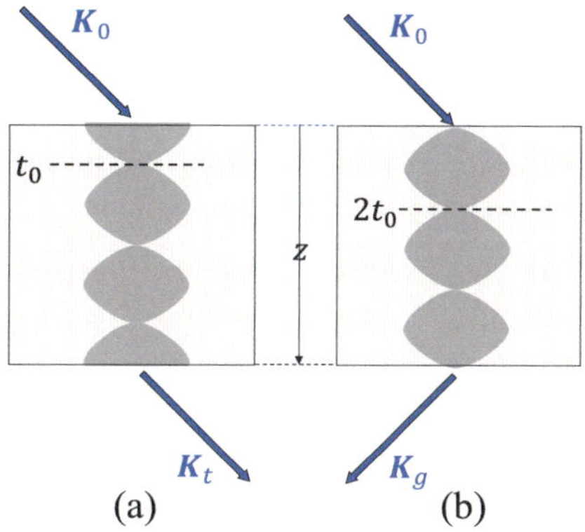

Fig. 6.7 Schematic drawings of Pendellösung fringes for $W = 0$ of the primary wave (**a**) and the secondary wave (**b**)

$$t_0 = \frac{\pi}{2s} = |\cos\theta_1 \cos\theta_2|^{1/2} \frac{\pi}{\kappa_0 P|\chi_g|}. \tag{6.26}$$

$I_0(z)$ and $I_g(z)$ increase and decrease with the period of $2t_0$ while keeping the sum of the intensities $I_g + I_0$ constant. This means that the primary wave and secondary wave propagate through the crystal while exchanging energy between the two waves. Since this situation is similar to the dynamics of vibrations of two-coupled pendulums, this solution is called the solution of a loosely coupled two-coupled pendulum (Pendellösung), and the interference fringes in the crystal shown in Figs. 6.4, 6.5 and 6.7 are called Pendellösung fringes. These Pendellösung fringes are also called the equal-thickness interference fringe because they occur depending on the depth from the crystal surface.

According to sin term (6.24), the period of the fringes is given by $s(W^2+1)^{1/2} l = \pi$ with l being the minimum distance from the surface given by

$$l = \frac{\pi}{s(W^2+1)^{\frac{1}{2}}} = \frac{2\pi |\cos\theta_1 \cos\theta_2|^{\frac{1}{2}}}{\kappa_0 P|\chi_g|(W^2+1)^{\frac{1}{2}}} = \frac{2\pi}{|\Delta \boldsymbol{k}|}. \tag{6.27}$$

From (6.16), $\Delta \boldsymbol{k}$ is given as

$$\Delta \boldsymbol{k} = \boldsymbol{k}_0^{(2)} - \boldsymbol{k}_0^{(1)} = \boldsymbol{k}_g^{(2)} - \boldsymbol{k}_g^{(1)} = \overline{\mathrm{A}_2\mathrm{A}_1} = Y_0^{(2)} - Y_0^{(1)}. \tag{6.28}$$

6.1.2.2 Standing Wave in a Crystal

We consider the interference between the primary wave and the secondary wave excited at the same tie-point. The interference is given by

$$
\begin{aligned}
I^{(j)}(\boldsymbol{r}) &= \left|D^{(j)}(\boldsymbol{r})\right|^2 = \left|D_0^{(j)}\exp\left(-i\boldsymbol{k}_0^{(j)}\cdot\boldsymbol{r}\right) + D_g^{(j)}\exp\left(-i\boldsymbol{k}_g^{(j)}\cdot\boldsymbol{r}\right)\right|^2 \\
&= \left|D_0^{(j)}\right|^2\left|1 + \left|\frac{D_g^{(j)}}{D_0^{(j)}}\right|\exp\left(i\omega^{(j)}\right)\exp(-i\boldsymbol{g}\cdot\boldsymbol{r})\right|^2 \\
&= \left|D_0^{(j)}\right|^2\left[1 + |r^{(j)}|^2 + 2|r^{(j)}|\cos\left(\boldsymbol{g}\cdot\boldsymbol{r} - \omega^{(j)}\right)\right].
\end{aligned} \tag{6.29}
$$

Here, we use $\boldsymbol{k}_g^{(j)} - \boldsymbol{k}_0^{(j)} = \boldsymbol{g}$ as shown in Fig. 6.2a. $r^{(j)}$ is given in (6.7) and (6.8) and rewritten as

$$
r^{(j)} = \exp\left(i\alpha_g\right)\left(\frac{\cos\theta_1}{\cos\theta_2}\right)^{1/2}\left[W + (-1)^j(W^2+1)^{\frac{1}{2}}\right], \tag{6.30}
$$

by using (6.14). We express this as

$$
r^{(j)} = \frac{D_g^{(j)}}{D_0^{(j)}} = |r^{(j)}|\exp\left(i\omega^{(j)}\right). \tag{6.31}
$$

with $\omega^{(j)}$ being the phase of $r^{(j)}$.

Since χ_0 is negative and χ_g is also generally negative for monatomic crystals, we put $\alpha_g = \pi$ in (6.30) and examine (6.29) in the case of symmetric reflection. As $\exp\left(i\alpha_g\right) = -1$, the term inside the square bracket in (6.30) is negative for $j = 1$, and $r^{(1)}$ is always positive and $\omega^{(1)} = 0$. For $j = 2$, the term inside the square bracket is positive, $r^{(2)}$ is always negative, and $\omega^{(2)} = \pi$. When $W = 0$, $|r^{(1)}| = |r^{(2)}| = 1$, and the term in the square bracket in (6.29) becomes

$$
2\left\{1 + \cos\left(\frac{2\pi x}{d} - \omega^{(j)}\right)\right\} \tag{6.32}
$$

by taking x-axis parallel to $\boldsymbol{g}$ and using $\boldsymbol{g}\cdot\boldsymbol{r} = 2\pi x/d$. $I^{(j)}(\boldsymbol{r})$ in (6.29) represents a standing wave with a period of d as shown in Fig. 6.8b and c. The standing wave of branch (1) has its antinode on each lattice plane, and the standing wave of branch (2) has its node on each lattice plane. As will be described in detail later, the standing wave of branch (1) has a strong electric wavefield on the lattice planes, so that it undergoes absorption stronger than the average absorption by the atoms on the planes. This effect is called the anomalous absorption. On the other hand, the standing wave of branch (2) has a weaker electric wavefield on the lattice planes, so that it undergoes almost no absorption by atoms on the planes. This effect is called anomalous transmission. It is also called the Borrmann effect, after Borrmann [1], who experimentally discovered the anomalous absorption and the anomalous transmission. However, it is noted that we are dealing with a non-absorbing crystal here, and there is no absorption by atoms on the planes and the Borrmann effect does not appear. The Borrmann effect will be discussed in detail in Chap. 7.

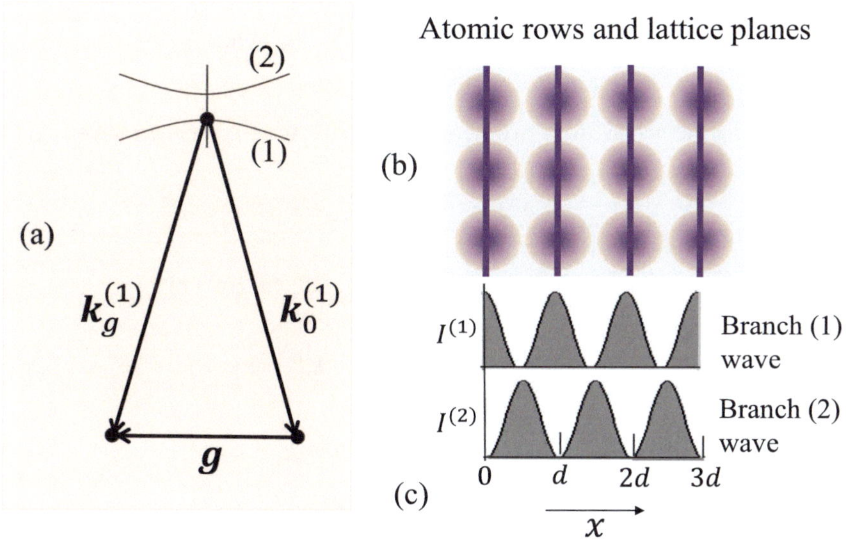

Fig. 6.8 Standing waves in a crystal in the symmetric Laue case. **a** Dispersion surface and the excited branch (1) waves. **b** Arrangement of atomic rows and lattice planes perpendicular to x–z plane. **c** Standing waves of $I^{(1)}$ and $I^{(2)}$

6.1.3 Energy Flow—In Terms of Poynting Vector

In the two-wave approximation, the primary and secondary waves in a crystal excited on the same branch cannot exist independently as given by (6.6). The direction of X-ray energy flow in the crystal is neither the direction of $\boldsymbol{k}_0^{(j)}$ nor the direction of $\boldsymbol{k}_g^{(j)}$. As the propagation direction of an electromagnetic wave is given by its Poynting vector, the flow of X-ray energy within the crystal is also given by the Poynting vector.

Since the Poynting vector $\boldsymbol{S}$ fluctuates with the frequency of X-ray, the temporal average Poynting vector $\boldsymbol{S}$ is

$$\boldsymbol{S} \approx \frac{\varepsilon_0}{2}\mathrm{Re}\left(\boldsymbol{D} \times \boldsymbol{H}^*\right), \tag{6.33}$$

where the asterisk denotes the complex conjugate. Strictly speaking, $\boldsymbol{S}$ is defined by

$$\boldsymbol{S} = \frac{1}{2}\mathrm{Re}\left(\boldsymbol{E} \times \boldsymbol{H}^*\right). \tag{6.34}$$

However, since $\boldsymbol{D} = \varepsilon_0\left(1 + \chi_g\right)\boldsymbol{E}$ and χ_g is very small, (6.33) holds as a good approximation. From (5.8) and (5.9), $\boldsymbol{D}$ and $\boldsymbol{H}$ are expressed by

$$\boldsymbol{D} = \exp(i\omega t)\sum_j \exp\left(-i\boldsymbol{k}_0^{(j)} \cdot \boldsymbol{r}\right)\sum_g \boldsymbol{D}_g \exp(-i\boldsymbol{g} \cdot \boldsymbol{r}), \tag{6.35}$$

$$\boldsymbol{H} = \exp(i\omega t)\sum_j \exp\left(-i\boldsymbol{k}_0^{(j)}\cdot\boldsymbol{r}\right)\sum_g \boldsymbol{H}_g \exp(-i\boldsymbol{g}\cdot\boldsymbol{r}) \tag{6.36}$$

by considering the waves from two tie-points ($j = 1$ and 2). We then have

$$\begin{aligned} \boldsymbol{D}\times\boldsymbol{H}^* &= \sum_j \exp\left(-i\boldsymbol{k}_0^{(j)}\cdot\boldsymbol{r}\right)\sum_g \boldsymbol{D}_g\exp(-i\boldsymbol{g}\cdot\boldsymbol{r}) \\ &\quad\times\sum_{j'} \exp\left(i\boldsymbol{k}_0^{*(j')}\cdot\boldsymbol{r}\right)\sum_{g'} \boldsymbol{H}_{g'}^*\exp(i\boldsymbol{g}'\cdot\boldsymbol{r}) \\ &= \sum_j\sum_{j'} \exp\left[i\left(\boldsymbol{k}_0^{*(j')} - \boldsymbol{k}_0^{(j)}\right)\cdot\boldsymbol{r}\right] \\ &\quad\sum_g\sum_{g'} (\boldsymbol{D}_g^{(j)}\times\boldsymbol{H}_g^{*(j')})\exp[i(\boldsymbol{g}'-\boldsymbol{g})\cdot\boldsymbol{r})]. \end{aligned} \tag{6.37}$$

Using (6.28), we can write the first exponential factor at depth z as

$$(\boldsymbol{k}_0^{*(j')} - \boldsymbol{k}_0^{(j)})\cdot\boldsymbol{r} = \Delta k_z^{(j,j')}z,\ \Delta k_z^{(j,j')} = \boldsymbol{k}_0^{*(j')} - \boldsymbol{k}_0^{(j)}. \tag{6.38}$$

The spatial average of $\exp[i(\boldsymbol{g}'-\boldsymbol{g})\cdot\boldsymbol{r})]$ must be zero for $\boldsymbol{g}' \neq \boldsymbol{g}$ and becomes 1 for $\boldsymbol{g}' = \boldsymbol{g}$, as it is a periodic function of the lattice. The time and spatial average of $\boldsymbol{D}\times\boldsymbol{H}^*$ is given by

$$\boldsymbol{D}\times\boldsymbol{H}^* = \sum_j\sum_{j'} \exp\left(i\Delta k_z^{(j,j')}z\right)\sum_g (\boldsymbol{D}_g^{(j)}\times\boldsymbol{H}_g^{*(j')}). \tag{6.39}$$

By taking the vector product of $\boldsymbol{k}_g$ from the left of both sides of $\omega\boldsymbol{D}_g = \boldsymbol{H}_g\times\boldsymbol{k}_g$ of (5.14), and using $\boldsymbol{k}_g^2\boldsymbol{H}_g = K_0(\boldsymbol{k}_g\times\boldsymbol{D}_g)$, $\omega = cK_0$ and $|\boldsymbol{k}_g| = K_0$, we get

$$\boldsymbol{H}_g^* = \frac{1}{|\boldsymbol{k}_g|}\left(\boldsymbol{k}_g\times\boldsymbol{D}_g^*\right). \tag{6.40}$$

Then we rewrite $\boldsymbol{D}_g^{(j)}\times\boldsymbol{H}_g^{*(j')}$ in (6.39) as

$$\begin{aligned} \boldsymbol{D}_g^{(j)}\times\boldsymbol{H}_g^{*(j')} &= \frac{c}{|\boldsymbol{k}_g|}(\boldsymbol{D}_g^{(j)}\times\left(\boldsymbol{k}_g\times\boldsymbol{D}_g^{*(j')}\right)) \\ &= cs_g(\boldsymbol{D}_g^{(j)}\cdot\boldsymbol{D}_g^{*(j')}) \end{aligned} \tag{6.41}$$

using the relation $\boldsymbol{D}_g\times\left(\boldsymbol{k}_g\times\boldsymbol{D}_g^*\right) = \boldsymbol{k}_g\left(\boldsymbol{D}_g\cdot\boldsymbol{D}_g^*\right) - \boldsymbol{D}_g^*(\boldsymbol{D}_g\cdot\boldsymbol{k}_g) = \boldsymbol{k}_g\left(\boldsymbol{D}_g\cdot\boldsymbol{D}_g^*\right)$. We now get temporally and spatially averaged Poynting vector as

$$\boldsymbol{S} = \frac{c}{2\varepsilon_0}\sum_g\left[\sum_j\left|\boldsymbol{D}_g^{(j)}\right|^2\boldsymbol{s}_g + \sum_j\sum_{j'}\cos\left(\Delta k_z^{(j,j')}z\right)\left(\boldsymbol{D}_g^{(j)}\cdot\boldsymbol{D}_g^{(j')}\right)\boldsymbol{s}_g\right]. \tag{6.42}$$

As shown by (6.27), $\Delta k_z^{(j,j\prime)}$ is related to the period of Pendellösung fringes l as

$$l = \frac{2\pi}{\Delta k_z^{(j,j\prime)}}.$$

This period l is on the order of micrometers. When spatial averaging is applied to the components of the Pendellösung fringes, the Poynting vector has only the first term in (6.42) remaining,

$$\boldsymbol{S} = \frac{c}{2\varepsilon_0}\sum_g\sum_j\boldsymbol{s}_g\left|\boldsymbol{D}_g^{(j)}\right|^2. \tag{6.43}$$

The Poynting vector corresponding to each branch $\boldsymbol{S}^{(j)}$ is expressed by

$$\boldsymbol{S}^{(j)} = \frac{c}{2\varepsilon_0}\left(\boldsymbol{s}_0\left|\boldsymbol{D}_0^{(j)}\right|^2 + \boldsymbol{s}_g\left|\boldsymbol{D}_g^{(j)}\right|^2\right). \tag{6.44}$$

The direction of Poynting vector $\boldsymbol{S}^{(j)}$ in the symmetric Laue case can be found as follows. From Fig. 6.9a, the surface normal component is given by

$$\boldsymbol{S}^{(j)}\cdot\boldsymbol{s} = S^{(j)}\cos\varphi^{(j)} = \frac{c}{2\varepsilon_0}\left(\left|\boldsymbol{D}_0^{(j)}\right|^2 + \left|\boldsymbol{D}_g^{(j)}\right|^2\right)\cos\theta_B'. \tag{6.45}$$

Here, $\boldsymbol{s}$ is the unit vector of surface normal direction and $\varphi^{(j)}$ is the angle between $\boldsymbol{S}^{(j)}$ and $\boldsymbol{s}$.

The component in the reciprocal lattice vector direction is given by

$$S^{(j)}\cdot g = S^{(j)}\sin\varphi^{(j)} = \frac{c}{2\varepsilon_0}\left(\left|D_0^{(j)}\right|^2 - \left|D_g^{(j)}\right|^2\right)\sin\theta_B'. \tag{6.46}$$

The angle $\varphi^{(j)}$ is obtained as

$$\tan\varphi^{(j)} = \frac{\left|\boldsymbol{D}_0^{(j)}\right|^2 - \left|\boldsymbol{D}_g^{(j)}\right|^2}{\left|\boldsymbol{D}_0^{(j)}\right|^2 + \left|\boldsymbol{D}_g^{(j)}\right|^2}\tan\theta_B'. \tag{6.47}$$

In Fig. 6.9b, $\overline{\mathrm{O'P}} = \overline{\mathrm{OO'}}\tan\varphi^{(2)}$ and $\overline{\mathrm{O'Q}} = \overline{\mathrm{OO'}}\tan\theta_B'$ and similarly, we get $\tan\varphi^{(j)}$. The ratio $\overline{\mathrm{O'P}}/\overline{\mathrm{O'Q}} = p^{(2)}$ and similarly the ratio $p^{(j)}$ are obtained as

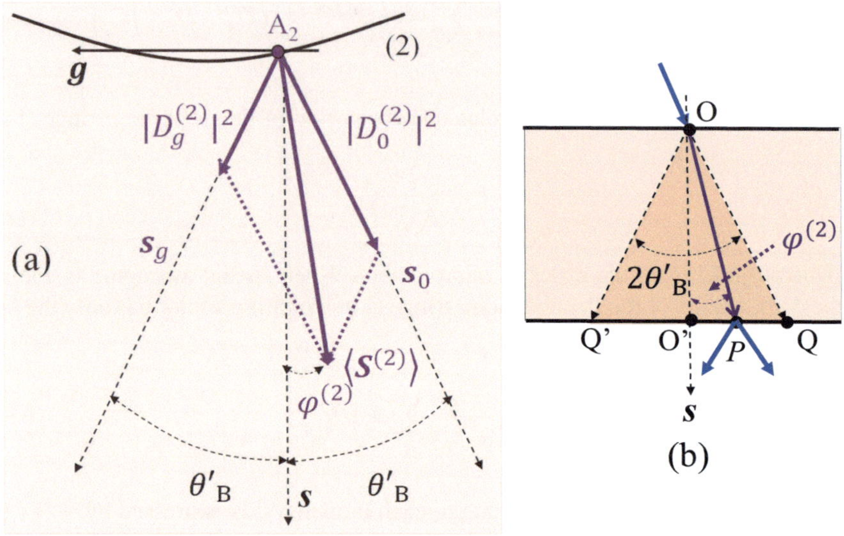

Fig. 6.9 Poynting vector of branch (2) wave. **a** Direction of the poynting vector $S^{(2)}$, **b** Trajectory of energy flow in purple. The dark brown area indicates the Bormann fan

$$p^{(j)} = \frac{\tan\varphi^{(j)}}{\tan\theta'_B} = \frac{\left|D_0^{(j)}\right|^2 - \left|D_g^{(j)}\right|^2}{\left|D_0^{(j)}\right|^2 + \left|D_g^{(j)}\right|^2} = \frac{1 - \left|r^{(j)}\right|^2}{1 + \left|r^{(j)}\right|^2}, \tag{6.48}$$

where $r^{(j)} = \left|D_g^{(j)}\right| / \left|D_0^{(j)}\right|$. Using (6.47), we get

$$\left|r^{(j)}\right|^2 = \frac{1 - p^{(j)}}{1 + p^{(j)}} = \frac{\sin\left(\theta'_B - \varphi^{(j)}\right)}{\sin\left(\theta'_B + \varphi^{(j)}\right)}. \tag{6.49}$$

From the above relations, we can obtain the magnitude and direction of the Poynting vector.

6.1.4 Poynting Vector and Dispersion Surface

When the tie-point is at the top of the dispersion surface ($W = 0$), $\left|D_0^{(2)}\right| = \left|D_g^{(2)}\right|$ so that $\langle S^{(2)} \rangle$ is parallel to the lattice planes and perpendicular to the dispersion surface at the tie-point. On the other hand, at $W = \infty$, the dispersion surface coincides with the asymptote and $\left|D_g^2\right| \approx 0$, $\langle S^{(2)} \rangle$ is parallel to the direction of the incident beam

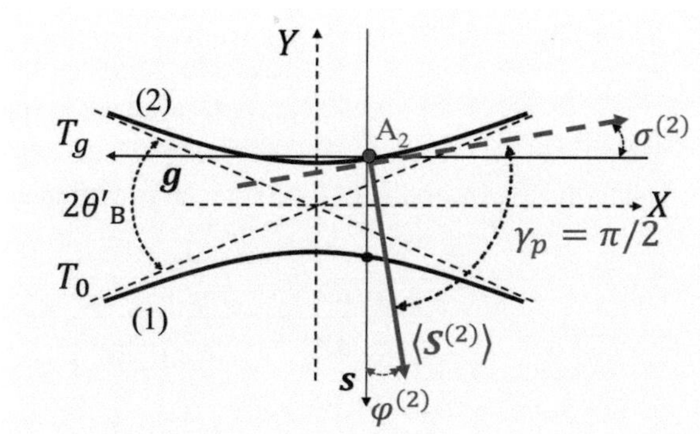

Fig. 6.10 The direction of temporally and spatially averaged poynting vector $S^{(2)}$ from tie-point A_2 on the dispersion surface (thick black lines)

and perpendicular to the dispersion surface. This means that $\varphi^{(2)} = 0$ at $W = 0$ and $\varphi^{(2)} = \theta'_B$ at $W = \infty$. That is, $\varphi^{(2)}$ changes from 0 to θ'_B when W changes from 0 to ∞.

As shown in Fig. 6.10, set the angle between the tangent of the dispersion surface and the X axis at tie-point A_2 to $\sigma^{(2)}$. In the symmetric Laue case, by putting $\beta = \pi/2$ in (5.61) and using $\theta_B = -\theta_1 = \theta_2$, the dispersion equation is obtained as

$$Y_0^2\cos^2\theta'_B - X_0^2\sin^2\theta'_B = \frac{1}{4}\kappa_0^2P^2\left|\chi_g\right|^2. \tag{6.50}$$

The asymptote of this hyperbola is given as

$$\frac{dY}{dX} = \frac{X_0}{Y_0}\tan^2\theta'_B = \tan\sigma^{(2)}. \tag{6.51}$$

From (5.68) and (5.69), ξ_0 and ξ_g are obtained as

$$\xi_0 = -X_0\,\sin\theta'_B + Y_0\,\cos\theta'_B, \tag{6.52}$$

$$\xi_g = X_0\,\sin\theta'_B + Y_0\,\cos\theta'_B. \tag{6.53}$$

By using (6.52) and (6.53), Y_0 and X_0 are expressed as

$$Y_0 = \frac{\xi_0 + \xi_g}{2\cos\theta'_B},\ X_0 = \frac{\xi_0 - \xi_g}{2\sin\theta'_B}. \tag{6.54}$$

Substituting these into (6.51) and express the results for each branch wave, we get

$$\frac{dY_0^{(j)}}{dX_0^{(j)}} = \frac{\xi_g^{(j)} - \xi_0^{(j)}}{\xi_h^{(j)} + \xi_0^{(j)}}\tan\theta'_B. \tag{6.55}$$

According to (5.70) and (5.71), we have relations

$$\xi_0^{(j)} = (-1)^j |\xi_0|, \xi_g^{(j)} = (-1)^j |\xi_g|.$$

Using these expressions, (6.55) can be rewritten as

$$\frac{dY_0^{(j)}}{dX_0^{(j)}} = \frac{\left|\xi_g^{(j)}\right| - |\xi_0^{(j)}|}{|\xi_g^{(j)}\left|+|\xi_0^{(j)}\right|} \tan\theta_B'. \tag{6.56}$$

By using the relation between ξ_0, ξ_g and D_0, D_g in (5.56), we get

$$\frac{dY_0^{(j)}}{dX_0^{(j)}} = \frac{\left|D_0^{(j)}|/|D_g^{(j)}|-|D_g^{(j)}|/|D_0^{(j)}\right|}{\left|D_g^{(j)}|/|D_0^{(j)}|+|D_0^{(j)}|/|D_g^{(j)}\right|} \tan\theta_B', \tag{6.57}$$

$$\tan\sigma^{(j)} = \frac{\left|D_0^{(j)}\right|^2 - \left|D_g^{(j)}\right|^2}{\left|D_0^{(j)}\right|^2 + \left|D_g^{(j)}\right|^2} \tan\theta_B'. \tag{6.58}$$

(6.57) is equal to (6.55), we get

$$\tan\varphi^{(j)} = \tan\sigma^{(j)} \tag{6.59}$$

and thus

$$\varphi^{(j)} = \sigma^{(j)} \text{ and } \gamma_p = \frac{\pi}{2}, \tag{6.60}$$

where γ_p is the angle between $\langle \boldsymbol{S}^{(j)} \rangle$ and the tangent at the tie-point $A_j (j = 1, 2)$. γ_p is shown that the temporal and spatial averaged Poynting vector $\langle \boldsymbol{S}^{(j)} \rangle$ is orthogonal to the dispersion surface of hyperbola. This relation is very useful in determining the direction of energy flow within the crystal and the trajectory of the beam by using the dispersion surface. $\varphi^{(j)}$ changes from $+\theta_B$ to $-\theta_B$ when W changes from $-\infty$ to $+\infty$ as can be seen in Fig. 6.9b. The region of the triangle OQ'Q where this Poynting vector moves is called the Borrmann fan or Borrmann triangle.

Figure 6.11a shows the directions of the temporal and spatial averaged Poynting vector around $W = 0$, and Fig. 6.11b shows the trajectories of the energy flow (the Poynting vector). When $W = 0$, $\boldsymbol{S}^{(1)}$ and $\boldsymbol{S}^{(2)}$ point in the same direction and parallel to the lattice plane. When $W = 2$, $\boldsymbol{S}^{(1)}$ reaches B_1 via $\overline{AB_1}$, while $\boldsymbol{S}^{(2)}$ reaches B_2 via $\overline{AB_2}$. Then, if we put a slit that allows the positions of B_1 and B_2 to be distinguished below the crystal, we can observe the X-rays of $\boldsymbol{S}^{(1)}$ and $\boldsymbol{S}^{(2)}$ separately.

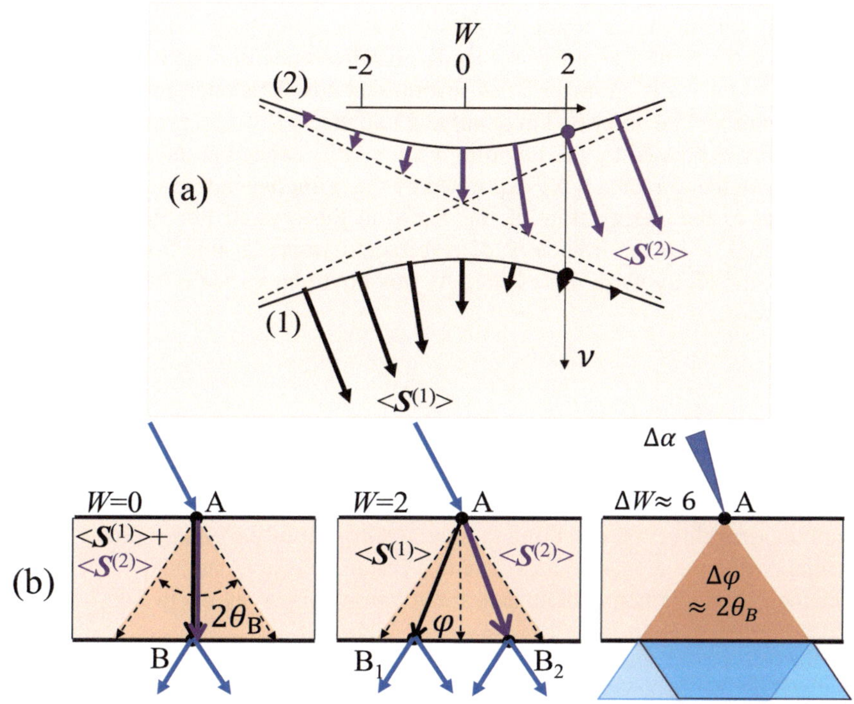

Fig. 6.11 The direction and the amplitude of temporally and spatially averaged Poynting vector $\boldsymbol{S}^{(j)}$ of branch (j) wave near $W = 0$ **a**. Poynting vectors $\boldsymbol{S}^{(j)}$ from tie-points corresponding to $W = 0, \pm1, \pm2, \pm3$ in purple. **b** The directions and the trajectories of the Poynting vectors for $W = 0$, $W = 2$ and $\Delta W = 6$. The shaded triangular regions in **b** indicate Borrmann triangles

6.1.5 Angular Amplification Effect and Pseudo-Refracted Wave

If we take the range of deviation parameter W from -3 to $+\,3$ ($\Delta W = 6$), the range of refracted angle $\varphi^{(j)}$ of the Poynting vector becomes from $-\theta_\mathrm{B}$ to $+\theta_\mathrm{B}$, and its width is $\Delta\varphi^{(j)} \approx 2\theta_\mathrm{B}$ as shown in Fig. 6.11b. By using the width of divergence angle of the incident wave $\Delta\alpha$ in (5.83), we get $\Delta\alpha \approx \Delta W\left|\chi_g\right|/\sin 2\theta_\mathrm{B}$. The angle $\Delta\alpha$ corresponding to the width of $\Delta W = 6$ is approximately several seconds, and the ratio $\Delta\varphi^{(j)}/\Delta\alpha$ is

$$\frac{\Delta\varphi^{(j)}}{\Delta\alpha} \approx 10^4 \sim 10^6.$$

A very large angular amplification rate is obtained, indicating that this dynamical diffraction has a lens effect. This angular amplification effect will be described in detail in Chap. 8. Due to this effect, the trajectory of the X-ray in a crystal change significantly due to lattice strain. Its effects will be described in detail in Chap. 9. As shown on the right side of Fig. 6.11b, X-rays diverge within the Borrmann fan in the crystal when $\Delta\alpha$ is on the order of several seconds. This interference within the Borrmann fan will be described in detail in Chap. 8.

The X-ray expressed by the Poynting vector $\langle \boldsymbol{S}^{(j)} \rangle$ excited at the j-th tie-point is different from the refracted wave expressed by the refractive index, and its trajectory corresponds to the energy flow of the X-ray in the crystal. For this reason, this trajectory looks like the trajectory of a refracted wave. In this book, as already mentioned, the X-ray wave represented by this Poynting vector is called a pseudo-refraction wave.

6.1.6 Integrated Reflection Power

We consider the total energy E reflected by a certain lattice plane when X-rays with a fixed incident direction enter a plane-parallel crystal as the energy flow of P_0 per unit time, while the crystal rotates around a certain axis at an angular velocity ω. If the incident X-ray is perpendicular to the rotation axis as shown in Fig. 6.12b, we get

$$\omega = \frac{d\alpha_0}{dt}, \tag{6.61}$$

where α_0 is the incident glancing angle. If we write the energy flow rate for the diffraction as P_g, E is given as

$$E = \frac{1}{\omega}\int P_g d\alpha_0. \tag{6.62}$$

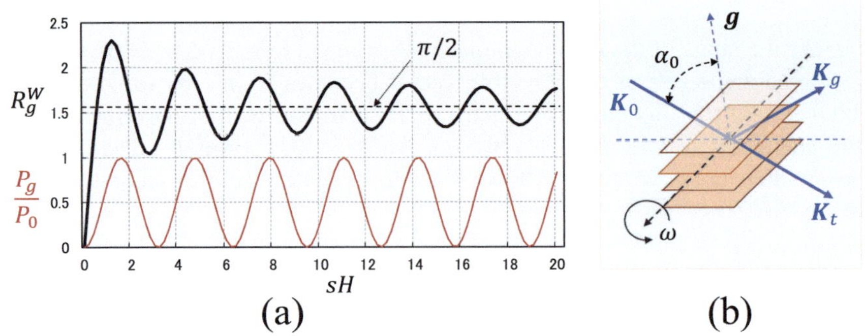

Fig. 6.12 **a** Integrated reflection intensity R_g^W (σ-polarization) in the symmetric Laue case (black line) and $P_g(W = 0)/P_0$ (brown line). **b** Integrated intensity is obtained by rotating the plane-parallel crystal around an axis with angular velocity ω. $\boldsymbol{g}$ is the reciprocal vector

This is called the integrated reflection intensity. As a more general quantity not related to either angular velocity ω or P_0, we get the integrated reflection power given by

$$R_g^{\alpha_0} = \frac{E\omega}{P_0} = \int \frac{P_g}{P_0} d\alpha_0. \tag{6.63}$$

P_g/P_0 is given by (6.20) as a function of W. Since there is also a relation between W and α_0 as given by (5.84), we can obtain

$$\frac{dW}{d\alpha_0} = \left(\frac{\cos\theta_1}{\cos\theta_2}\right)^{1/2} \frac{\sin 2\theta_B}{P|\chi_g|}. \tag{6.64}$$

Then, $R_g^{\alpha_0}$ is expressed as

$$R_g^{\alpha_0} = \left(\frac{\cos\theta_2}{\cos\theta_1}\right)^{1/2} \frac{P|\chi_g|}{\sin 2\theta_B} R_g^W. \tag{6.65}$$

Here R_g^W is the integrated reflection power in terms of parameter W is given as

$$R_g^W = \int_{-\infty}^{+\infty} \frac{P_g(W)}{P_0} dW. \tag{6.66}$$

In the Laue case, as P_g/P_0 is given by (6.20),

$$\begin{aligned} R_g^W(sH) &= \int \frac{\sin^2\left[sH\left(1+W^2\right)^{1/2}\right]}{1+W^2} dW \\ &= \frac{\pi}{2} \int_0^{2sH} J_0(x)dx, \end{aligned} \tag{6.67}$$

according to Waller [2]. Here, $J_0(x)$ is the zero-order Bessel function. From (6.65), $R_g^{\alpha_0}$ is written by

$$R_g^{\alpha_0}(sH) = \left(\frac{\cos\theta_2}{\cos\theta_1}\right)^{\frac{1}{2}} \frac{\pi P|\chi_g|}{2\sin 2\theta_B} \int_0^{2sH} J_0(x)dx. \tag{6.68}$$

Figure 6.12a shows R_g^W as a function of sH. R_g^W increases rapidly as sH increases from 0, and shows its maximum value at $sH = \pi/2$. As sH increases further beyond $\pi/2$, R_g^W shows repeating oscillations and eventually converges to $\pi/2$. For reference,

$P_g(W = 0)/P_0$ is also shown in the figure. The oscillation seen in R_g^W strongly reflects the oscillation of $P_g(W = 0)/P_0$ and is a Pendellösung fringe.

6.2 Bragg Case ($\cos\theta_2 < 0$)

We consider dynamical diffraction in the Bragg case using a plane-parallel crystal as an example. Figure 6.13a shows is the beam geometry in real space, (b, c) shows the beam geometry in the reciprocal space. In Fig. 6.13b and c, the line T_0' is the dispersion surface in vacuum, point A is the tie-point for the incident wave $\boldsymbol{K}_0$. As the incident angle α_0 increases, W increases and A moves to upper right. As shown in Fig. 6.13b, when A comes to the Laue point $\mathrm{L_a}$ on the X-axis, $\boldsymbol{K}_0+\boldsymbol{g} = \boldsymbol{K}_g$ and the Bragg condition in vacuum is exactly satisfied. The Bragg angle is denoted as θ_B. At this condition, the incident wave in the crystal $\boldsymbol{k}_0$ does not satisfy the Bragg condition exactly. The Bragg condition in the crystal is satisfied when $W = 0$ and A comes to L_o'. The Bragg angle at this condition is denoted as θ_B', and the corresponding incident angle as α_B. From Fig. 6.13b, we have a relation

$$\theta_\mathrm{B} < \theta_\mathrm{B}' = \alpha_\mathrm{B}. \tag{6.69}$$

Figure 6.13a shows the geometry when $\varphi \neq 0$ and $|W| \geq 1$. At this condition, a part of the incident wave is reflected at the point of incidence, and the rest enters the crystal as a pseudo-refracted wave. In the following, we will consider the tie-point and the trajectory of the pseudo-refracted wave by dividing into three regions $W \leq -1$, $-1 \leq W \leq 1$, and $W \geq 1$.

Region of $W \leq -1$: A pseudo-refraction wave $\boldsymbol{S}^{(1)}$ can be excited at the tie-point $\mathrm{A_1}$ on the dispersion surface in Fig. 6.13c. This branch is labeled (1) ($Y_r \leq 0$). When the wave $\boldsymbol{S}^{(1)}$ reaches the bottom surface as shown in Fig. 6.13a, a part of it becomes a transmitted wave ($\boldsymbol{K}_t$), and the rest is reflected and becomes the wave $\boldsymbol{S}^{(2)}$. At this time, the tie-point changes to $\mathrm{A_2}$ on branch (2) ($Y_r \geq 0$). The wave $\boldsymbol{S}^{(2)}$ returns to the incident plane, and a part of it becomes a diffracted wave $\boldsymbol{K'}_g$, and the rest is reflected to become a pseudo-refracted wave and propagates into the crystal. It is better to explain how to label a branch. In Fig. 6.13c, the dispersion surface through which tie-points $\mathrm{A_1}$ and $\mathrm{A_2}$ pass looks continuous. However, as will be explained later, when the absorption effect is considered, it is reasonable to label the dispersion surfaces where tie-points $\mathrm{A_1}$ and $\mathrm{A_2}$ to pass through as (1) and (2) separately. Here, a branch with tie-points that excite downward Poynting vectors is labeled (1) (thick black line), and a branch with tie-points that excite upward Poynting vectors is labeled (2) (thick purple line).

Region of $-1 \leq W \leq 1$: The dispersion surface in this region is between $\mathrm{a_1}$ and $\mathrm{a_2}$ in Fig. 6.13c, and the value of Y is pure imaginary as seen in (5.53). In Fig. 6.13c is shown the real part of Y-axis (Y_r), which is zero when $-1 \leq W \leq 1$ (cf. 6.15b and

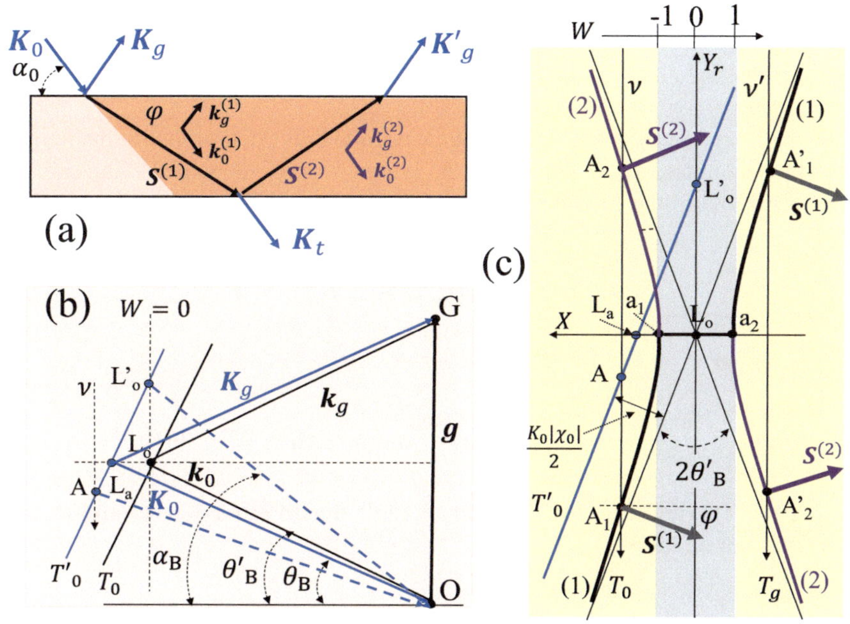

Fig. 6.13 Diffraction geometries in the symmetric Bragg case. **a** Geometry of the incident, diffracted and transmitted waves in the real space. The dark brown contrast shows the Borrmann fan. **b** Geometry of the incident and diffracted waves in the reciprocal space with $\boldsymbol{g}$ being the reciprocal vector of the diffraction. **c** Relation between a tie-point on the dispersion surface and a diffraction condition

c). The pseudo-refracted wave is parallel to the crystal surface (evanescent wave), and the diffracted wave becomes a totally reflected wave.

Conventionally, the straight line between a_1 and a_2 is not depicted as the real part of the dispersion surface (for example, Batterman & Cole [3], Pinsker [4], Authier [5]), but physically, it is preferable to depict it [6].

Region of $W \geq 1$: When a pseudo-refraction wave $\boldsymbol{S}^{(1)}$ is excited at a tie-point on branch (1) of $Y_r \geq 0$, it propagates downward and when a pseudo-refraction wave $\boldsymbol{S}^{(2)}$ is excited at a tie-point on branch (2) of $Y_r \leq 0$, it propagates upward. The pseudo-refracted wave excites the diffracted wave and the transmitted wave while it propagates in the crystal.

In the Bragg case, ξ_0 and ξ_g are given in (5.76) and (5.77). By using them and (5.56), we get

$$r^{(1)} = \frac{D_g^{(1)}}{D_0^{(1)}} = \frac{|\chi_g|}{\chi_{-g}}\left(\frac{\cos\theta_1}{\cos\theta_2}\right)^{1/2}\left[-W(\pm)\left(W^2-1\right)^{1/2}\right], \tag{6.70}$$

$$r^{(2)} = \frac{D_g^{(2)}}{D_0^{(2)}} = \frac{|\chi_g|}{\chi_{-g}}\left(\frac{\cos\theta_1}{\cos\theta_2}\right)^{1/2}\left[-W(\mp)\left(W^2-1\right)^{1/2}\right]. \tag{6.71}$$

6.2.1 Deviation Parameter W and Complex Dispersion Surface

In Fig. 6.14, let A_0 be the intersection of the normal line ν determined at the tie-point A and the asymptote T_0. If we denote the intersection of the straight line extended from A_0 parallel to the crystal surface and the asymptote T_g as A'$_0$, from the figure we get

$$\overline{A_0A'_0}\cos\theta_B = \xi_0 + \xi_g. \tag{6.72}$$

Then we get

$$\overline{A_0A'_0} = \frac{\xi_0 + \xi_g}{\cos\theta_B} = 2\overline{L'_oA}\sin\theta_B = 2K_0(\alpha_0 - \theta_B - \Delta\theta_0)\sin\theta_B. \tag{6.73}$$

Here, $\Delta\theta_0$ is the deviation from the Bragg angle due to refraction. From Fig. 6.14, we get another relation

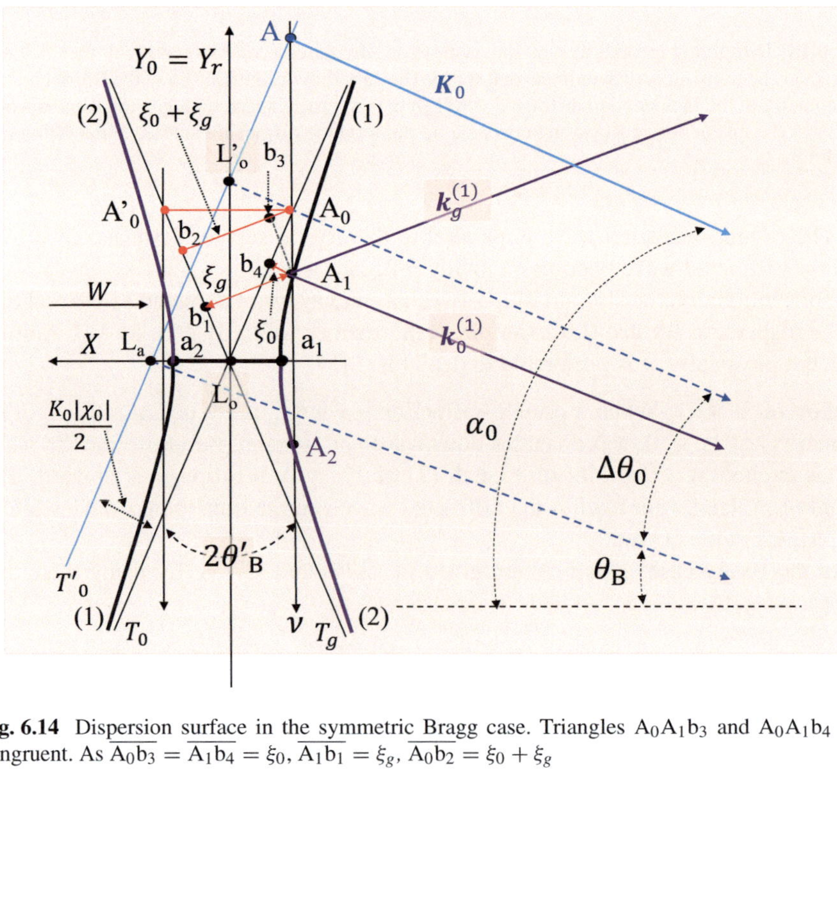

Fig. 6.14 Dispersion surface in the symmetric Bragg case. Triangles $A_0A_1b_3$ and $A_0A_1b_4$ are congruent. As $\overline{A_0b_3} = \overline{A_1b_4} = \xi_0$, $\overline{A_1b_1} = \xi_g$, $\overline{A_0b_2} = \xi_0 + \xi_g$

$$\overline{L'_o L_a}\sin\theta_B = \frac{K_0|\chi_0|}{2\cos\theta_B} \tag{6.74}$$

Using (6.74), we get

$$\Delta\theta_0 = \frac{\overline{L'_o L_a}}{K_0} = \frac{|\chi_0|}{\sin 2\theta_B} \tag{6.75}$$

$\Delta\theta_0$ is zero in the symmetric Laue case, while in the Bragg case, $\Delta\theta_0$ is not equal to zero, regardless of whether it is symmetric or asymmetric Bragg case. The deviation parameter W in the symmetric Bragg case is

$$W = \frac{\overline{A_0A'_0}}{\overline{a_1a_2}} = \frac{\overline{A_0A'_0}}{G_B} = \frac{2K_0(\alpha_0 - \theta_B - \Delta\theta_0)\sin\theta_B}{G_B} \tag{6.76}$$

$$= \frac{2K_0\sin\theta_B}{G_B}\left\{\alpha_0 - \left(\theta_B + \frac{|\chi_0|}{\sin 2\theta_B}\right)\right\}. \tag{6.77}$$

From (6.76), when $W = 0$, $\alpha_0 = \theta_B + \Delta\theta_0$ and the tie-point in vacuum is at L'_o which is located at the angle higher than the Bragg angle by $\Delta\theta_0$.

In the symmetric Bragg case, by putting $\beta = 0$ in Eq. (5.67), we get $\cos\theta_1 = |\cos\theta_2| = \sin\theta'_B$. Y_0 given by (5.67) becomes

$$Y_0 = \frac{G_B}{2\tan\theta'_B}\left(W^2 - 1\right)^{1/2}. \tag{6.78}$$

Y_0 is real when $|W| \geq 1$, but pure imaginary when $|W| < 1$. When $|W| < 1$, the normal line ν does not intersect with the hyperbola in Fig. 6.14. We then write Y_0 as

$$Y_0 = Y_{0r} + iY_{0i} \tag{6.79}$$

with Y_{0r} and Y_{0i} being the real and imaginary parts. Then, for $|W| \geq 1$,

$$Y_{0r} = \frac{G_B}{2\tan\theta'_B}\left(W^2 - 1\right)^{1/2} \tag{6.80}$$

and $Y_{0i} = 0$. For $|W| \leq 1$, $Y_{0r} = 0$ and

$$Y_{0i} = \frac{G_B}{2\tan\theta'_B}\left(1 - W^2\right)^{1/2}. \tag{6.81}$$

As shown in Fig. 6.15b and c, Y_{0r} represents a hyperbola with an east–west opening as a function of W when $|W| \geq 1$ and is zero for $|W| \leq 1$. The imaginary part Y_{0i} represents an ellipse as a function of W from (6.81) for $|W| < 1$. When Y_0 becomes

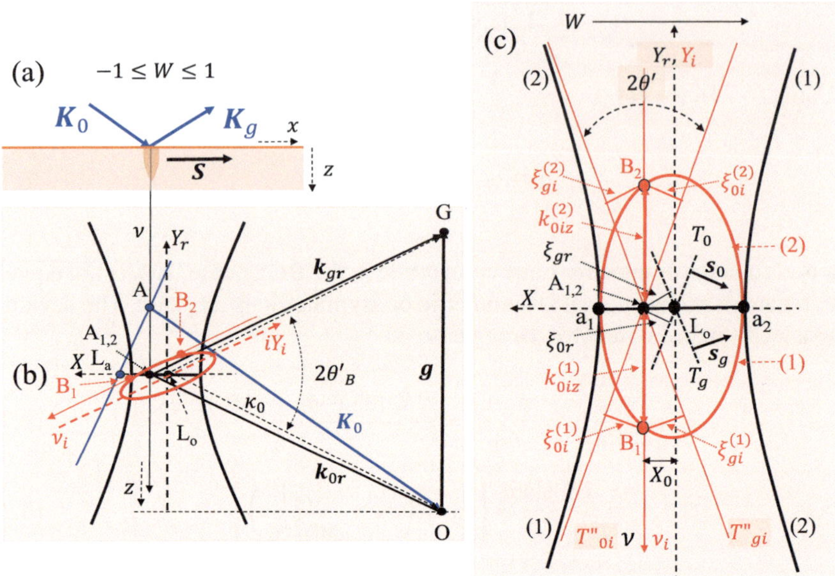

Fig. 6.15 Complex dispersion surfaces. **a** Geometry of the incident and diffracted waves in real space. **b** and **c** shows the dispersion surface in reciprocal space. The black line shows the real part and the red line the imaginary part. **c** Y_i is shown in the same plane with Y_r

a complex number, both ξ_0 and ξ_g in (5.70) and (5.71) become complex numbers. We write as

$$\xi_0 = \xi_{0r} + i\xi_{0i}, \xi_g = \xi_{gr} + i\xi_{gi}. \tag{6.82}$$

From Fig. 6.15, we obtain the relations

$$\xi_{0r} = \xi_{gr} = X_0 \cos\theta'_B = \frac{G_B \cos\theta'_B}{2} W, \tag{6.83}$$

$$\xi_{0i} = -\xi_{gi} = Y_{0i} \sin\theta'_B = \frac{G_B \cos\theta'_B}{2} \left(1 - W^2\right)^{1/2}. \tag{6.84}$$

The tie-point of the real part of Y_0 is at $Y_{0r} = 0$ on the dispersion surface. Then $\boldsymbol{k}_0$ in (4.44) becomes complex, and we write $\boldsymbol{k}_0 = \boldsymbol{k}_{0r} + \boldsymbol{i}\boldsymbol{k}_{0i}$. As $|\boldsymbol{k}_{0r}| \gg |\boldsymbol{k}_{0i}|$, we can write

$$\sqrt{\boldsymbol{k}_0} = |\boldsymbol{k}_{0r}| + \boldsymbol{i}\boldsymbol{s}_0 \cdot \boldsymbol{k}_{0i}, \sqrt{\boldsymbol{k}_g} = \left|\boldsymbol{k}_{gr}\right| + \boldsymbol{i}\boldsymbol{s}_g \cdot \boldsymbol{k}_{gi}. \tag{6.85}$$

By writing the average wavenumber in the crystal as $\kappa_0 \left(= \overline{\mathrm{L_0 O}}\right)$, and writing

$$\sqrt{\boldsymbol{k}_0} - \kappa_0 = |\boldsymbol{k}_{0r}| - \kappa_0 + i\boldsymbol{s}_0 \cdot \boldsymbol{k}_{0i}, = \xi_{0r} + i\xi_{0i}, \tag{6.86}$$

$$\sqrt{\boldsymbol{k}_g} - \kappa_0 = \left|\boldsymbol{k}_{gr}\right| - \kappa_0 + i\mathbf{s}_g \cdot \boldsymbol{k}_{gi}, = \xi_{gr} + i\xi_{gi}, \tag{6.87}$$

then we get

$$\begin{aligned} \xi_{0r} &= |\boldsymbol{k}_{0r}| - \kappa_0, \xi_{0i} = \boldsymbol{s}_0 \cdot \boldsymbol{k}_{0i}, \\ \xi_{gr} &= \left|\boldsymbol{k}_{gr}\right| - \kappa_0, \xi_{gi} = \mathbf{s}_g \cdot \boldsymbol{k}_{gi}. \end{aligned} \tag{6.88}$$

Now, using Fig. 6.15, we consider the dispersion surface in the region $-1 \leq W \leq 1$. Diffracted waves are totally reflected and pseudo-refracted waves propagate parallel to the crystal surface, resulting in evanescent waves in the depth direction as shown in Fig. 6.15a, which will be described in detail in the next subsection. This range of dispersion surface corresponds to the forbidden band in the band theory [7]. X-rays cannot exist within the crystal, so they are monotonically attenuated and the incident waves are totally reflected. The penetration depth ranges from several to several tens of micrometers.

In Fig. 6.15a, x-direction is parallel to the crystal surface, and z-direction (depth direction) is perpendicular to the surface and downward. Figure 6.15b shows the complex dispersion surface. The real part of dispersion surface (hyperbola in black) and the imaginary part (ellipse in red) are shown. In the dispersion surface, X-axis is parallel to x in real space. Y_r-axis is parallel to z-direction of real space. Y_i-axis is orthogonal to Y_r-axis. The imaginary part of the dispersion surface is in the XY_i-plane, which is perpendicular to XY_r-plane. Let A_{12} be the intersection of the real part of the dispersion surface with the normal line ν, which is parallel to z-axis, defined by the diffraction condition. The normal line of the imaginary part ν_i defined by the diffraction conditions is parallel to Y_i and perpendicular to ν, and intersects with ν at A_{12} on X-axis. In the imaginary part of the dispersion surface, the negative side of the branch is labeled (1) and the positive side is labeled (2). Tie-point B_1 is on branch (1) and B_2 is on branch (2).

As stated in 4.2.1, the components of wave vectors parallel to the crystal surface must have the same magnitude inside and outside the crystal. Since the wave vector $\boldsymbol{K}_0$ in vacuum is a real vector, the imaginary part $\boldsymbol{k}_{0i}$ of the wave vector inside the crystal exists only in the component perpendicular to the crystal surface $(\boldsymbol{k}_{0i})_z = k_{0iz}$. From the z component of $\boldsymbol{k}_{0i}$ and Fig. 6.15b, we get

$$k_{0z} = k_{0rz} + ik_{0iz} = \kappa_0 \sin\theta'_{\mathrm{B}} + (Y_{0r} = 0) + iY_{0i} \tag{6.89}$$

and $Y_{0i} = k_{0iz}$.

In Fig. 6.15c, the Y_r-axis and Y_i-axis are shown in the same plane. The real part of the dispersion surface is shown by a solid black line, and the imaginary part is shown by a solid red line. Figure 6.15c has the advantage of displaying the shape of the imaginary part of the dispersion surface without distorting it. Using this figure, we consider $\xi_{0i} = \boldsymbol{s}_0 \cdot \boldsymbol{k}_{0i}$. As described in Fig. 4.5, $\boldsymbol{s}_0 \cdot \boldsymbol{k}_{0i}$ gives a plane perpendicular to

$\boldsymbol{s}_0$, indicating a plane with constant absorption. In Fig. 6.15c, the cross section of the plane perpendicular to $\boldsymbol{s}_0$ is the line segment T''_{0i}, which is parallel to T_0. $\xi_{0i}^{(1)}$ is the length of a line drawn perpendicular to T''_{0i} from B_1. The same goes for $\xi_{gi}^{(1)} = \boldsymbol{s}_g \cdot \boldsymbol{k}_{gi}$. There are relations of $\xi_{0i}^{(2)} = -\xi_{0i}^{(1)}$ and $\xi_{gi}^{(2)} = \xi_{0i}^{(1)} = -\xi_{gi}^{(1)}$. This imaginary part of the dispersion surface is useful for interpreting the extinction effect, anomalous absorption, and anomalous transmission, as will be described later.

The tie-point $\mathrm{A}_{1,2}$ of the real part is on the line $\overline{\mathrm{a}_1\mathrm{a}_2}$, and the values of the real part are the same for branches (1) and (2). From Fig. 6.15c, ξ_{0r} and ξ_{gr} are the lengths of the perpendicular lines drawn from $\mathrm{A}_{1,2}$ to T_0 and T_g, respectively. Then we obtain $\xi_{0r}^{(1)} = \xi_{0r}^{(2)}$ and $\xi_{gr}^{(1)} = \xi_{gr}^{(2)}$.

For $|W| \leq 1$, the dispersion equation of (5.45) becomes

$$(\xi_{0r} + i\xi_{0i})(\xi_{gr} + i\xi_{gi}) = (\xi_{0r}\xi_{gr} - \xi_{0i}\xi_{gi}) + i(\xi_{0r}\xi_{gi} + \xi_{gr}\xi_{0i}) = \left(\frac{P\kappa_0}{2}\right)^2 |\chi_g|^2. \tag{6.90}$$

Since the right-hand side is a real number, the imaginary part $\xi_{0r}\xi_{gi} + \xi_{gr}\xi_{0i}$ must be zero, which is satisfied because of $\xi_{0r} = \xi_{gr}$ and $\xi_{0\mathrm{i}} = -\xi_{gi}$. From the real part $\xi_{0r}\xi_{gr} - \xi_{0i}\xi_{gi}$, we can derive the result on the right side by using (6.83) and (6.84).

6.2.2 *Diffraction Intensity*

We first describe the diffraction intensity from a sufficiently thick crystal, i.e., a semi-infinite crystal, and then that from a thin crystal.

6.2.2.1 Diffraction Intensity from Semi-infinite Crystal

In a semi-infinite crystal, there is no reflected wave corresponding to the pseudo-refracted wave within the crystal because there is no bottom surface. Let us try to find the intensity of the diffracted wave in such a case.

As already mentioned, the pseudo-refracted wave $\boldsymbol{S}^{(1)}$ is excited on the branch (1) dispersion surface to propagates downward, as shown in Fig. 6.16a. From Fig. 6.16b, the boundary condition at the crystal surface is given as

$$E_0^a = D_0^{(1)}, E_g^a = D_g^{(1)} = r^{(1)} E_0^a. \tag{6.91}$$

From (6.70), we obtain $r^{(1)}$ as

$$r^{(1)} = \exp(i\alpha_g)\left(\frac{\cos\theta_1}{|\cos\theta_2|}\right)^{1/2}\left[-W(\pm)(W^2-1)^{\frac{1}{2}}\right]. \tag{6.92}$$

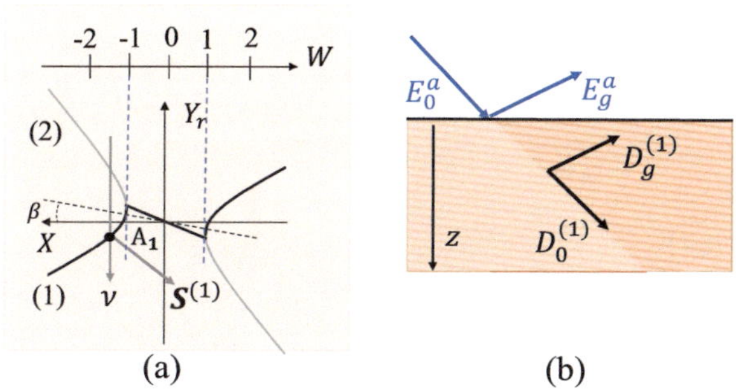

Fig. 6.16 Diffraction condition for a semi-infinite crystal (asymmetric reflection). **a** Dispersion surface. **b** Electric field vectors of the incident $\boldsymbol{E}_0^a$ and diffracted waves $\boldsymbol{E}_g^a$ in real space, and the corresponding electric flux density vectors $\boldsymbol{D}_0^{(1)}$ and $\boldsymbol{D}_g^{(1)}$ in the crystal

For $|W| \geq 1$, the term in the square bracket is real and the diffraction intensity P_g is given by

$$\frac{P_g}{P_0} = \left|\frac{E_g^a}{E_0^a}\right|^2 \frac{|\cos\theta_2|}{\cos\theta_1} = \left|r^{(1)}\right|^2 \frac{|\cos\theta_2|}{\cos\theta_1}$$
$$= \left[|W| - \left(W^2 - 1\right)^{1/2}\right]^2 = \frac{1}{\left[|W| + \left(W^2 - 1\right)^{1/2}\right]^2}. \tag{6.93}$$

On the other hand, for $|W| < 1$, the term in the square bracket is complex and

$$\frac{P_g}{P_0} = 1. \tag{6.94}$$

The diffraction intensity of (6.93) and (6.94) is called Darwin's reflection intensity formula. The rocking curve based on it is called the Darwin curve, which shows the reflection intensity as a function of W as shown in Fig. 6.17. In the Darwin curve, total reflection is observed for $|W| \leq 1$ and the FWHM of the peak $\Delta W = 2.12$.

We consider the effect of refractive index when observing rocking curves. In the symmetric Bragg case, the Bragg angle θ_B' in the crystal is related to the Bragg angle θ_B in vacuum as

$$\theta_B' = \theta_B + \Delta\theta_0 = \theta_B + \frac{|\chi_0|}{\sin 2\theta_B} \tag{6.95}$$

by using (6.76) and (6.77) and writing the incident angle α_0 at $W = 0$ as θ_B'. It is noted that the maximum diffraction intensity is not observed at the Bragg angle θ_B in vacuum. It is obtained at the higher angle than θ_B by $\Delta\theta_0 = |\chi_0|/\sin 2\theta_B$ as shown in Fig. 6.17.

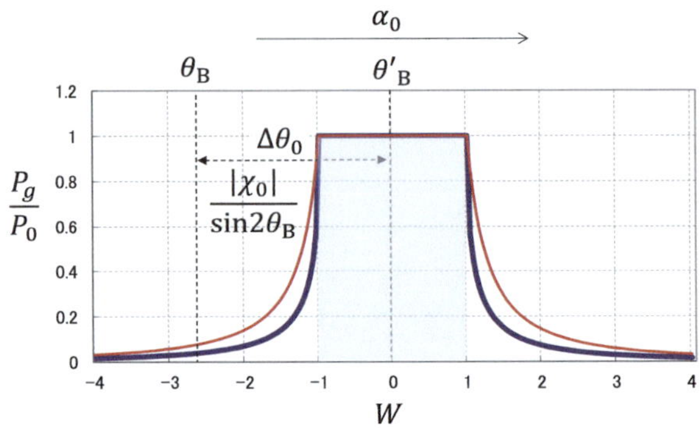

Fig. 6.17 Reflection intensity curves from a semi-infinite crystal. The Darwin curve is shown in blue and the Ewald's curve in red. The Bragg angle in the crystal is θ'_B and that in vacuum is θ_B

Extinction Effect

Let's examine the behavior of the primary wave inside the crystal in the total reflection region ($|W| < 1$). When $|W| \geq 1$, $\boldsymbol{k}_0$ is a real vector and the pseudo-refraction wave is not absorbed within the crystal. On the other hand, when $|W| < 1$, it becomes a complex vector and undergoes attenuation expressed by its imaginary part. $\boldsymbol{S}^{(1)}$ of the pseudo-refraction wave for $|W| < 1$ is shown in Fig. 6.18. From (6.70), $\left|r^{(1)}\right|^2 = \cos\theta_1/|\cos\theta_2|$, and

$$|D_0|^2\cos\theta_1 = |D_g|^2|\cos\theta_2| = \left|r^{(1)}\right|^2|D_0|^2|\cos\theta_2|. \tag{6.96}$$

The direction of $\boldsymbol{S}^{(1)}$ is parallel to the crystal surface, although it is not parallel to the lattice plane in an asymmetric reflection.

Since $\boldsymbol{S}^{(1)}$ is parallel to the crystal surface, the pseudo-refracted wave is an evanescent wave and localized near the crystal surface, which is called extinction effect. The z component of the imaginary part is given by $k_{0iz} = Y_i$ and $E_0^a = D_0^{(1)}$ at $z = 0$, then the intensity of the X-ray inside the crystal is given by

$$\left|\frac{D_0^{(1)}(z)}{E_0^a}\right|^2 = \exp(-2|Y_i|z) = \exp\left[-2s\left(1 - W^2\right)^{1/2}z\right]. \tag{6.97}$$

Here, s is given as

$$s = \frac{K_0P\left(\chi_g\chi_{-g}\right)^{1/2}}{2|\cos\theta_1\cos\theta_2|^{1/2}}.$$

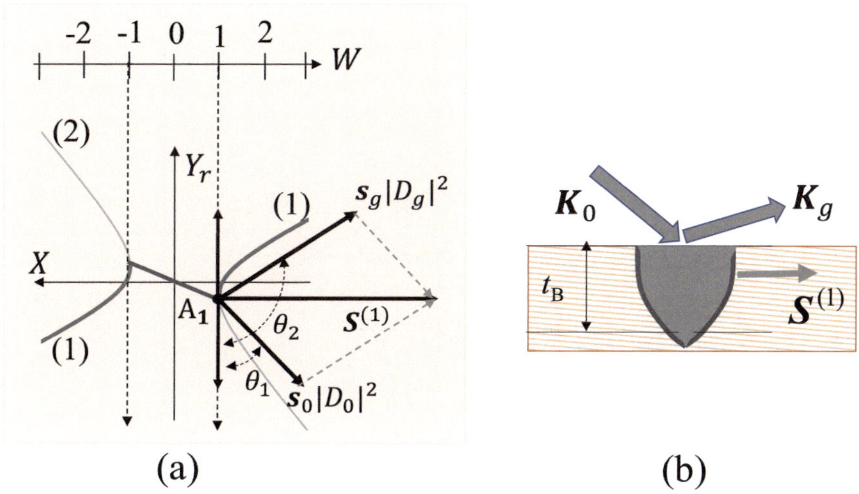

Fig. 6.18 The direction of pseudo-refracted wave $S^{(1)}$ in reciprocal space (**a**) real space (**b**)

The primary wave attenuates rapidly as the depth z increases. This total reflection has something in common with the total internal reflection in the one-wave approximation. If the depth at which the intensity of the primary wave attenuates to $\exp(-1)$ from the surface at $W = 0$ is expressed as t_B (see Fig. 6.18b), then

$$t_B = \frac{1}{2s} = |\cos\theta_1 \cos\theta_2|^{1/2} \frac{1}{K_0 P \left(\chi_g \chi_{-g}\right)^{1/2}}. \tag{6.98}$$

This depth t_B is called the extinction distance in the Bragg case. t_B is on the order of micrometers, and is about 100 times larger than the penetration distance l in the one-wave approximation (4.37).

Standing Wave Inside Crystal

In the Bragg case of the semi-infinite crystal, there is only one branch (1), and Pendellösung fringes do not appear. There is, however, an interference between the primary and secondary waves of the same branch. The strength of the standing wave excited at a tie-point of branch (1) $I^{(1)}(z)$ is given by (6.29), and $r^{(1)}$ is given by (6.92). Then, $I^{(1)}(z)$ is obtained as

$$I^{(1)}(z) = \left|D_0^{(1)}\right|^2 \left[1 + |r^{(1)}|^2 + 2|r^{(1)}|\cos\left(\frac{2\pi z}{d} - \omega^{(1)}\right)\right]. \tag{6.99}$$

with $\omega^{(1)}$ being the phase of $r^{(1)}$. When $\exp(i\alpha_h) = -1$, $r^{(1)}$ given by (6.92) is positive for $W \geq 1$, and we get $\omega^{(1)} = 0$. For $W \leq -1$, $r^{(1)}$ is negative and $\omega^{(1)} = \pi$. For

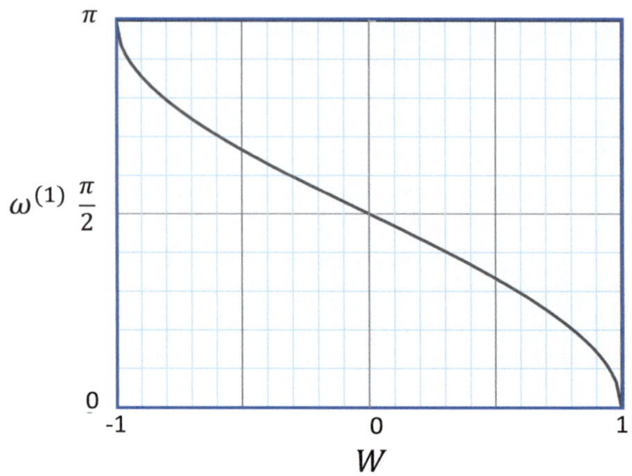

Fig. 6.19 The phase $\omega^{(1)}$ of standing wave in a crystal as a function of W

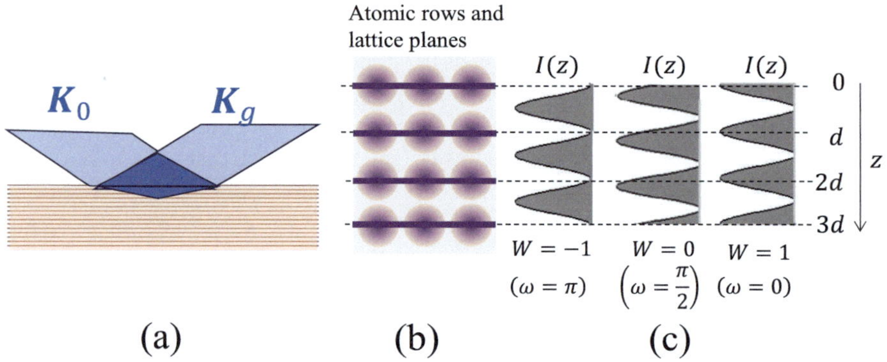

Fig. 6.20 Standing wave in a crystal in the Bragg case. **a** Standing wave is formed in the dark shaded region inside the crystal and above the surface. **b** Arrangement of atomic rows and lattice planes perpendicular to x–z plane. **c** Standing waves of $I(z)$ for $W = -1, 0$ and 1

$-1 < W < 1$, $\omega^{(1)}$ varies from π to 0. For $W = 0$, $\omega^{(1)} = \pi/2$. Figure 6.19 shows the change in $\omega^{(1)}$ as a function of W.

Figure 6.20 shows $I(\mathrm{z})$ for $W = -1, 0$ and 1. For $W \geq 1$, $I(\mathrm{z})$ shows its maximum on the lattice plane, while for $W \leq -1$, it shows its minimum on the lattice plane. This effect of X-ray standing wave will be described in detail in the case of absorbing crystals; when $W \geq 1$, the strong interaction with atoms in the lattice plane results in anomalous absorption of X-rays, while for $W \leq -1$, the weak interaction with atoms results in anomalous transmission.

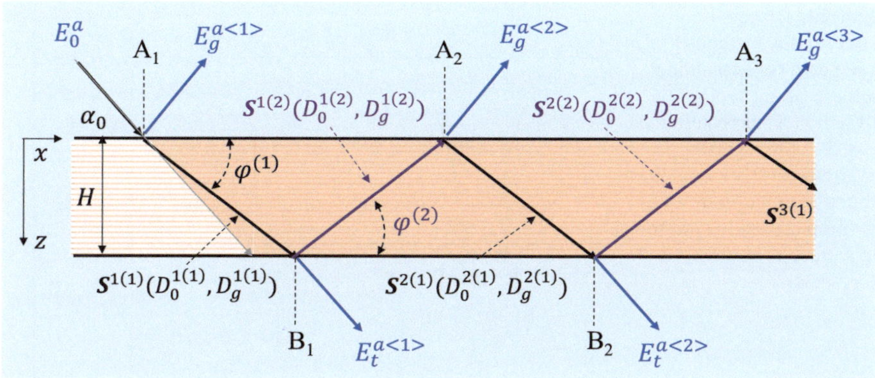

Fig. 6.21 A schematic diagram of multiple Bragg type diffractions. The origin is taken at the incident point A_1, x-axis is parallel to the crystal surface and z-axis points downward and perpendicular to the surface. The crystal thickness is H

6.2.2.2 Diffraction Intensity from Thin Crystal

Partial Wave Method (Wagner's Method)

We describe the diffraction intensity from a thin crystal in the symmetric Bragg case. Figure 6.21 shows a schematic diagram to explain the beam geometry. This was proposed by Wagner [8]. The partial wave reflection approach in a single layer film mentioned in 4.6 can be used as a reference. At the incident point A_1, a part of the incident beam E_0^a is diffracted from the crystal as $E_g^{a\langle 1\rangle}$ (here, the number in the superscript indicates the number of reflections), and the rest of it propagates into the crystal as a pseudo-refraction wave $\boldsymbol{S}^{1(1)}$. When the wave $\boldsymbol{S}^{1(1)}$ reaches the bottom surface at B_1, a part of it is transmitted as $E_t^{a\langle 1\rangle}$ and the rest is reflected from the bottom to propagate as a pseudo-refraction wave $\boldsymbol{S}^{1(2)}$. Here, the first number in superscript of $\boldsymbol{S}$ indicates the number of pairs of reflections at the top and bottom surfaces in the crystal and the second number in the round bracket indicates the branch number. When the pseudo-refracted wave $\boldsymbol{S}^{1(2)}$ returns to the top surface, a part of it is emitted outside the crystal as the diffracted wave $E_g^{a\langle 2\rangle}$ and the rest is reflected as a pseudo-refracted wave $\boldsymbol{S}^{2(1)}$. In this way, a pseudo-refracted wave is repeatedly reflected at the bottom and top surfaces of the crystal, and propagates within the crystal while exciting diffracted waves and transmitted waves. We consider the diffraction intensity based on this process.

The boundary condition at the incident point A_1 is written as

$$E_0^a = D_0^{1(1)},\ E_g^{a\langle 1\rangle} = D_g^{1(1)}. \tag{6.100}$$

By using $r^{(1)} = D_g^{(1)}/D_0^{(1)}$, the second equation can be written as

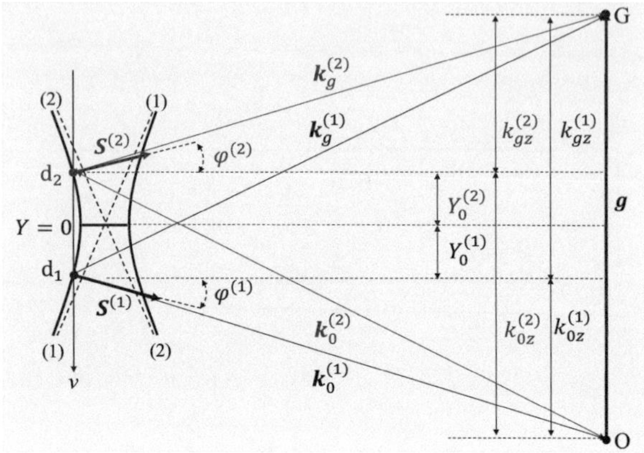

Fig. 6.22 Dispersion surface in the symmetric Bragg case for explaining multiple Bragg-type diffraction. Pseudo-refracted wave $\boldsymbol{S}^{(1)}$ propagating toward the bottom surface is excited at tie-point d_1 and $\boldsymbol{S}^{(2)}$ propagating toward the top surface is excited at tie-point d_2

$$E_g^{a\langle 1\rangle} = D_g^{1(1)} = r^{(1)} D_0^{1(1)} = r^{(1)} E_0^a. \tag{6.101}$$

This agrees with (6.91) for a semi-infinite crystal. In other words, there is no influence of the reflected wave from the bottom surface. The boundary condition at B_1 on the bottom surface is written as

$$D_0^{1(1)} \exp\left(-ik_{0z}^{(1)} H\right) + D_0^{1(2)} \exp\left(-ik_{0z}^{(2)} H\right) = E_t^{a\langle 1\rangle} \exp(-iK_{0z} H), \tag{6.102}$$

$$D_g^{1(1)} \exp\left(-ik_{gz}^{(1)} H\right) + D_g^{1(2)} \exp\left(-ik_{gz}^{(2)} H\right) = 0. \tag{6.103}$$

From (6.103), we get

$$D_g^{1(2)} = -D_g^{1(1)} \frac{\exp\left(-ik_{gz}^{(1)} H\right)}{\exp\left(-ik_{gz}^{(2)} H\right)} = -r^{(1)} \exp\left(-i2sH\sqrt{B}\right) E_0^a. \tag{6.104}$$

From Fig. 6.22, we get

$$k_{gz}^{(1)} - k_{gz}^{(2)} = k_{0z}^{(2)} - k_{0z}^{(1)} = Y_0^{(2)} - Y_0^{(1)} = 2s\sqrt{B}, \tag{6.105}$$

where s and B in the symmetric Bragg case are

$$s = \frac{G}{2\tan\theta_B}, \sqrt{B} = (W^2 - 1)^{1/2}. \tag{6.106}$$

Using $D_g^{1(2)} = r^{(2)} D_0^{1(2)}$ and (6.104), we get

$$D_0^{1(2)} = \frac{D_g^{1(2)}}{r^{(2)}} = -\frac{r^{(1)}}{r^{(2)}} \exp\left(-i2sH\sqrt{B}\right) E_0^a. \tag{6.107}$$

Substituting (6.101) and (6.107), we write (6.102) as

$$E_0^a \exp\left(-ik_{0z}^{(1)}H\right) - \frac{r^{(1)}}{r^{(2)}} \exp\left(-i2sH\sqrt{B}\right) \exp\left(-ik_{0z}^{(2)}H\right) E_0^a = E_t^{a\langle 1\rangle} \exp(-iK_{0z}H)$$

and modify this to obtain

$$E_t^{a\langle 1\rangle} \exp(-iK_{0z}H) = \left(1 - \frac{r^{(1)}}{r^{(2)}}\right) \exp\left(-ik_{0z}^{(1)}H\right) E_0^a. \tag{6.108}$$

Next, the boundary conditions at A_2 are given as

$$D_0^{2(1)} + D_0^{1(2)} = 0, \tag{6.109}$$

$$D_g^{2(1)} + D_g^{1(2)} = E_g^{a\langle 2\rangle}. \tag{6.110}$$

From (6.109), we get

$$D_0^{2(1)} = -D_0^{1(2)} = \frac{r^{(1)}}{r^{(2)}} \exp\left(-i2sH\sqrt{B}\right) E_0^a \tag{6.111}$$

by using $D_0^{1(2)}$ in (6.107). From (6.111), we get

$$D_g^{2(1)} = r^{(1)} D_0^{2(1)} = \frac{\left(r^{(1)}\right)^2}{r^{(2)}} \exp\left(-i2sH\sqrt{B}\right) E_0^a. \tag{6.112}$$

From the boundary condition of (6.110), we obtain

$$E_g^{a\langle 2\rangle} = r^{(1)} \left(\frac{r^{(1)}}{r^{(2)}} - 1\right) \exp\left(-i2sH\sqrt{B}\right) E_0^a. \tag{6.113}$$

By writing

$$\gamma = 2sH\sqrt{B}, \tag{6.114}$$

we get for $n > 1$,

$$D_0^{n(1)} = \left(\frac{r^{(1)}}{r^{(2)}}\right)^{n-1} \exp[-i(n-1)\gamma] E_0^a,\ D_0^{n(2)} = -\left(\frac{r^{(1)}}{r^{(2)}}\right) \exp(-i\gamma) D_0^{n(1)} \tag{6.115}$$

and

$$D_g^{n(1)} = r^{(1)} D_0^{n(1)},\ D_g^{n(2)} = -\exp(-i\gamma) D_0^{n(1)}. \tag{6.116}$$

Then the electric field of diffracted wave $E_g^{a\langle n\rangle}$ after $(n-1)$-times $(n \geq 1)$ reflections from the bottom surface is given by

$$E_g^{a\langle n\rangle} = r^{(1)}\left(\frac{r^{(1)}}{r^{(2)}} - 1\right)\left(\frac{r^{(1)}}{r^{(2)}}\right)^{n-1} \exp\left[i(n-1)\gamma\right]E_0^a \tag{6.117}$$

and that of transmitted wave $E_t^{\langle n\rangle}$ after $(n-1)$-times reflections form the top surface is given by

$$E_t^{\langle n\rangle}\exp(-iK_{0z}H) = \left(\frac{r^{(1)}}{r^{(2)}}\right)^{n-1}\left(1 - \frac{r^{(1)}}{r^{(2)}}\right)\exp\left[i(n-1)\gamma\right]\exp\left(-ik_{0z}^{(1)}H\right)E_0^a. \tag{6.118}$$

The electric field of the diffracted wave obtained from an infinitely wide crystal surface E_g^a must be given by

$$E_g^a = E_g^{\langle 1\rangle} + E_g^{\langle 2\rangle} + \cdots\cdots + E_g^{\langle\infty\rangle}. \tag{6.119}$$

By writing $R_g = E_g^a/E_0^a$, we obtain

$$R_g = r^{(1)} + r^{(1)}\left(\frac{r^{(1)}}{r^{(2)}} - 1\right)\exp(-i\gamma)$$
$$\left[1 + \frac{r^{(1)}}{r^{(2)}}\exp(-i\gamma) + \left(\frac{r^{(1)}}{r^{(2)}}\right)^2\exp(-i2\gamma) + \cdots\right]. \tag{6.120}$$

The term in the square bracket of (6.120) is a geometric series and R_g can be modified as

$$R_g = r^{(1)} + \frac{r^{(1)}\left(\frac{r^{(1)}}{r^{(2)}} - 1\right)\exp(-i\gamma)}{1 - \frac{r^{(1)}}{r^{(2)}}\exp(-i\gamma)} = \frac{r^{(1)}\left[1 - \exp(-i\gamma)\right]}{1 - \frac{r^{(1)}}{r^{(2)}}\exp(-i\gamma)}$$
$$= \frac{r^{(1)}r^{(2)}\left[1 - \exp(-i\gamma)\right]}{r^{(2)} - r^{(1)}\exp(-i\gamma)}. \tag{6.121}$$

The ratio of the diffraction intensity P_g to the incident wave intensity P_0 is given by

$$\frac{P_g}{P_0} = \left|\frac{E_g^a}{E_0^a}\right|^2 = \left|R_g\right|^2 = \left|\frac{r^{(1)}r^{(2)}\left[1 - \exp(-i\gamma)\right]}{r^{(2)} - r^{(1)}\exp(-i\gamma)}\right|^2$$
$$= \frac{\sin^2\left[sH\left(W^2 - 1\right)^{1/2}\right]}{W^2 - 1 + \sin^2\left[sH\left(W^2 - 1\right)^{1/2}\right]}. \tag{6.122}$$

For $|W| \leq 1$, $\sin\left[isH\left(1 - W^2\right)^{1/2}\right]$ is to be replaced by $i\sinh\left(sH\left(1 - W^2\right)^{1/2}\right)$.

Let's find the intensity of the transmitted wave using the same procedure. The electric field of the transmitted wave is given by

$$E_t^a \exp(-iK_{0z}H)$$
$$= E_0^a \exp\left(-ik_{0z}^{(1)}H\right)\left(1 - \frac{r^{(1)}}{r^{(2)}}\right)$$
$$\left[1 + \frac{r^{(1)}}{r^{(2)}}\exp(-i\gamma) + \left(\frac{r^{(1)}}{r^{(2)}}\right)^2 \exp(-i2\gamma) + \cdot\cdot\right]. \tag{6.123}$$

As the transmission coefficient T is expressed as

$$T = \frac{E_t^a \exp(-iK_{0z}H)}{E_0^a \exp\left(-ik_{0z}^{(1)}H\right)} = \frac{1 - \frac{r^{(1)}}{r^{(2)}}}{1 - \frac{r^{(1)}}{r^{(2)}}\exp(-i\gamma)} = \frac{r^{(2)} - r^{(1)}}{r^{(2)} - r^{(1)}\exp(-i\gamma)}, \tag{6.124}$$

the transmitted intensity is given as

$$\frac{P_t}{P_0} = \left|\frac{E_t^a}{E_0^a}\right|^2 = |T|^2\left|\exp\left(-ik_{0z}^{(1)}H\right)\right| = \left|\frac{r^{(2)} - r^{(1)}}{r^{(2)} - r^{(1)}\exp(-i\gamma)}\right|^2 \exp[\mathrm{Im}(-\gamma)]. \tag{6.125}$$

Here, $k_{0z}^{(1)}$ is complex for $|W| \leq 1$. By referring to Fig. 6.22, we get

$$\mathrm{Im}(k_{0z}^{(1)})H = \mathrm{Im}\left(\frac{h}{2} + Y^{(1)}\right)H = sH\left(1 - W^2\right)^{1/2} = \mathrm{Im}(\gamma\ /2)\ . \tag{6.126}$$

The approach using partial waves described here can be applied to a wide range of applications, including crystals of finite size, as described later.

Ideal Plane Wave Method (Ewald-Laue Method)

When an ideal plane wave is incident on an infinitely wide and flat crystal surface with a crystal thickness of H, there is always a reflected wave due to the pseudo-refracted wave from the bottom surface and there are tie-points on two branches of the dispersion surface. The boundary conditions at the crystal surface are

$$D_0^{(1)} + D_0^{(2)} = E_0^a, \tag{6.127}$$

$$D_g^{(1)} + D_g^{(2)} = E_g^a. \tag{6.128}$$

As for the boundary conditions at the bottom surface, one is obtained from (6.103) as

$$D_g^{(2)} = -D_g^{(1)} \exp(-i\gamma) = -r^{(1)} D_0^{(1)} \exp(-i\gamma), \tag{6.129}$$

by replacing

$$D_0^{(1)} = D_0^{1(1)}, D_0^{(2)} = D_0^{1(2)}, E_t^a = E_t^{a\langle 1\rangle}$$

in (6.102) and by replacing,

$$D_g^{(1)} = D_g^{1(1)}, D_g^{(2)} = D_g^{1(2)}$$

in (6.103). The other boundary condition is similarly obtained from (6.103) as

$$D_0^{(2)} = \frac{D_g^{(2)}}{r^{(2)}} = -\frac{r^{(1)}}{r^{(2)}} \exp(-i\gamma) D_0^{(1)}. \tag{6.130}$$

Substituting (6.130) into (6.127), we get $D_0^{(1)}$ as

$$D_0^{(1)} = \frac{E_0^a}{1 - \frac{r^{(1)}}{r^{(2)}} \exp(-i\gamma)}. \tag{6.131}$$

Then, the reflection coefficient is given by

$$R_g = \frac{E_g^a}{E_0^a} = \frac{r^{(1)} D_0^{(1)} + r^{(2)} D_0^{(2)}}{E_0^a} = \frac{r^{(1)}[1 - \exp(-i\gamma)]}{1 - \frac{r^{(1)}}{r^{(2)}} \exp(-i\gamma)} = \frac{r^{(1)} r^{(2)}[1 - \exp(-i\gamma)]}{r^{(2)} - r^{(1)} \exp(-i\gamma)}. \tag{6.132}$$

This result is the same as (6.121) obtained by the partial wave method.

In the case of transmitted waves, we first obtain E_t by substituting $D_0^{(1)}$ of (6.131) and $D_0^{(2)}$ of Eq. (6.130) into (6.102), then we obtain

$$T = \frac{E_t^a \exp\left(-ik_{0z}^{(2)} H\right)}{E_0^a \exp\left(-ik_{0z}^{(1)} H\right)} = \frac{1 - \frac{r^{(1)}}{r^{(2)}}}{1 - \frac{r^{(1)}}{r^{(2)}} \exp(-i\gamma)} = \frac{r^{(2)} - r^{(1)}}{r^{(2)} - r^{(1)} \exp(-i\gamma)}. \tag{6.133}$$

This agrees with (6.124).

Calculation of Diffraction Intensity

Figure 6.23 shows the calculated results of P_g/P_0 and P_t/P_0 as a function of W for sH being $\pi/4$, $\pi/2$ and π. P_g/P_0 always shows the maximum at $W = 0$ and

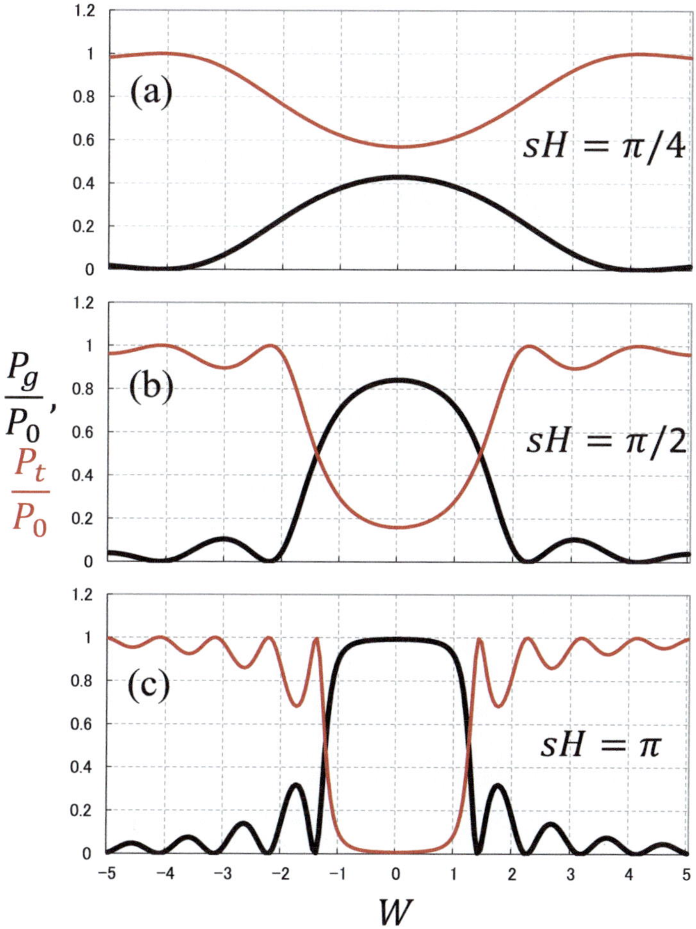

Fig. 6.23 Rocking curves from a thin crystal in the Bragg case. P_g/P_0 is shown by black line and P_t/P_0 by brown line

approaches 1 as sH increases up to π. In addition, interference fringes that are not seen from semi-infinite crystals are seen in P_g/P_0 and P_t/P_0. These interference fringes are due to the phase difference between branch (1) and (2) waves and are Pendellösung fringes. The phase difference between P_g/P_0 and P_t/P_0 is π.

The rocking curve from a sufficiently thick crystal is called the Ewald's curve, which is shown in Fig. 6.17 together with the Darwin's curve. For $|W| > 1$, the Ewald's curve is somewhat larger than the Darwin's curve. This is due to the influence of reflected waves from the bottom of an infinitely thick crystal. The question sometimes arises as to whether the experimental rocking curve is an Ewald's curve or a Darwin's curve. If the bottom reflection can be ignored, it will be a Darwin's curve, and if it cannot be ignored, it will be an Ewald's curve. On the other hand, in mirage diffraction, which will be described in Chap. 9, the rocking curve becomes an Ewald's curve when the pseudo-refracted wave returns to the incident surface without reaching the bottom surface even if the crystal is thin.

6.2.3 Integrated Reflection Power

Let us consider the integrated reflection power in the Bragg case with reference to the Laue case in 6.1.6. The integrated reflection power R_g^W in terms of parameter W for a thin crystal is given by Waller [2] as

$$R_g^W = \int_{-\infty}^{\infty} \frac{P_g(W)}{P_0} dW = \int_{-\infty}^{\infty} \frac{\sin^2\left[sH\left(W^2-1\right)^{1/2}\right]}{W^2 - 1 + \sin^2\left[sH\left(W^2-1\right)^{1/2}\right]} dW = \pi \tanh(sH). \tag{6.134}$$

Figure 6.24 shows R_g^W as a function of sH. R_g^W increases rapidly as sH increases from 0 to around 2. Then it converges to π as sH increases further and $\tanh(sH)$ converges to 1. The integrated reflection power $R_g^{\alpha_0}$ of the Ewald's curve from a sufficiently thick crystal is obtained by using (6.64) as

$$R_g^{\alpha_0} = \left(\frac{|\cos\theta_2|}{\cos\theta_1}\right)^{1/2} \frac{\pi P|\chi_g|}{\sin 2\theta_B}. \tag{6.135}$$

The value of $R_g^{\alpha_0}$ in the Bragg case is twice as large as that in the Laue case, because R_g^W in the Laue case converges to $\pi/2$ as sH becomes infinitely large. $R_g^{\alpha_0}$ of the Darwin's curve from a semi-infinite crystal is obtained as

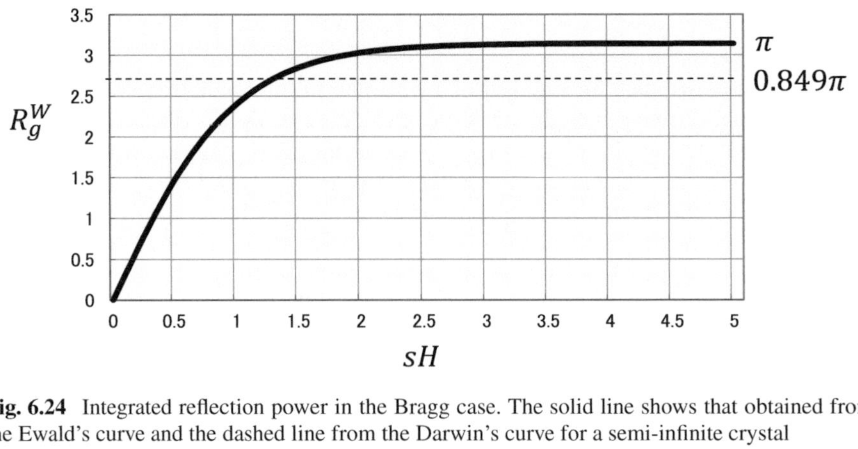

Fig. 6.24 Integrated reflection power in the Bragg case. The solid line shows that obtained from the Ewald's curve and the dashed line from the Darwin's curve for a semi-infinite crystal

$$R_g^{\alpha_0} = \left(\frac{|\cos\theta_2|}{\cos\theta_1}\right)^{1/2} \frac{8P|\chi_g|}{3\sin 2\theta_B}. \tag{6.136}$$

This value is $8/(3\pi) \approx 0.849$ times of that from the Ewald's curve. Warren [9] describes the detail about the integration of the Ewald's curve.

6.3 Divergence Angle in Asymmetric Reflection

We consider the variation of the divergence angle of diffracted waves for an asymmetric reflection. Figure 6.25a shows the arrangement of asymmetric reflections in the Bragg case, and Fig. 6.25b shows its dispersion surface. We focus on the dispersion surfaces in vacuum, T_0' and T_g'. Since total reflection occurs for W from -1 to 1, the corresponding angular region of the incident wave is the range where the tie-point lies within the region of ΔX. Let $\Delta\theta_g$ be the angular range in the diffracted wave direction that causes total reflection, and $\Delta\theta_0$ be the angular range in the transmitted wave direction. $\overline{\mathrm{AB}}$ on T_0' is the total reflection area of the incident wave, and $\overline{\mathrm{CD}}$ on T_g' is that of the reflected wave. Since $\overline{\mathrm{AO}} = \overline{\mathrm{BO}} = \overline{\mathrm{CG}} = \overline{\mathrm{DG}} = K_0$ (K_0 is the wave number in vacuum), the ratio of the angular ranges $b = \Delta\theta_g/\Delta\theta_0$ is given as

$$b = \frac{\Delta\theta_g}{\Delta\theta_0} = \frac{\overline{\mathrm{CD}}/K_0}{\overline{\mathrm{AB}}/K_0} = \frac{\frac{\Delta X}{|\cos\theta_2|}}{\frac{\Delta X}{\cos\theta_1}} = \frac{\cos\theta_1}{|\cos\theta_2|}. \tag{6.137}$$

In Fig. 6.25a, by using the width l_0 of the parallel incident beam of X-ray and that l_g of the diffracted beam, the ratio of these widths is given as

$$b = \frac{l_0}{l_g}. \tag{6.138}$$

This can be rewritten as

$$b = \frac{\sin(\theta_B - \beta)}{\sin(\theta_B + \beta)}. \tag{6.139}$$

For symmetric reflection, $\beta = 0$, then $b = 1$. In the example of Fig. 6.25, $b < 1$, and $\Delta\theta_g$ is smaller than $\Delta\theta_0$. This means that the beam parallelism of the diffracted wave is higher than that of the incident wave. If asymmetric reflection is repeated using n pieces of crystals, the angular width can be reduced by a factor of b^n. This asymmetric reflection is often used as a monochromator for intrinsic rocking curve measurements. An example is given in 7.10.1. In the beam arrangement in Fig. 6.25, if the incident and diffracted beams are swapped, $b > 1$.

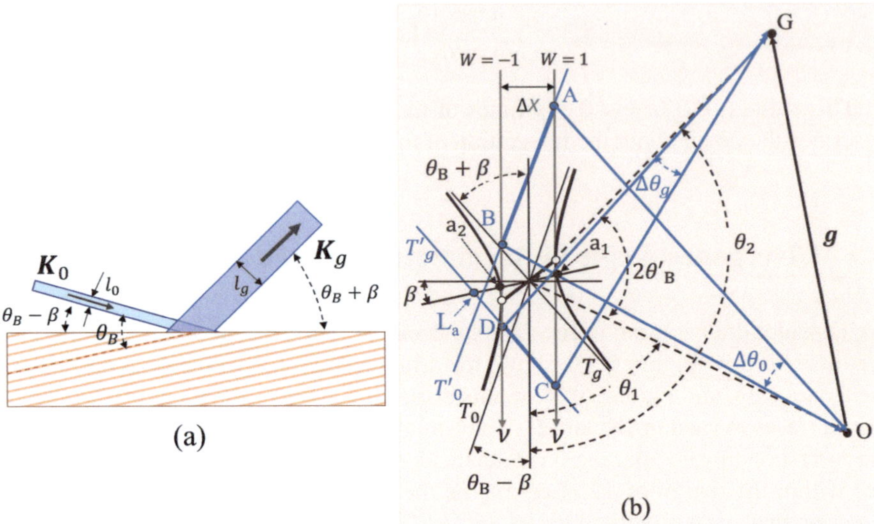

Fig. 6.25 **a** Beam geometry to explain amplification of beam width in case of asymmetric reflection. Oblique lines in the crystal indicate the direction of lattice planes. **b** Dispersion surface for the asymmetric reflection

6.4 Trajectory of Pseudo-Refracted Wave in Crystal

We will consider the trajectory of the pseudo-refracted wave, first in the symmetric Laue case, then in the symmetric Bragg case.

6.4.1 Trajectory of Pseudo-Refracted Wave in the Laue Case

From (6.47), the angle $\varphi^{(j)}$ between the reflecting plane and the pseudo-refracted wave of branch j is obtained as

$$\tan\varphi^{(j)} = \frac{1 - \left|r^{(j)}\right|^2}{1 + \left|r^{(j)}\right|^2}\tan\theta_{\mathrm{B}}. \tag{6.140}$$

with $r^{(j)} = D_h^{(j)}/D_0^{(j)}$. This can be rewritten as

$$\frac{\tan\varphi^{(j)}}{\tan\theta_{\mathrm{B}}} = \frac{1 - \left|W(\mp)\left(W^2+1\right)^{1/2}\right|^2}{1 + \left|W(\mp)\left(W^2+1\right)^{1/2}\right|^2} = \pm\frac{W}{\left(W^2+1\right)^{1/2}}. \tag{6.141}$$

The direction of $\varphi^{(j)}$ is taken to be counterclockwise. The sign of the last term is positive for branch (1) ($j = 1$) and negative for branch (2) ($j = 2$). When $W = 0$, $\varphi^{(j)} = 0$, and the direction of the pseudo-refracted wave is parallel to the reflecting plane. On the other hand, when $|W| = \infty$, $\left|\varphi^{(j)}\right| = \theta_B$.

In Fig. 6.26, pseudo-refracted waves are excited at tie-points d_1 and d_2 on the dispersion surface, when a beam with small width is incident at point A on a crystal of thickness H. As for the pseudo-refracted wave $\boldsymbol{S}^{(1)}$, the angle $\varphi^{(1)}$ changes from θ_B to $-\theta_B$ when W changes from $-\infty$ to ∞, so that the trajectory is within the Borrmann fan of ΔAE′E. The direction of $\boldsymbol{S}^{(1)}$ is $\varphi^{(1)}$ from the normal. $\boldsymbol{S}^{(1)}$ reaches point B on the exit surface. Let O be the midpoint of E'E, and take the ratio p of $\overline{\mathrm{OB}}$ and $\overline{\mathrm{OE}}$, then

$$p = \frac{\overline{\mathrm{OB}}}{\overline{\mathrm{OE}}} = \frac{\tan\varphi^{(j)}}{\tan\theta_B} = \pm\frac{W}{\left(W^2 + 1\right)^{1/2}}. \tag{6.142}$$

Once W is given, the beam emission position B on the exit surface can be obtained. Conversely, (6.142) can be rewritten as

$$W = \pm\frac{p}{\left(1 - p^2\right)^{\frac{1}{2}}}. \tag{6.143}$$

This means that the value of W can be obtained by measuring the emission point B of the pseudo-refraction wave.

Figure 6.27 shows the trajectories of the pseudo-refracted waves for $W = 0$ (a), $W \neq 0$ (b) and (c) when the incident wave has a divergence angle of $\Delta\alpha$. For $W = 0$ in (a), both $\boldsymbol{S}^{(1)}$ and $\boldsymbol{S}^{(2)}$ are parallel to the lattice plane, and the transmitted wave ($\boldsymbol{E}_t^a$) and diffracted wave ($\boldsymbol{E}_g^a$) are emitted in two separate directions from point O. In (b), when $W \neq 0$, $\boldsymbol{S}^{(1)}$ and $\boldsymbol{S}^{(2)}$ propagate in different directions. $\boldsymbol{S}^{(1)}$ reaches point B_1, and the waves $E_t^{a(1)}$ and $E_g^{a(1)}$ are emitted into two separate directions. $\boldsymbol{S}^{(2)}$ reaches point B_2 and $E_t^{a(2)}$ and $E_g^{a(2)}$ are emitted into two separate directions. When the incident wave has a divergence of angle $\Delta\alpha$, and as stated in 6.1.5, W has a finite

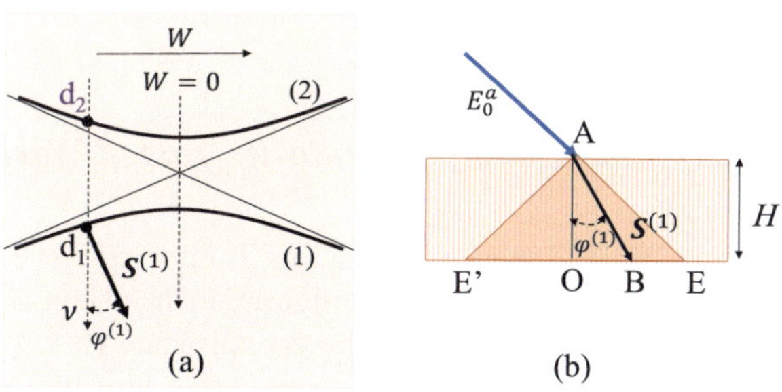

Fig. 6.26 Trajectory of a pseud-refracted wave in the symmetric Laue case.
a Dispersion surface and a pseud-refraction wave $\boldsymbol{S}^{(1)}$ from tie-point d_1.
b Trajectory of a pseud-refracted wave in crystal. A triangular region AE'E represents the Borrmann fan

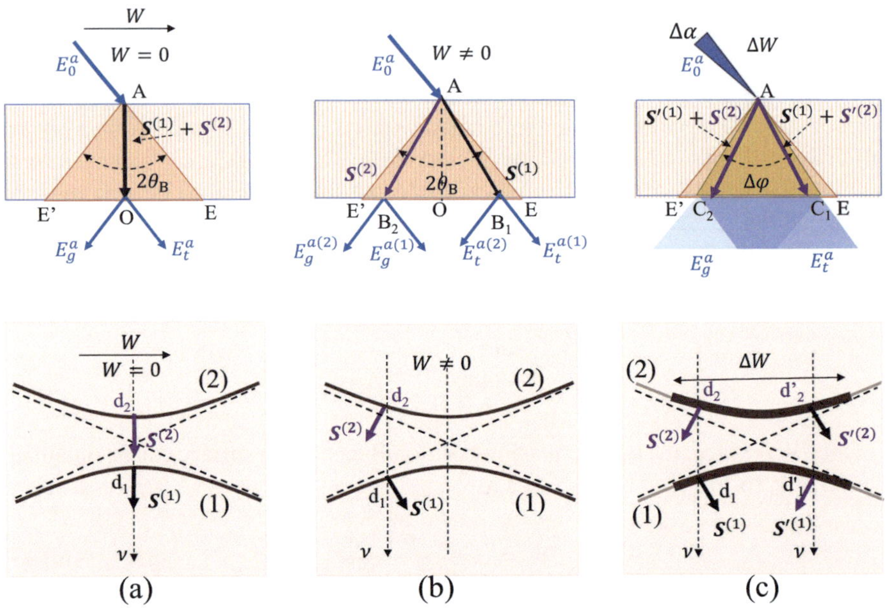

Fig. 6.27 The upper part shows trajectories of pseudo-refraction waves for $W = 0$ (**a**), $W \neq 0$ (**b**) and for $W = 0$ with a width of ΔW (**c**). The lower part shows the corresponding dispersion surfaces and the tie-points from which pseudo-refracted waves are excited

width ΔW on the dispersion surface. (In Fig. 6.27c, the entire parts of the dispersion surface indicated by the thick lines are tie-points.) $\Delta\varphi$ becomes large due to the angular amplification effect to cover most of the Borrmann fan. In this case, $\boldsymbol{S}^{(1)}$ of branch (1) and $\boldsymbol{S}'^{(2)}$ of branch (2) are in the same direction, and at the same time, $\boldsymbol{S}'^{(1)}$ and $\boldsymbol{S}^{(2)}$ are in the same direction. Generally, it is not easy to experimentally obtain ideal plane waves; the X-rays used in experiments are wave-packets, and the beam obtained by an X-ray monochromator has a slight divergence angle. Figure 6.27c shows trajectories in such an actual condition. This issue will be described in detail in Chap. 8.

6.4.2 Trajectory of Pseudo-Refracted Wave in the Bragg Case

We consider the trajectory of a quasi-refracted wave in a crystal in the symmetric Bragg case. In the Bragg configuration in a non-absorbing crystal, the pseudo-refracted wave cannot enter the crystal when $|W| \leq 1$ as described in 6.2. When $|W| \geq 1$, pseudo-refracted waves propagate into the crystal. We consider the trajectory of pseudo-refracted waves in this case.

6.4.2.1 Bragg-Laue Case

Figure 6.28a shows the beam geometry in the symmetric Bragg-Laue case for a thick crystal. Figure 6.28b shows the dispersion surface in this case. The diffraction condition on the entrance surface is the Bragg case, but that on the exit plane on the lateral surface of the crystal is the Laue case. We call this arrangement as the Bragg-Laue (BL) case. A part of the incident wave is reflected at the incident point A and the diffracted wave E_g^a is emitted, and the rest becomes a pseudo-refracted wave $\boldsymbol{S}^{(1)}$. $\boldsymbol{S}^{(1)}$ is excited at the tie-point d_1 on branch (1) on the dispersion surface. When $\boldsymbol{S}^{(1)}$ reaches point B on the lateral surface of the crystal, the X-rays are separated into the transmission direction (E_{lt}^a) and the diffraction direction (E_{lg}^a) and emitted from the crystal. If the angle between the reflecting lattice plane and $\boldsymbol{S}^{(1)}$ is $\varphi^{(1)}$, then from (6.47), we get

$$\tan\varphi^{(1)} = \frac{1 - \left|r^{(1)}\right|^2}{1 + \left|r^{(1)}\right|^2}\tan\theta_{\mathrm{B}} \tag{6.144}$$

If we write (6.144) using the deviation parameter W, we get

$$\frac{\left|\tan\varphi^{(1)}\right|}{\tan\theta_{\mathrm{B}}} = \left|\frac{1 - \left|W(\mp)\left(W^2 - 1\right)^{1/2}\right|^2}{1 + \left|W(\mp)\left(W^2 - 1\right)^{1/2}\right|^2}\right| = \frac{\left(W^2 - 1\right)^{1/2}}{W} \tag{6.145}$$

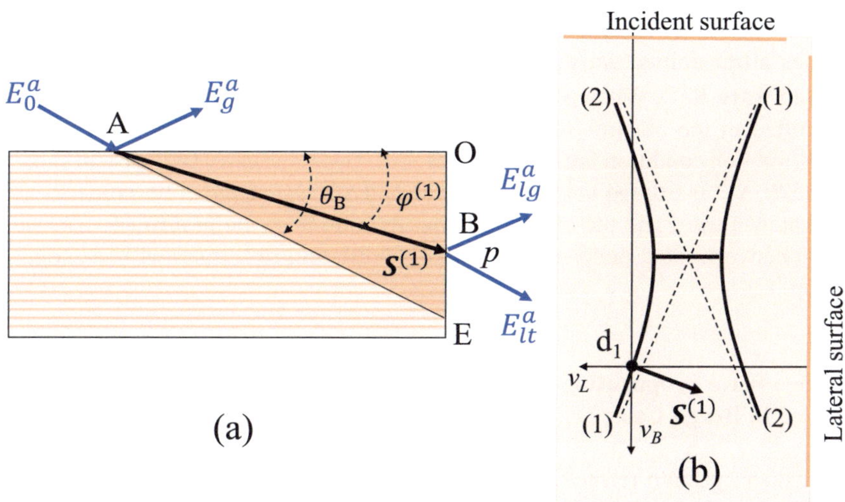

Fig. 6.28 **a** Trajectory of pseudo-refracted wave in the Bragg-Laue case in real space. Dark brown contrast shows the Borrmann fan. **b** Dispersion surface and a tie-point corresponding to the pseudo-refracted wave

And the ratio p is given by

$$p = \frac{\overline{\mathrm{OB}}}{\overline{\mathrm{OE}}} = \frac{|\tan\varphi|}{\tan\theta_{\mathrm{B}}} = \frac{\left(W^2 - 1\right)^{1/2}}{|W|} \tag{6.146}$$

Once W is given, the beam emission point B can be obtained. Conversely, we can rewrite (6.146) as

$$W = \pm \frac{1}{\left(1 - p^2\right)^{1/2}} \tag{6.147}$$

Once the emission point B (p) is given, we can get the deviation parameter W of the pseudo-refracted wave.

6.4.2.2 Bragg-Bragg-Laue Case

Another example of the trajectory of pseudo-refracted waves in a parallel plate crystal of finite length is shown in Fig. 6.29a, when the incident wave E_0^a has a slight divergent angle. Figure 6.29b shows the dispersion surface and pseudo-refracted waves in the reciprocal space. Pseudo-refracted waves are excited approximately within the Borrmann fan. $\boldsymbol{S}^{(1)}$ corresponds to a component of E_0^a, which is incident on the crystal at a shallow angle to the surface. $\boldsymbol{S}^{(1)}$ is a pseudo-refracted wave in the BL case, reaching directly point B on the lateral surface.

$\boldsymbol{S}'^{(1)}$ corresponds to a component of $\boldsymbol{E}_0^a$ that enters at a larger glancing angle than the above incident wave, and reaches point C on the bottom surface. A part of it becomes a transmitted wave E_t^a, and the rest is reflected and becomes a pseudo-refracted wave $\boldsymbol{S}'^{(2)}$. When $\boldsymbol{S}^{(1)}$ and $\boldsymbol{S}'^{(2)}$ reach point B on the side surface, X-rays are emitted in the transmission direction (E_{lt}^a) and the diffraction direction (E_{lg}^a). This diffraction condition is called the Bragg-Bragg-Laue case, BbL case in short. In Fig. 6.29b, $\boldsymbol{S}^{(1)}$ is excited at the tie-point d_1 and $\boldsymbol{S}'^{(1)}$ is excited at tie-point d'_1. $\boldsymbol{S}'^{(2)}$ propagates upward and the corresponding tie-point is d'_2 on branch (2). The interference between the pseudo-refracted waves in BL and BbL cases will be explained in Chap. 8.

6.4.2.3 Wave Propagating in Thin Crystal with Multiple Reflections in Bragg Case

We consider a pseudo-refracted wave excited by a narrow incident beam, which propagates in a thin crystal while reflected from the top and bottom surfaces repeatedly as shown in Fig. 6.30. We restrict the number of reflections three in the following. Let the electric field of the diffracted wave after ($n-1$)-times reflections be $E_g^{a\langle n\rangle}$, and that of the transmitted wave be $E_t^{a\langle n\rangle}$.

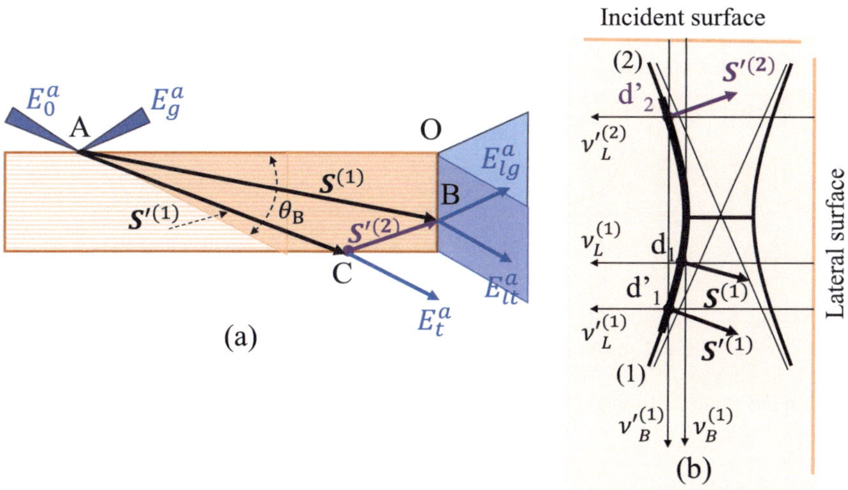

Fig. 6.29 **a** Trajectories of pseudo-refracted waves $S^{(1)}$, $S'^{(1)}$, and $S'^{(2)}$ in Bragg-Bragg-Laue case in real space. Dark brown contrast shows the Borrmann fan region. **b** Dispersion surface and tie-points corresponding to the pseudo-refracted waves

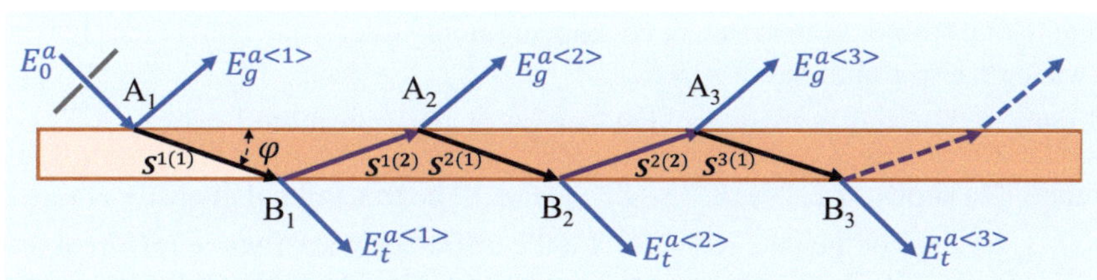

Fig. 6.30 Trajectory of a pseudo-refracted wave in an infinitely long flat parallel crystal and the diffracted and transmitted waves emitted from either top or bottom surface of the crystal in the symmetric Bragg case. Crystal thickness is H

The closer $|W|$ approaches 1, the smaller the angle φ between the crystal surface and the pseudo-refracted wave $\boldsymbol{S}$ becomes; for $|W| = 1$, $\varphi = 0$, and $\boldsymbol{S}^{1(1)}$ is parallel to the surface. In other words, as $|W|$ approaches 1, the distance between the incident point A_1 and the reflection point B_1 on the bottom surface increases. (At the same time, the distance between B_1 and A_2 increases.) $E_t^{\boldsymbol{a}\langle n\rangle}$ and $E_g^{a\langle n\rangle}$ can be observed separately. The calculated results of $\left|E_g^{a\langle n\rangle}/E_0^a\right|^2$ are shown in Fig. 6.31a and b. The calculated $|(E_g^{a\langle 1\rangle} + E_g^{a\langle 2\rangle} + E_g^{a\langle 3\rangle})/E_0^a|^2$ are shown in Fig. 6.31c. The rocking curve for $n = 1$ is the Darwin's curve.

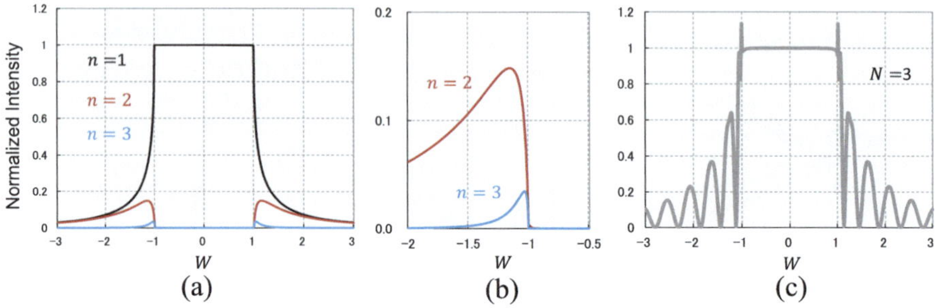

Fig. 6.31 Rocking curves of the diffracted waves. **a** Rocking curves for $n = 1, 2$ and 3. **b** Magnified rocking curves for $n = 2$ and 3 in **a** in the range of $-2 < W < -0.5$. **c** Rocking curve of sum of three diffracted waves of $n = 1, 2$ and 3. The value of the intensities around $W = \pm 1$ is larger than 1 because the number of waves in the sum is limited. $sH = 6$ is assumed

For $n = 2$ and 3, the diffracted intensity is zero when $|W| < 1$, because the total reflection occurs for the diffracted wave for $n = 1$. The diffracted intensity for $n = 2$ shows the peak of height 0.15 at $|W|$=1.05, and its FWHM $\Delta W_g^{\langle 2\rangle}$ is 0.8. The diffracted intensity for $n = 3$ shows its peak of 0.03 high at $|W|$=1.02, and its FWHM $\Delta W_g^{\langle 3\rangle}$ is 0.1, which is 1/8 of FWHM of the peak for $n = 2$. In this way, the peak height decreases and the width becomes narrower as the number of reflections increases. Pendelösung fringes seen in Fig. 6.31c appear mainly due to the interference between the wave of $n = 1$ and that of $n = 2$.

Figure 6.32a and b show rocking curves of the transmitted waves $\left|E_t^{\langle n\rangle}/E_0\right|^2$. Figure 6.32c shows $\left|\left(E_t^{\langle 1\rangle} + E_t^{\langle 2\rangle} + E_t^{\langle 3\rangle}\right)/E_0\right|^2$. The transmitted intensity is zero for $|W| < 1$, because of the total reflection for the diffracted wave for $n = 1$.The curve for $n = 2$ in Fig. 6.32b shows a peak of 0.06 at $|W|$=1.05 with its FWHM $\Delta W_t^{\langle 2\rangle} = 0.3$. The curve for $n = 3$ shows the maximum value of 0.02 at $|W|$=1.01 with its FWHM $\Delta W_g^{\langle 3\rangle} = 0.1$. The transmitted intensity for $n = 3$ from B_3 on the bottom surface which is farthest from the incident point A_1 is 0.02. In terms of the magnitude of the electric field, $|E_t^{\langle 3\rangle}/E_0|$ is 0.14, which is still not so small. This means that even if the internal reflections are repeated, the decrease in the amplitude of the electric field of the pseudo-refracted wave is small. The interference effects due to $\Sigma E_g^{\langle n\rangle}$ and $\Sigma E_t^{\langle n\rangle}$ will be described in Chap. 8.

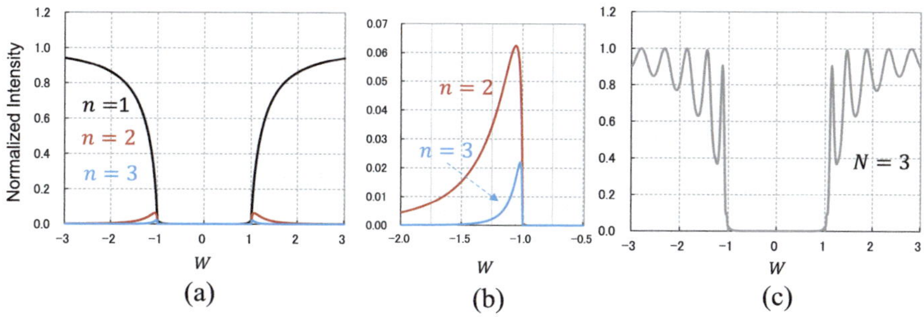

Fig. 6.32 Rocking curves of transmitted waves from the bottom surface. **a** Rocking curves for $n =$ 1, 2 and 3. **b** Magnified rocking curves for $n = 2$ and 3 in **a** in the range of $-2 < W < -0.5$. **c** Rocking curve of sum of three diffracted waves of $n = 1$, 2 and 3. $sH = 6$ is assumed

References

1. G. Borrmann, Phys. Z. **42**, 157–162 (1941)
2. I. Waller, Ann. Phys. **79**, 261 (1928)
3. B.W. Batterman, H. Cole, Rev. Modern Phys. **36**, 681–717 (1964)
4. Z.G. Pinsker, *Dynamical Scattering of X-Rays in Crystals* (Springer-Verlag, 1978)
5. A. Authier, *Dynamical Theory of X-ray Diffraction* (Oxford University Press, 2001)
6. A. Authier, Acta Cryst. **A64**, 337–340 (2008)
7. P.H. Dederichs, Solid State Phys. **27**, 135–236 (1972)
8. H. Wagner, Z. Phys. **146**, 127–168 (1956)
9. B.E. Warren, *X-ray Diffraction* (Addison-Wiesley, 1969)

Chapter 7
Resonant Scattering and Dynamical Theory of Diffraction

In Chaps. 5 and 6, we considered dynamical theories of diffraction, that is, multiple scattering theories of diffraction due to Thomson (normal) scattering. In real crystals, diffraction occurs not only by normal scattering but also by resonant (anomalous) scattering. Here, we will consider dynamical diffraction for real crystals. The most dynamical theories of diffraction including resonant scattering are called "dynamical theories of diffraction considering absorption" or "dynamical theory of diffraction for absorbing crystals", which considers resonant scattering as a correction term, assuming that it is smaller than the normal atomic scattering factor. In contrast, we will consider resonant scattering and normal scattering equally and consider a dynamic theory of diffraction in this chapter.

The normal atomic scattering factor is a real positive number. When resonant scattering is involved, the atomic scattering factor f becomes a complex number as described in Chap. 3, and from (3.83), we have

$$f = f^0(k_s) + f'(\omega) + if''(\omega),$$

where $f^0(k_s)$ is the normal atomic scattering factor, k_s is the magnitude of the scattering vector. $f'(\omega) + if''(\omega)$ is called anomalous scattering factor due to resonant scattering and ω is the angular frequency of the incident X-ray. The anomalous scattering factor shows large change near the absorption edge of the constituent atom in a crystal. The real part $f^0(k_s) + f'(\omega)$ can be either positive or negative. When the real part becomes zero, only the imaginary part $f''(\omega)$ has non-zero value. $f''(\omega)$ is mainly due to photoelectric absorption and has a large value on the high energy side of the absorption edge. The anomalous scattering factor is not a correction term at all.

T. Fukamachi and T. Kawamura, *Resonant Dynamical X-ray Diffraction*, Springer Series in Solid-State Sciences 208, https://doi.org/10.1007/978-981-96-8406-9_7

7.1 X-ray Susceptibility Including Resonant Scattering

The X-ray susceptibility $\chi(\boldsymbol{r})$ due to normal scattering is a real number, and its Fourier coefficient χ_g is a complex number satisfying

$$\chi_g = \chi_{-g}^*. \tag{7.1}$$

When resonant scattering is included, $\chi(\boldsymbol{r})$ becomes a complex number and can be written as

$$\chi(\boldsymbol{r}) = \chi_r(\boldsymbol{r}) + i\chi_i(\boldsymbol{r}) \tag{7.2}$$

from (4.39). By writing the Fourier coefficients of $\chi_r(\boldsymbol{r})$ and $\chi_i(\boldsymbol{r})$ as χ_{gr} and χ_{gi} respectively, the Fourier coefficient χ_g of $\chi(\boldsymbol{r})$ is written by

$$\chi_g = \chi_{gr} + i\chi_{gi}. \tag{7.3}$$

Since we have relations

$$\chi_{gr} = \chi_{-gr}^*, \ \chi_{gi} = \chi_{-gi}^*, \tag{7.4}$$

we obtain

$$\chi_g \neq \chi_{-g}^*. \tag{7.5}$$

We write these Fourier coefficients as

$$\chi_{gr} = |\chi_{gr}| \exp(i\alpha_{gr}), \ \chi_{gi} = |\chi_{gi}| \exp(i\alpha_{gi}) \tag{7.6}$$

to examine the relation of the atomic scattering factor f with these χ_{gr} and χ_{gi}. In (5.6), by replacing the normal atomic scattering factor of j-th atom f_j^0 with the atomic scattering factor f_j including resonant scattering, the crystal structure factor F_g can be written as

$$F_g(\omega) = \sum_j \left[f_j^0(\boldsymbol{g}) + f_j'(\omega) + if_j''(\omega)\right] \exp(i\boldsymbol{g}\cdot\boldsymbol{r}) \exp(-M_j), \tag{7.7}$$

where $\exp(-M_j)$ is the Debye–Waller factor used in (2.51). From (5.5), we get

$$\chi_g = \chi_{gr} + i\chi_{gi} = -r_e \frac{\lambda^2}{\pi V_c} F_g(\omega) = -r_e \frac{\lambda^2}{\pi V_c}\left[F_{gr}(\omega) + iF_{gi}(\omega)\right]. \tag{7.8}$$

In (7.8), χ_{gr} and χ_{gi} are given by

$$\chi_{gr} = -r_e \frac{\lambda^2}{\pi V_c} F_{gr}(\omega) = -r_e \frac{\lambda^2}{\pi V_c} \sum_{j=1}^{n} \left[f_j^0(\boldsymbol{g}) + f_j'(\omega) \right] \exp(i\boldsymbol{g} \cdot \boldsymbol{r}_j) \exp(-M_j), \tag{7.9}$$

$$\chi_{gi} = -r_e \frac{\lambda^2}{\pi V_c} F_{gi}(\omega) = -r_e \frac{\lambda^2}{\pi V_c} \sum_{j=1}^{n} f_j''(\omega) \exp(i\boldsymbol{g} \cdot \boldsymbol{r}_j) \exp(-M_j). \tag{7.10}$$

7.2 One-Wave Approximation and Complex Refractive Index

In the one-wave approximation, as mentioned in Chap. 5, only one Fourier component (g = 0) corresponding to the incident wave exists and it can be written as

$$\chi_0 = \frac{1}{V_c} \int_{\text{u.c.}} \chi(\boldsymbol{r}) d\boldsymbol{r}.$$

Since the permittivity is given by

$$\varepsilon(\boldsymbol{r}) = \varepsilon_0[1 + \chi(\boldsymbol{r})] = \varepsilon_0(1 + \chi_0) \ ,$$

its ratio to the permittivity in vacuum is expressed as

$$\frac{\varepsilon}{\varepsilon_0} = 1 + \chi_0 = 1 + \chi_{0r} + i\chi_{0i}. \tag{7.11}$$

From (7.9) and (7.10), we obtain

$$\chi_{0r} = -r_e \frac{\lambda^2}{\pi V_c} \sum_{j=1}^{n} \left[Z_j + f_j'(\omega) \right], \quad \chi_{0i} = -r_e \frac{\lambda^2}{\pi V_c} \sum_{j=1}^{n} f_j''(\omega). \tag{7.12}$$

Using (4.19) and writing the average wave number in the crystal as κ_0, the complex refractive index n_0 can be expressed as

$$n_0 = \frac{\kappa_0}{K_0} = \sqrt{\frac{\varepsilon}{\varepsilon_0}} = (1 + \chi_{0r} + i\chi_{0i})^{1/2} \approx 1 + \frac{1}{2}\chi_{0r} + i\frac{1}{2}\chi_{0i}. \tag{7.13}$$

K_0 is the wave number in vacuum and is a real number. κ_0 is a complex number and can be written as

$$\kappa_0 = \kappa_{0r} + i\kappa_{0i}. \tag{7.14}$$

The real and the imaginary parts are respectively written as

$$\kappa_{0r} = \left(1 + \frac{\chi_{0r}}{2}\right)K_0 \ , \ \kappa_{0i} = \frac{\chi_{0i}}{2}K_0. \tag{7.15}$$

The dispersion relation in the one-wave approximation is given by (5.21) as

$$\boldsymbol{k}_0^2 = \frac{\varepsilon}{\varepsilon_0}K_0^2.$$

When $\boldsymbol{k}_0$ is a complex vector, we write

$$\boldsymbol{k}_0 = \boldsymbol{k}_{0r} + i\boldsymbol{k}_{0i} \tag{7.16}$$

and obtain the dispersion relation as

$$\boldsymbol{k}_0^2 = (\boldsymbol{k}_{0r} + i\boldsymbol{k}_{0i})^2 = (1 + \chi_{0r} + i\chi_{0i})K_0^2. \tag{7.17}$$

From the real part of (7.17), we get

$$\boldsymbol{k}_{0r}^2 - \boldsymbol{k}_{0i}^2 = (1 + \chi_{0r})K_0^2. \tag{7.18}$$

Usually, we have a relation $\boldsymbol{k}_{0r}^2 >> \boldsymbol{k}_{0i}^2$ and can write (7.18) as

$$|\boldsymbol{k}_{0r}|^2 \approx K_0^2(1 + \chi_{0r}) = \kappa_{0r}^2. \tag{7.19}$$

From the imaginary part of (7.17), we get

$$2\boldsymbol{k}_{0r}\boldsymbol{k}_{0i} = \chi_{0i}K_0^2. \tag{7.20}$$

By writing the unit vector in the direction of $\boldsymbol{k}_{0r}$ as $\boldsymbol{s}_0$, we can approximately write as $\boldsymbol{s}_0 = \boldsymbol{k}_{0r}/K_0$ and get

$$\boldsymbol{s}_0 \cdot \boldsymbol{k}_{0i} = \frac{\chi_{0i}}{2}K_0. \tag{7.21}$$

This equation expresses a plane perpendicular to $\boldsymbol{s}_0$ in the $\boldsymbol{k}_{0i}$ space, and that the magnitude of $\boldsymbol{k}_{0i}$ in the direction of $\boldsymbol{s}_0$ is $\chi_{0i}K_0/2$.

The dispersion relation of (7.17) is the same as the that of (4.46). The optical property of X-rays within the one-wave approximation is the same as described in Chap. 4.

7.3 Two-Wave Approximation

When resonant scattering is involved, both n_0 and κ_0 are complex numbers. We will consider the dispersion surface in the two-wave approximation under this condition.

7.3.1 Basic Equations

The dispersion equations given in (5.39) and (5.40) in the two-wave approximation are extended as

$$\frac{\boldsymbol{k}_0^2 - \kappa_0^2}{\boldsymbol{k}_0^2} D_0 = \frac{(\boldsymbol{k}_{0r} + i\boldsymbol{k}_{0i})^2 - (\kappa_{0r} + i\kappa_{0i})^2}{\boldsymbol{k}_0^2} D_0 = P\big(\chi_{-gr} + i\chi_{-gi}\big) D_g, \quad (7.22)$$

$$\frac{\boldsymbol{k}_g^2 - \kappa_0^2}{\boldsymbol{k}_g^2} D_g = \frac{\big(\boldsymbol{k}_{gr} + i\boldsymbol{k}_{gi}\big)^2 - (\kappa_{0r} + i\kappa_{0i})^2}{\boldsymbol{k}_g^2} D_g = P\big(\chi_{gr} + i\chi_{gi}\big) D_0. \quad (7.23)$$

The term $(\boldsymbol{k}_{0r} + i\boldsymbol{k}_{0i})^2$ can be approximated as

$$(\boldsymbol{k}_{0r} + i\boldsymbol{k}_{0i})^2 = \boldsymbol{k}_{0r}^2 - \boldsymbol{k}_{0i}^2 + 2i\boldsymbol{k}_{0r} \cdot \boldsymbol{k}_{0i}, \quad (7.24)$$

$$\boldsymbol{k}_0^2 = \boldsymbol{k}_{0r}^2 + 2i\boldsymbol{k}_{0r} \cdot \boldsymbol{k}_{0i}. \quad (7.25)$$

Then, we get

$$\sqrt{\boldsymbol{k}_0^2} = |\boldsymbol{k}_{0r}| + i\boldsymbol{s}_0 \cdot \boldsymbol{k}_{0i}. \quad (7.26)$$

In (7.23), $\boldsymbol{k}_g$ can be written as

$$\boldsymbol{k}_g = \boldsymbol{k}_{gr} + i\boldsymbol{k}_{gi}, \quad (7.27)$$

with

$$\boldsymbol{k}_{gr} = \boldsymbol{k}_{0r} + \boldsymbol{g}, \quad (7.28)$$

because the reciprocal lattice vector $\boldsymbol{g}$ is real. Then we can write

$$\boldsymbol{k}_g = \boldsymbol{k}_0 + \boldsymbol{g} = \boldsymbol{k}_{gr} + i\boldsymbol{k}_{0i} \quad (7.29)$$

and we obtain

$$\boldsymbol{k}_{gi} = \boldsymbol{k}_{0i}. \quad (7.30)$$

Similar to (7.26), we get

$$\sqrt{\boldsymbol{k}_g^2} = |\boldsymbol{k}_{gr}| + i\boldsymbol{s}_g \cdot \boldsymbol{k}_{gi}. \tag{7.31}$$

We define the resonance error ξ_0 and ξ_g according to (5.41), as

$$\xi_0 = \sqrt{\boldsymbol{k}_0^2} - \kappa_0, \quad \xi_g = \sqrt{\boldsymbol{k}_g^2} - \kappa_0. \tag{7.32}$$

ξ_0 and ξ_g are given as

$$\xi_0 = |\boldsymbol{k}_{0r}| - \kappa_{0r} + i(\boldsymbol{s}_0 \cdot \boldsymbol{k}_{0i} - \kappa_{0i}), \quad \xi_g = |\boldsymbol{k}_{gr}| - \kappa_{0r} + i(\boldsymbol{s}_g \cdot \boldsymbol{k}_{gi} - \kappa_{0i}) \tag{7.33}$$

and become complex. We can express ξ_0 and ξ_g as

$$\xi_0 = \xi_{0r} + i\xi_{0i}, \quad \xi_g = \xi_{gr} + i\xi_{gi}. \tag{7.34}$$

From (7.33), we can write

$$\xi_{0r} = |\boldsymbol{k}_{0r}| - \kappa_{0r}, \xi_{0i} = \boldsymbol{s}_0 \cdot \boldsymbol{k}_{0i} - \kappa_{0i}. \tag{7.35}$$

By rewriting (7.32) as $\kappa_0^2 = \boldsymbol{k}_0^2 - 2\sqrt{\boldsymbol{k}_0^2}\xi_0 + \xi_0^2$, we get

$$\boldsymbol{k}_0^2 - \kappa_0^2 \approx \boldsymbol{k}_0^2 - \left(\boldsymbol{k}_0^2 - 2\sqrt{\boldsymbol{k}_0^2}\xi_0\right) = 2\sqrt{\boldsymbol{k}_0^2}\xi_0 \tag{7.36}$$

by ignoring the term ξ_0^2. Equation (7.22) is rewritten as

$$\frac{\boldsymbol{k}_0^2 - \kappa_0^2}{\boldsymbol{k}_0^2} = \frac{2\sqrt{\boldsymbol{k}_0^2}\xi_0}{\boldsymbol{k}_0^2} = \frac{2\xi_0}{\kappa_{0r}} = \frac{2(\xi_{0r} + i\xi_{0i})}{\kappa_{0r}} \tag{7.37}$$

by using $\boldsymbol{k}_0^2 \approx \boldsymbol{k}_{0r}^2 \approx \kappa_{0r}^2$. Similarly, (7.23) is rewritten as

$$\frac{\boldsymbol{k}_g^2 - \kappa_0^2}{\boldsymbol{k}_g^2} = \frac{2(\xi_{gr} + i\xi_{gi})}{\kappa_{0r}}. \tag{7.38}$$

The basic equations in the two-wave approximation are given by

$$\frac{2(\xi_{0r} + i\xi_{0i})}{\kappa_{0r}} D_0 - P(\chi_{-gr} + i\chi_{-gi})D_g = 0, \tag{7.39}$$

$$P(\chi_{gr} + i\chi_{gi})D_0 - \frac{2(\xi_{gr} + i\xi_{gi})}{\kappa_{0r}}D_g = 0. \tag{7.40}$$

For finding a solution satisfying the condition that D_0 and D_g are nonzero, it is required to satisfy the following equation

$$(\xi_{0r} + i\xi_{0i})(\xi_{gr} + i\xi_{gi}) = \frac{\kappa_{0r}^2 P^2}{4}(\chi_{gr} + i\chi_{gi})(\chi_{-gr} + i\chi_{-gi}). \tag{7.41}$$

This gives the relation between ξ_0 and ξ_g, that is, the dispersion relation. From (7.39) and (7.40), we get

$$\frac{D_g}{D_0} = \frac{2(\xi_{0r} + i\xi_{0i})}{\kappa_{0r}P(\chi_{-gr} + i\chi_{-gi})} = \frac{\kappa_{0r}P(\chi_{gr} + i\chi_{gi})}{2(\xi_{gr} + i\xi_{gi})}. \tag{7.42}$$

The magnitude of the electric flux density D_g of the diffracted wave with respect to the magnitude of the electric flux density D_0 of the incident wave is obtained.

7.3.2 Relation Between Resonance Circles and Resonant Dynamical Diffraction

The right side of (7.41) can be rewritten as

$$\begin{aligned}
\chi_g\chi_{-g} &= (\chi_{gr} + i\chi_{gi})(\chi_{-gr} + i\chi_{-gi}) \\
&= [|\chi_{gr}|\exp(i\alpha_{gr}) + i|\chi_{gi}|\exp(i\alpha_{gi})][|\chi_{gr}|\exp(-i\alpha_{gr}) + i|\chi_{gi}|\exp(-i\alpha_{gi})] \\
&= |\chi_{gr}|^2 - |\chi_{gi}|^2 + i2|\chi_{gr}||\chi_{gi}|\cos\delta.
\end{aligned} \tag{7.43}$$

The parameter δ is defined as

$$\delta = \alpha_{gi} - \alpha_{gr}, \tag{7.44}$$

which is the difference between the phase α_{gi} of the imaginary part of χ_g and the phase α_{gr} of the real part. Dividing both sides of (7.43) by $|\chi_{gr}|^2 + |\chi_{gi}|^2$, we get

$$\frac{\chi_g\chi_{-g}}{|\chi_{gr}|^2 + |\chi_{gi}|^2} = \left(\frac{|\chi_{gr}|^2 - |\chi_{gi}|^2}{|\chi_{gr}|^2 + |\chi_{gi}|^2} + i\frac{2|\chi_{gr}||\chi_{gi}|\cos\delta}{|\chi_{gr}|^2 + |\chi_{gi}|^2}\right). \tag{7.45}$$

By writing the magnitude of $\overline{\chi}_g$ as

$$\overline{\chi}_g = \left(\left|\chi_{gr}\right|^2 + \left|\chi_{gi}\right|^2\right)^{1/2} \tag{7.46}$$

and defining two parameters u and v as

$$u = \frac{\left|\chi_{gr}\right|^2 - \left|\chi_{gi}\right|^2}{\overline{\chi}_g^2}, \ v = \frac{2\left|\chi_{gr}\right|\left|\chi_{gi}\right|\cos\delta}{\overline{\chi}_g^2} = \left(1 - u^2\right)^{1/2}\cos\delta, \tag{7.47}$$

we rewrite the right side of (7.45) as

$$\frac{\chi_g \chi_{-g}}{\overline{\chi}_g^2} = (u + iv). \tag{7.48}$$

Equation (7.48) can be also written as

$$\frac{\chi_g \chi_{-g}}{\overline{\chi}_g^2} = \left(u^2 + v^2\right)^{1/2} \exp(i\phi), \ \tan\phi = \frac{v}{u}. \tag{7.49}$$

By above definition, u and v each vary within the range $[-1, 1]$. In the complex (u, v)-plane with u in the real axis and v in the imaginary axis, $\chi_g \chi_{-g}/\overline{\chi}_g^2$ takes all values within the unit circle as shown in Fig. 7.1. We call this circle as the resonance circle. If the crystal has a center of symmetry, δ should be zero or π and $\cos\delta = \pm 1$, $\chi_g \chi_{-g}/\overline{\chi}_g^2$ takes the value on the unit circle. If resonant scattering is ignored, $\chi_{gi} = 0$, which corresponds to one point on this unit circle at $u = 1$ ($\phi = 0$). The values of u and v for some typical values of χ_{gr} and χ_{gi} are shown in Table 7.1. The listed values from ① to ⑤ are indicated in Fig. 7.1.

For expressing the relative magnitude of resonant scattering to that of normal scattering, we define a parameter

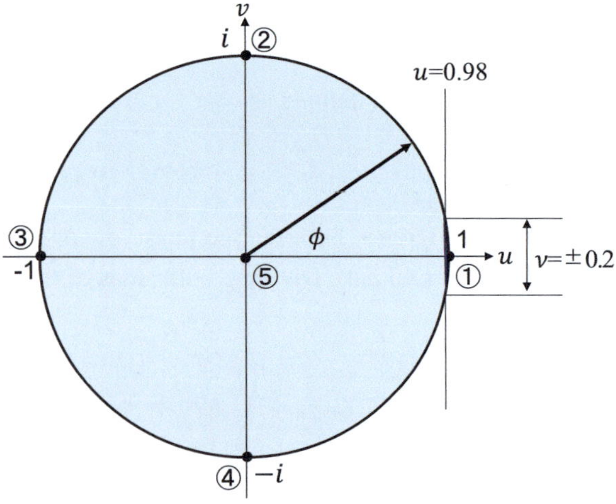

Fig. 7.1 Resonance circle expressing (7.49)

Table 7.1 Values of u, v, ϕ and k_a for some typical values of χ_{gr} and χ_{gi}

	u	v	ϕ	k_a	
①	1	0	0	0	$\lvert\chi_{gr}\rvert \neq 0, \lvert\chi_{gi}\rvert = 0$
②	0	1	$\pi/2$	1	$\lvert\chi_{gr}\rvert = \lvert\chi_{gi}\rvert, \cos\delta = 1$
③	-1	0	π	∞	$\lvert\chi_{gr}\rvert = 0, \lvert\chi_{gi}\rvert \neq 0$
④	0	-1	$-\pi/2$	1	$\lvert\chi_{gr}\rvert = \lvert\chi_{gi}\rvert, \cos\delta = -1$
⑤	0	0	–	1	$\chi_g\chi_{-g} = 0, \cos\delta = 0$

$$k_a = \frac{\left|\chi_{gi}\right|}{\left|\chi_{gr}\right|}. \tag{7.50}$$

Using this parameter, we can write $\chi_g\chi_{-g}$ as

$$\begin{aligned} \chi_g\chi_{-g} &= \left|\chi_{gr}\right|^2\left(1 - \frac{\left|\chi_{gi}\right|^2}{\left|\chi_{gr}\right|^2} + i2\frac{\left|\chi_{gi}\right|}{\left|\chi_{gr}\right|}\cos\delta\right) \\ &= \left|\chi_{gr}\right|^2\left(1 - k_a^2 + i2k_a\cos\delta\right). \end{aligned} \tag{7.51}$$

In many previous works, X-ray energy in experiment is assumed to be far enough away from the absorption edge and $\left|\chi_{gi}\right|$ is much smaller than $\left|\chi_{gr}\right|$. The anomalous scattering factor is regarded as a correction term for the normal scattering factor, and k_a is much smaller than 1. If $k_a < 0.1$, k_a^2 can be ignored and (7.51) can be written by

$$\chi_g\chi_{-g} \approx \left|\chi_{gr}\right|^2(1 + i2k_a\cos\delta). \tag{7.52}$$

This region ($k_a < 0.1$) corresponds to the range of $u = 0.98 \sim 1.0$ and $v = -0.2 \sim 0.2$ on the resonance circle, as shown in Fig. 7.1. In this region, small absorption approximation works.

At a typical value of $(u, v) = (1, 0)$ ① on the unit circle, χ_{gi} is zero and the anomalous scattering factor is ignored (even if resonant scattering is included, χ_{gi} can be zero as will be explained later). When χ_{gi} is close to zero, small absorption approximation can be applied and the conventional dynamical diffraction theory works. On the other hand, in cases from ② to ⑤, the resonant scattering dynamic diffraction theory must be applied and new diffraction phenomena peculiar to it can be expected. Especially in case of ③ $(u, v) = (-1, 0)$, $\chi_{gr} = 0$ and dynamical diffraction only due to χ_{gi}, that is, only resonant scattering, is expected. In cases of ② and ④ $(u, v) = (0, \pm 1)$ $\left|\chi_{gr}\right| = \left|\chi_{gi}\right|$, and some characteristics for polar crystals can be expected. In case of ⑤ $(u, v) = (0, 0)$, $\chi_g\chi_{-g}$ is zero and dynamical diffraction cannot be observed.

The parameters u and v are related to k_a as

$$u = \frac{1 - k_a^2}{1 + k_a^2}, \quad v = \frac{2k_a \cos\delta}{1 + k_a^2}. \tag{7.53}$$

The first equation can be rewritten as

$$k_a^2 = \frac{1 - u}{1 + u}. \tag{7.54}$$

χ_g can be expressed as

$$\begin{aligned} \chi_g &= \chi_{gr} + i\chi_{gr} = |\chi_{gr}| \exp(i\alpha_{gr}) + i|\chi_{gi}| \exp(i\alpha_{gi}) \\ &= |\chi_{gr}| \left(1 - 2k_a \sin\delta + k_a^2\right)^{1/2} \exp[i(\alpha_{gr} + \theta)], \end{aligned} \tag{7.55}$$

where θ is

$$\theta = \tan^{-1}\left(\frac{k_a \cos\delta}{1 - k_a \sin\delta}\right). \tag{7.56}$$

This indicates that θ is a function of δ. For polar crystals, the sign of δ is reversed between χ_g and χ_{-g}. Then (7.55) is useful for determining the phase of crystal structure factors and the polarity of polar crystals, as will be explained later. $|\chi_{gr}|$ and $|\chi_{gi}|$ are related to $\overline{\chi}_g$ and u as

$$|\chi_{gr}| = \overline{\chi}_g \left(\frac{1 + u}{2}\right)^{1/2}, \quad |\chi_{gi}| = \overline{\chi}_g \left(\frac{1 - u}{2}\right)^{1/2}. \tag{7.57}$$

For reference, Fig. 7.2 shows the relations between f, χ_g, $\chi_g \chi_{-g}$ and θ in a monoatomic crystal.

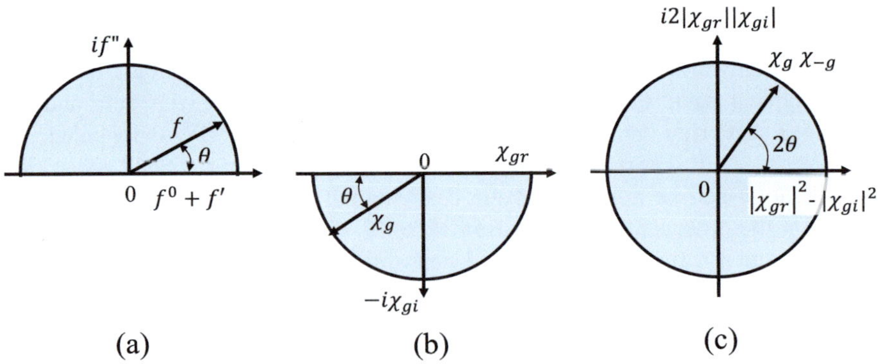

Fig. 7.2 Relations between f and θ (**a**), χ_g and θ (**b**), $\chi_g \chi_{-g}$ and θ (**c**)

7.3.3 Complex Dispersion Surface

The basic dispersion equation of (7.41) is now rewritten as

$$(\xi_{0r} + i\xi_{0i})(\xi_{gr} + i\xi_{gi}) = \frac{\kappa_{0r}^2 P^2}{4}\overline{\chi}_g^2 (u + iv). \tag{7.58}$$

The representation of this equation in the reciprocal space of $\boldsymbol{k}$ is called the complex dispersion surface. Separating (7.58) into the real and imaginary parts, we get

$$\xi_{0r}\xi_{gr} - \xi_{0i}\xi_{gi} = \frac{\kappa_{0r}^2 P^2}{4}\overline{\chi}_g^2 u, \tag{7.59}$$

$$\xi_{0r}\xi_{gi} + \xi_{gr}\xi_{0i} = \frac{\kappa_{0r}^2 P^2}{4}\overline{\chi}_g^2 v. \tag{7.60}$$

A similar expression of dispersion equation can also be seen in the paper by Batterman et al. [1], where the approximation of $|\xi_{0r}\xi_{gr}| >> |\xi_{0i}\xi_{gi}|$ is used and diffraction phenomena within the small absorption approximation are studied.

Even if resonant scattering is ignored, the dispersion surface becomes complex in the total reflection region in the Bragg case, and the complex dispersion equation is derived in (6.90). The right and left sides of (6.90) are real, but they are complex in (7.58). Since $\xi_{0r}\xi_{gr} = 0$ and $\xi_{0i}\xi_{gi} \neq 0$ when $W = 0$ in (6.90), we must be careful about adopting the conventional approximation of $|\xi_{0r}\xi_{gr}| >> |\xi_{0i}\xi_{gi}|$ [2, 3].

We consider the complex dispersion surface based on a Lorentz sphere whose radius is the real component κ_{0r} of wave number according to Miyake [4] by considering comparison with experimental results. By writing

$$\sqrt{\boldsymbol{k}_0^2} - \kappa_{0r} = \xi_0', \quad \sqrt{\boldsymbol{k}_g^2} - \kappa_{0r} = \xi_g', \tag{7.61}$$

the relations between (ξ_0', ξ_g') and (ξ_0, ξ_g) are obtained

$$\xi_0 = \xi_0' - i\kappa_{0i}, \quad \xi_g = \xi_g' - i\kappa_{0i}. \tag{7.62}$$

When κ_{0r} is used as a reference, (7.58) becomes

$$\left(\xi_0' - i\kappa_{0i}\right)\left(\xi_g' - i\kappa_{0i}\right) = \frac{\kappa_{0r}^2 P^2}{4}\overline{\chi}_g^2 (u + iv). \tag{7.63}$$

By rewriting in this way, it is possible to expand the dispersion relation using the X–Y orthogonal coordinate system used in Chap. 5.

As shown in (5.47) and (5.48), the left side of (7.63) can be expressed in the $X' - Y'$ coordinates for a symmetric reflection as

$$\left(\xi_0' - i\kappa_{0i}\right)\left(\xi_g' - i\kappa_{0i}\right) = \left(Y_0'\sin\theta_B' + X_0'\cos\theta_B' - i\kappa_{0i}\right)\left(X_0'\cos\theta_B' - Y_0'\sin\theta_B' - i\kappa_{0i}\right)$$
$$= X_0'^2\cos^2\theta_B' - Y_0'^2\sin^2\theta_B' - i2\kappa_{0i}X_0'\cos\theta_B' - \kappa_{0i}^2. \tag{7.64}$$

Using (7.64), (7.63) can be written as

$$X_0'^2\cos^2\theta_B' - Y_0'^2\sin^2\theta_B' - i2\kappa_{0i}X_0'\cos\theta_B' - \kappa_{0i}^2 = \frac{\kappa_{0r}^2P^2}{4}\overline{\chi}_g^2(u+iv). \tag{7.65}$$

In the X–Y coordinate of (5.63), which is rotated this $X' - Y'$ coordinate counterclockwise by β, (7.65) becomes

$$X_0^2\sin\theta_1\sin\theta_2 + Y_0^2\cos\theta_1\cos\theta_2 + X_0Y_0\sin2\beta - i2\kappa_{0i}(X_0\cos\beta + Y_0\sin\beta)\cos\theta_B'$$
$$- \kappa_{0i}^2 = \frac{\kappa_{0r}^2P^2}{4}\overline{\chi}_g^2(u+iv). \tag{7.66}$$

As described in Chap. 6, we take X_0 as a real number and Y_0 as a complex number, which is written as

$$Y_0 = Y_{0r} + iY_{0i}. \tag{7.67}$$

Substituting (7.67) into (7.66), we get

$$X_0^2\sin\theta_1\sin\theta_2 + \left(Y_{0r}^2 - Y_{0i}^2\right)\cos\theta_1\cos\theta_2 + X_0Y_{0r}\sin2\beta$$
$$+2\kappa_{0i}Y_{0i}\sin\beta\cos\theta_B' - \kappa_{0i}^2 = \frac{\kappa_{0r}^2P^2}{4}\overline{\chi}_g^2u \tag{7.68}$$

from the real part, and

$$2\left\{Y_{0r}Y_{0i}\cos\theta_1\cos\theta_2 + X_0Y_{0i}\cos\beta - \kappa_{0i}\cos\theta_B'(X_0\cos\beta + Y_{0r}\sin\beta)\right\} = \frac{\kappa_{0r}^2P^2}{4}\overline{\chi}_g^2v \tag{7.69}$$

from the imaginary part. When X_0 is given by the incident conditions, we can find Y_{0r} and Y_{0i}, that is the solution of dispersion Eq. (7.66).

By solving a quadratic Eq. (7.66) for Y_0, we get

$$Y_0 = \frac{1}{2\cos\theta_1\cos\theta_2}\Bigg[-X_0\sin2\beta + i2\kappa_{0i}\sin\beta\cos\theta_B'$$
$$\pm\left\{\left(-X_0\sin2\beta + i2\kappa_{0i}\sin\beta\cos\theta_B'\right)^2 - 4\cos\theta_1\cos\theta_2\left(X_0^2\sin\theta_1\sin\theta_2 - \kappa_{0i}^2 - \frac{\kappa_{0r}^2P^2}{4}\overline{\chi}_g^2(u+iv)\right)\right\}^{\frac{1}{2}}\Bigg]. \tag{7.70}$$

We expand the terms in the curly brackets in (7.70) in terms of X_0 and simplify the coefficients of X_0^2, X_0 and κ_{0i}^2. The coefficients of X_0^2 is modified as $\sin^2 2\beta - 4\sin\theta_1\sin\theta_2\cos\theta_1\cos\theta_2 = \sin^2 2\theta_B'$, that of X_0 is modified as

$$-i4\kappa_{0i}\left(\sin 2\beta\sin\beta\cos\theta_B' - 2\cos\theta_1\cos\theta_2\cos\beta\cos\theta_B'\right) = -4i\kappa_{0i}\sin 2\theta_B'\sin\theta_B'\cos\beta$$

and that of κ_{0i}^2 as $-4\left(\sin^2\beta\cos^2\theta_B' - \cos\theta_1\cos\theta_2\right) = -4\sin^2\theta_B'\cos^2\beta$. Then (7.70) is rewritten as

$$Y_0 = \frac{1}{2\cos\theta_1\cos\theta_2}\Big[-X_0\sin 2\beta + i2\kappa_{0i}\sin\beta\cos\theta_B' \pm\left\{\left(X_0\sin 2\theta_B' - i2\kappa_{0i}\sin\theta_B'\cos\beta\right)^2 + \cos\theta_1\cos\theta_2\kappa_{0r}^2P^2\overline{\chi}_g^2(u+iv)\right\}^{\frac{1}{2}}\Big] \tag{7.71}$$

By introducing the deviation parameter W defined by

$$W = -\frac{X_0\sin 2\theta_B'}{|\cos\theta_1\cos\theta_2|^{\frac{1}{2}}}\frac{1}{\kappa_{0r}P\overline{\chi}_g}, \tag{7.72}$$

we rewrite (7.71) using W instead of X_0 as

$$Y_0 = \frac{(\pm)\kappa_{0r}P\overline{\chi}_g}{2|\cos\theta_1\cos\theta_2|^{\frac{1}{2}}}\left[\frac{\sin 2\beta}{\sin 2\theta_B'}W + ig\frac{\cos\theta_B'\sin\beta}{|\cos\theta_1\cos\theta_2|^{1/2}} \pm\left\{\left(W+ig'\right)^2(\pm)(u+iv)\right\}^{\frac{1}{2}}\right]. \tag{7.73}$$

In the double sign, the + sign corresponds to the Laue case ($\cos\theta_2 > 0$) and the − sign to the Bragg case ($\cos\theta_2 < 0$). g and g' are defined by

$$g = \frac{\chi_{0i}}{P\overline{\chi}_g},\quad g' = g\frac{\sin\theta_B'\cos\beta}{|\cos\theta_1\cos\theta_2|^{1/2}}. \tag{7.74}$$

Since χ_{0i} is negative, g is negative. From (7.73), the real and imaginary parts of Y_0 are expressed as

$$Y_{0r} = \mathrm{Re}(Y_0) = \frac{(\pm)\kappa_{0r}P\overline{\chi}_g}{2|\cos\theta_1\cos\theta_2|^{\frac{1}{2}}}\left[\frac{\sin 2\beta}{\sin 2\theta_B'}W \pm \mathrm{Re}\left\{\left(W+ig'\right)^2(\pm)(u+iv)\right\}^{\frac{1}{2}}\right], \tag{7.75}$$

$$Y_{0i} = \mathrm{Im}(Y_0) = \frac{(\pm)\kappa_{0r}P\overline{\chi}_g}{2|\cos\theta_1\cos\theta_2|^{\frac{1}{2}}}\left[g\frac{\cos\theta_B'\sin\beta}{|\cos\theta_1\cos\theta_2|^{\frac{1}{2}}} \pm \mathrm{Im}\left\{\left(W+ig'\right)^2(\pm)(u+iv)\right\}^{\frac{1}{2}}\right]. \tag{7.76}$$

The real part of the dispersion surface is given by Y_{0r} as a function of $X_0(W)$, and the imaginary part of the dispersion surface is given by Y_{0i} as a function of $X_0(W)$.

Resonance errors ξ_0' and ξ_g' are related to Y_{0r} and Y_{0i} as

$$\xi'_{0r} = \xi_{0r} = X_0 \sin\theta_1 + Y_{0r} \cos\theta_1, \quad \xi'_{0i} = Y_{0i} \cos\theta_1 = \xi_{0i} + \kappa_{0i}, \tag{7.77}$$

$$\xi'_{gr} = \xi_{gr} = X_0 \sin\theta_2 + Y_{0r} \cos\theta_2, \quad \xi'_{gi} = Y_{0i} \cos\theta_2 = \xi_{gi} + \kappa_{0i}. \tag{7.78}$$

Once ξ_0 and ξ_g are given, D_0 and D_g are given by

$$D_0 = \frac{\kappa_{0r}}{2\xi_0} P\chi_{-g} D_g = \frac{2\xi_g}{\kappa_{0r}} \frac{1}{P\chi_g} D_g.$$

Here, ξ_0 is expressed as

$$\xi_0 = X_0 \sin\theta_1 + Y_0 \cos\theta_1 - i\kappa_{0i}, \tag{7.79}$$

with κ_{0i} being given by

$$\kappa_{0i} = \frac{k_{0r}\chi_{0i}}{2} = \frac{k_{0r} g P \overline{\chi}_g}{2}. \tag{7.80}$$

ξ_0 is also expressed as a function of W as

$$\xi_0 = (\pm) \frac{\kappa_{0r} P \overline{\chi}_g}{2} \left(\frac{\cos\theta_1}{\cos\theta_2} \right)^{\frac{1}{2}} \left[(W + ig') \pm \sqrt{Q} \right]. \tag{7.81}$$

Here, we use a parameter Q given by

$$\sqrt{Q} = \left[(W + ig')^2 (\pm)(u + iv) \right]^{1/2} \tag{7.82}$$

and the relations $2\theta'_{\mathrm{B}} = \theta_2 - \theta_1$, $\beta - \theta'_{\mathrm{B}} = \theta_1$ from Fig. 7.3, and

$$\sin 2\beta = \sin\theta_2 \cos\theta_1 + \cos\theta_2 \sin\theta_1,$$

$$\cos\theta_2 = \sin\beta \cos\theta'_{\mathrm{B}} - \cos\beta \sin\theta'_{\mathrm{B}}.$$

Deviation parameter *W* for resonant dynamical theory

In the complex dispersion relations, the X coordinate is expressed as a real number and the Y coordinate as a complex number. Since the parameter W is the normalized X coordinate, W is related to the angle $\Delta\alpha = \alpha_0 - \alpha_{0L}$ in the same manner as (5.82) and (5.84) as

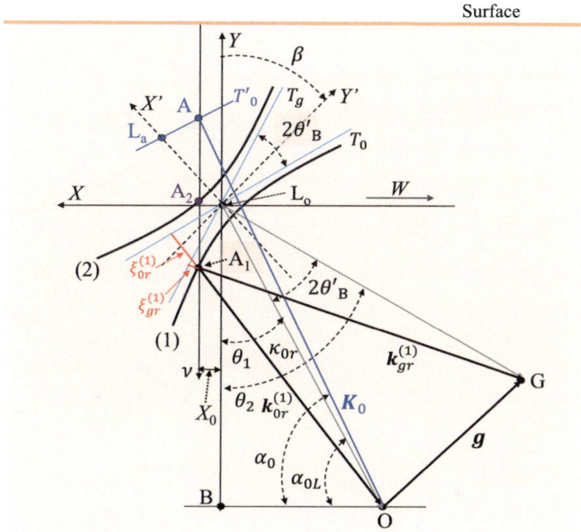

Fig. 7.3 The real part of dispersion surface based on κ_{0r}. L_a is Laue point and L_o Lorentz point. The straight line passing through AL_a (shown as a solid line in dark blue) is the dispersion surface T_0' in vacuum, and point A is the tie point. The intersections of the perpendicular line ν passing through point A with dispersion surfaces of branches (1) and (2) are tie-points A_1 and A_2 in the crystal

$$W = (\alpha_0 - \alpha_{0L})\left(\frac{\cos\theta_1}{|\cos\theta_2|}\right)^{\frac{1}{2}} \frac{\sin 2\theta_B'}{n_0 P \overline{\chi}_g} = \Delta\alpha\left(\frac{\cos\theta_1}{|\cos\theta_2|}\right)^{\frac{1}{2}} \frac{\sin 2\theta_B'}{\mathrm{n}_0 \mathrm{P} \overline{\chi}_g} \quad (7.83)$$

by replacing $\left(\chi_g \chi_{-g}\right)^{1/2}$ by $\overline{\chi}_g$ [5, 6]. As the glancing angle α_0 increases, W increases.

7.3.4 *Construction of Complex Dispersion Surface*

Let construct the dispersion surface corresponding to the dispersion equation given by (7.75) and (7.76). The X axis is a real number, and Y_r and Y_i represent points on the real and imaginary axes, respectively. The $X - Y_r$ plane and the $X - Y_i$ plane are orthogonal to each other (see Fig. 6.15b). For representing dispersion surfaces in $X - Y_i$ plane as well as $X - Y_r$ plane, we depict them separately as shown in Fig. 7.4 in the following.

Figure 7.4 shows the complex dispersion surface in the symmetric Laue case for $u = 0\,(\delta = 0 \text{ or } \pi)$. Figure 7.4a shows the real part and Fig. 7.4b shows the imaginary part. The real part shows a hyperbola opening north–south. The branch closer to the origin of the reciprocal lattice point is labeled as (1), and the other branch farther is label as (2). As for the tie-points, the intersection between the normal ν determined from the incident conditions and the branch (1) is tie-point A_1, and the intersection with the branch (2) is tie-point A_2. The interpretation of the real part of the dispersion surface is the same as described in Chap. 6. The length of the line segment from A_1 to its intersection with the asymptote T_g is $\xi_{gr}^{'(1)}$, and that to its intersection with the asymptote T_0 is $\xi_{0r}^{'(1)}$. As for the imaginary part of the dispersion surface shown in

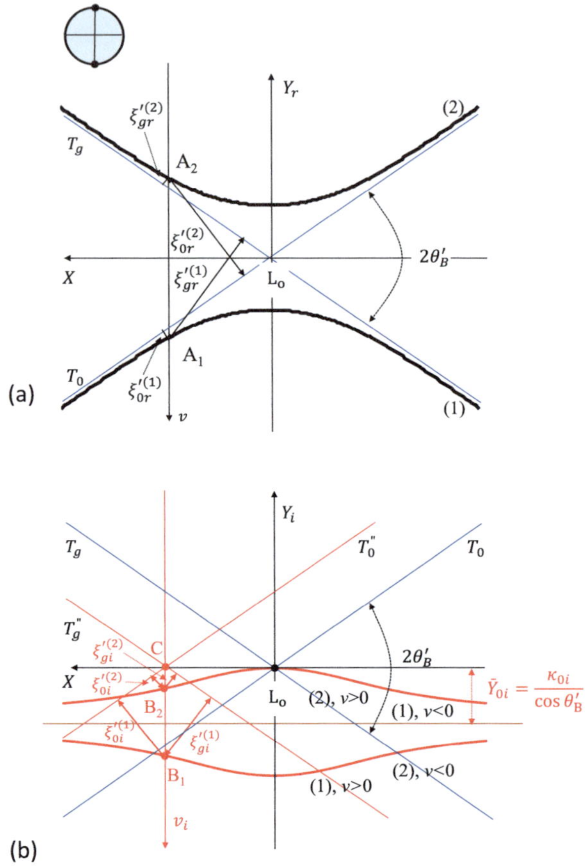

Fig. 7.4 **a** The real part and **b** the imaginary part of complex dispersion surface in the symmetric Laue case

Fig. 7.4b, the value of Y_{0i} is always negative. $|Y_{0i}|$ of branch (1) is maximum and that of branch (2) is minimum at the center ($X = 0$) when $v > 0$. The intersections of Y_{0i} with the perpendicular line v_i are tie-points B_1 of branch (1) and B_2 of branch (2). Suppose that the intersection of X-axis with v_i is C and the straight line parallel to T_0 passing through C is T_0''. The length of the perpendicular line drawn from B_1 to T_0'' is $\xi_{0i}^{'(1)}$. Since $\xi_{0i}^{\prime(1)} = \boldsymbol{s}_0 \cdot \boldsymbol{k}_{0i}$, T_0'' expresses a planc in which the absorption in reciprocal space for O wave is constant. However, the wave number K_0 of the incident beam has no imaginary component, only the perpendicular component to the surface has the imaginary part according to the boundary conditions at the surface. $\xi_{gi}^{\prime(1)} = \boldsymbol{s}_g \cdot \boldsymbol{k}_{gi}$ expresses the length of the perpendicular line drawn from B_1 to T_g'' which is parallel to T_g and passes through C.

7.3.5 Borrmann Effect and Extinction Effect

We now explain the meaning of the imaginary parts of the dispersion equation, Y_{0i} and ξ_i. We take the distance from the crystal surface as z, and denote the z component of the wave vector $\boldsymbol{k}_0$ as k_{0z}. In the symmetric Laue case,

$$k_{0z}^{(j)} = \kappa_{0r}\cos\theta'_{\mathrm{B}} + Y_{0r}^{(j)} + iY_{0i}^{(j)} = k_{0zr}^{(j)} - ik_{0zi}^{(j)}, \tag{7.84}$$

where $Y_{0i}^{(j)}$ is negative, and $k_{0zi}^{(j)} = -Y_{0i}^{(j)}$. The intensity of the X-ray at depth z is proportional to

$$\left|\exp\left(-k_{0zi}^{(j)}z\right)\right|^2 = \exp\left(2Y_{0i}^{(j)}z\right). \tag{7.85}$$

$Y_{0i}^{(j)}$ is related to the effective absorption coefficient $\mu^{(j)}$ as

$$Y_{0i}^{(j)} = -\frac{\mu^{(j)}}{2\cos\theta'_{\mathrm{B}}}. \tag{7.86}$$

We also know the relation

$$\left|k_{0i}^{(j)}\right| = \left|k_{gi}^{(j)}\right| = \left|\xi_{0i}^{\prime(j)}\right| = \left|\xi_{gi}^{\prime(j)}\right| = \frac{\mu^{(j)}}{2}. \tag{7.87}$$

If we write Y_{0i} when $|X_0| \to \infty$ as $\overline{Y}_{0i}$, we get

$$\overline{Y}_{0i} = Y_{0i}(|X_0| \to \infty) = \frac{\kappa_{0i}}{\cos\theta'_{\mathrm{B}}} = \frac{-\mu_a}{2\cos\theta'_{\mathrm{B}}}, \tag{7.88}$$

where μ_a is the average absorption coefficient in the crystal expressed by

$$\mu_a = \mu^{(j)}(|X_0| \to \infty) = -2\kappa_{0i}. \tag{7.89}$$

In the symmetric Laue case, $Y_{0i}^{(j)}$ in (7.76) is written as

$$Y_{0i}^{(j)} = \frac{-\mu^{(j)}}{2\cos\theta'_{\mathrm{B}}} = \frac{1}{2\cos\theta'_{\mathrm{B}}}\left[\kappa_{0r}P\overline{\chi}_g g \pm \kappa_{0r}P\overline{\chi}_g \mathrm{Im}\left\{W^2 + (u+iv)\right\}^{1/2}\right]. \tag{7.90}$$

Using $g = \chi_{0i}/\left(P\overline{\chi}_g\right)$ in (7.74) and $\mu_a = -\chi_{0i}K_0$ in (4.43), we obtain relations $\kappa_{0r}P\overline{\chi}_g g = \kappa_{0r}\chi_{0i} = -\mu_a$. Then, $\mu^{(j)}$ can be written as

$$\mu^{(j)} = \mu_a + (-1)^{j-1}\mu_d, \tag{7.91}$$

where μ_d is a parameter given by

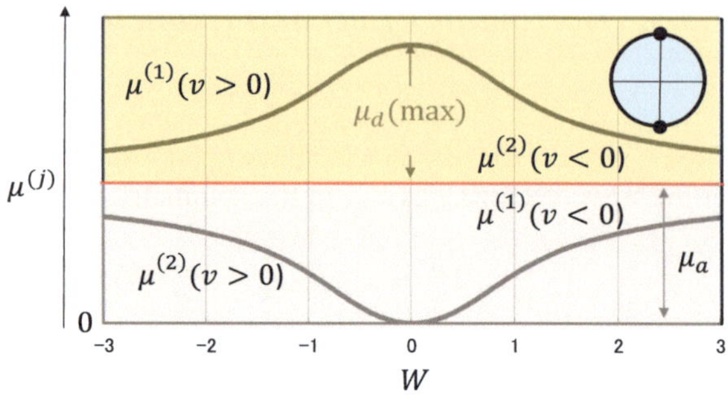

Fig. 7.5 Absorption coefficient $\mu^{(j)}$ in Laue case when $u = 0$ and $v = \pm 1$

$$\mu_d = \kappa_{0r} P \overline{\chi}_g \mathrm{Im}\{W^2 + (u + iv)\}^{1/2}. \tag{7.92}$$

The average absorption coefficient μ_a does not depend temperature, but μ_d depends on temperature due to the temperature dependence of $\overline{\chi}_g$. μ_d depends on the polarization factor P, too. The maximum value of μ_d is obtained for $W = 0$, which is denoted as μ_d(max). Under the conditions in Fig. 7.5, $(u, v) = (0, 1)$, then we have $|\chi_{gr}| = |\chi_{gi}|$. By using relations $\overline{\chi}_g = \sqrt{2}|\chi_{gi}|$ and $\mathrm{Im}(u + iv)^{1/2} = \mathrm{Im}\big((0 + i)^{1/2}\big) = 1/\sqrt{2}$ when $v > 0$, we can write

$$\mu_d(\max) = \kappa_{0r} P \overline{\chi}_g \mathrm{Im}(u + iv)^{1/2} = \kappa_{0r} P |\chi_{gi}|. \tag{7.93}$$

If the temperature effect is ignored, $|\chi_{gi}| = |\chi_{0i}|$, and from (7.93)

$$\mu_d(\max) = \mu_a$$

is obtained for $P = 1$. When $W = 0$, $Y_{0i}^{(2)} = 0$, and $\mu^{(2)} = 0$ $(\mu^{(1)} = 2\mu_a)$. The pseudo-refracted wave excited at branch (2) in Figs. 7.4 and 7.5 propagates in the crystal without absorption. When $v < 0$, $\mathrm{Im}\Big((0 - i)^{\frac{1}{2}}\Big) = -1/\sqrt{2}$, $\mu_d(\max) = -\mu_a$ and $\mu^{(1)} = 0$. The pseudo-refracted wave excited at branch (1) propagates in the crystal without absorption.

In Fig. 7.4b, $\left|Y_{0i}^{(2)}\right| < \left|\overline{Y}_{0i}\right|$ (i.e., $\mu^{(2)} < \mu_a$) for branch (2), while for branch (1), $\left|Y_{0i}^{(1)}\right| > \left|\overline{Y}_{0i}\right|$ (i.e. $\mu^{(1)} > \mu_a$). The X-ray wave of branch (2) suffers weaker absorption than the average and are hardly attenuated within the crystal, which is called anomalous transmission. Whereas the X-ray wave of branch (1) suffers stronger absorption than the average and are quickly attenuated in the crystal, which is called anomalous absorption. If μ_d is zero, the absorption will be only the average absorption. For $\mu_d > 0$, anomalous absorption will occur, and for $\mu_d < 0$, anomalous transmission will occur. The anomalous absorption and anomalous transmission can be directly interpreted by the sign of μ_d. There are two factors that cause such absorption in the dynamical theory of diffraction. These are the case of (1) the Borrmann effect in the Laue case and the case of (2) the Borrmann effect and the extinction effect in the Bragg case.

7.3.5.1 Borrmann Effect in the Laue Case

We explain an example of this case using Fig. 7.5. $|\mu_d|$ shows the maximum at $W = 0$, and the value is obtained by inserting $u = 0$, $P = 1$ and $\kappa_{0r} \approx K_0$ into (7.92) and (7.93), as

$$|\mu_d(\max)| = |\mu_a|.$$

The anomalous absorption and transmission due to χ_{gi} were experimentally discovered by Borrmann [7] and are therefore called the Borrmann effect. Zachariasen [8] and Laue [9] theoretically clarified that this effect is due to resonant scattering. Without resonant scattering, that is, with only Thomson scattering, $\chi_{0i} = \chi_{gi} = 0$, and the Borrmann effect does not occur. If resonant scattering is included in X-ray scattering, χ_{0i} is not zero, but it is possible to make $\chi_{gi} = 0$, in which case $\mu_d = 0$ and the Borrmann effect will disappear. Negishi et al. [10–12] observed this effect experimentally (see Fig. 7.23a). Since $|\chi_{gi}|$ is always smaller than $|\chi_{0i}|$, $\mu_d(\max) \leq \mu_a$. Furthermore, since $|\chi_{gi}|$ is proportional to the Debye–Waller factor as shown in (7.10), the Borrmann effect depends on temperature. It is better to lower the temperature of the crystal for enhancing anomalous transmission. X-rays propagate over long distances in the crystal due to anomalous transmission.

7.3.5.2 Borrmann Effect and Extinction Effect in the Bragg Case

An example of the W dependence of the absorption coefficients $\mu^{(1)}(v > 0)$ and $\mu^{(1)}(v < 0)$ in the Bragg case is shown in Fig. 7.6 when $u = 0.98$ ($k_a = 0.1$). $\mu^{(1)}(v > 0)$ and $\mu^{(1)}(v < 0)$ are inversion symmetric. The absorption coefficient $\mu^{(1)}(v > 0)$ shows minimum at $W = -0.995$ and maximum at $W \approx 0$. For discussing the variations of $\mu^{(1)}(v > 0)$, we divide W into three regions: (1) $W < -1$, (2) $-1 < W < 1$ and (3) $1 < W$ according to James [13].

Region (1): $\mu^{(1)}(v > 0)$ is always less than μ_a and anomalous transmission occurs. $\mu^{(1)}(v > 0)$ becomes zero when W is close to -1.

Region (2): $\mu^{(1)}(v > 0)$ is significantly large and becomes the maximum when W is close to 0 as

$$\mu_d(\max) = K_0|\chi_{hr}|. \tag{7.94}$$

In this figure, for example, $\mu_d(\max) \approx 10\mu_a$. Equation (7.94) holds even when $\chi_{0i} = 0$, that is, the large absorption coefficient of $\mu^{(j)}$ caused only by normal scattering and is due to the extinction effect described in Chap. 6. X-rays cannot

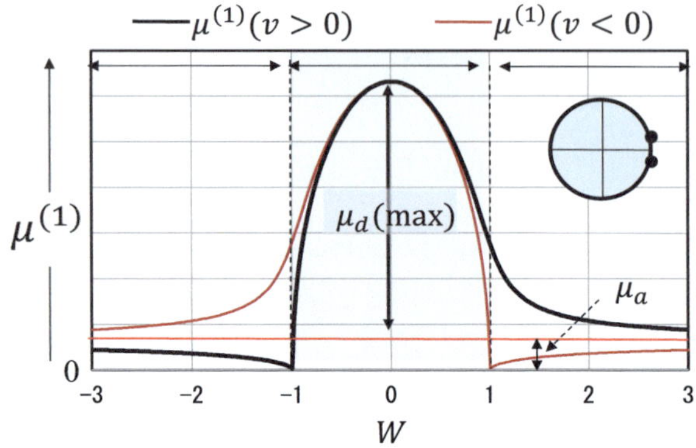

Fig. 7.6 Absorption coefficient $\mu^{(1)}$ in the Bragg case for $u = 0.98(k_a = 0.1)$. $\mu^{(1)}(v > 0)$ is zero at $W = -0.995$ and $\mu^{(1)}(v < 0)$ is zero at $W = 0.995$

penetrate in the crystal and are localized very close to the surface, resulting in an X-ray reflectance of almost 100%.

Region (3): $\mu^{(1)}(v > 0)$ is always larger than μ_a and anomalous absorption occurs.

It is not always true that the extinction effect occurs in the Bragg case, as will be explained in detail later. When $u = -1$, $\chi_{hr} = 0$, the extinction effect completely disappears and only anomalous transmission due to the Borrmann effect occurs, leaving only the region (1) while regions (2) and (3) disappear. (See Fig. 7.11b).

7.4 Diffraction Intensity in the Laue Case

We explain how to calculate the diffraction and transmission intensities from a plane parallel crystal with a thickness of H in the Laue case, by using the Ewald-Laue method and referring to the procedure described in 6.1.

There is a relation between the electric displacement densities D_0 and D_g in the crystal as

$$\frac{D_g}{D_0} = \frac{2\xi_0}{\kappa_{0r}} \frac{1}{P\chi_{-g}}.$$

Once the deviation parameter W is given, ξ_0 in the Laue case is obtained from (7.81) as

$$\xi_0 = (\pm)\frac{\kappa_{or}P\overline{\chi}_g}{2}\left(\frac{\cos\theta_1}{\cos\theta_2}\right)^{1/2}\left[\left(W + ig'\right) \pm \sqrt{L}\right], \tag{7.95}$$

where the parameter $\sqrt{L}$ is given according to (7.82) as,

$$\sqrt{L} = \left[\left(W + ig'\right)^2 + u + iv\right]^{1/2}. \tag{7.96}$$

The reflection coefficient $r^{(j)} = D_g^{(j)}/D_0^{(j)}$ is given by

$$r^{(1)} = \frac{2\xi_0^{(1)}}{\kappa_{0r}P\chi_{-g}} = \frac{\overline{\chi}_g}{\chi_{-g}}\left(\frac{\cos\theta_1}{\cos\theta_2}\right)^{1/2}\left[\left(W+ig'\right)-\sqrt{L}\right] \tag{7.97}$$

$$r^{(2)} = \frac{2\xi_0^{(2)}}{\kappa_{0r}P\chi_{-g}} = \frac{\overline{\chi}_g}{\chi_{-g}}\left(\frac{\cos\theta_1}{\cos\theta_2}\right)^{1/2}\left[\left(W+ig'\right)+\sqrt{L}\right]. \tag{7.98}$$

From the boundary conditions in the Laue case given in (6.9) and (6.10), we have

$$D_0^{(1)} = \frac{-r^{(2)}}{r^{(1)}-r^{(2)}}E_0^a, \quad D_0^{(2)} = \frac{r^{(1)}}{r^{(1)}-r^{(2)}}E_0^a.$$

Using these relations, we get

$$D_0^{(1)} = \frac{1}{2}\left(1+\frac{W+ig'}{\sqrt{L}}\right)E_0^a,\ D_0^{(2)} = \frac{1}{2}\left(1-\frac{W+ig'}{\sqrt{L}}\right)E_0^a. \tag{7.99}$$

We also obtain

$$D_g^{(1)} = \frac{1}{2}\frac{\overline{\chi}_g}{\chi_{-g}}\left(\frac{\cos\theta_1}{\cos\theta_2}\right)^{1/2}\frac{[-(u+iv)]}{\sqrt{L}}E_0^a, \quad D_g^{(2)} = \frac{1}{2}\frac{\overline{\chi}_g}{\chi_{-g}}\left(\frac{\cos\theta_1}{\cos\theta_2}\right)^{1/2}\frac{(u+iv)}{\sqrt{L}}E_0^a, \tag{7.100}$$

by using the relation

$$\left(W+ig'\right)^2 - L = -(u+iv). \tag{7.101}$$

The boundary condition at the bottom plane is given by (6.5) as

$$D_g^{(1)}\exp\left(-ik_{gz}^{(1)}H\right) + D_g^{(2)}\exp\left(-ik_{gz}^{(2)}H\right) = E_g^a\exp\left(-iK_{gz}H\right).$$

Using this condition, we get

$$\frac{E_g^a}{E_0^a} = \left(\frac{\cos\theta_1}{\cos\theta_2}\right)^{\frac{1}{2}}\frac{\overline{\chi}_g}{\chi_{-g}}\frac{(u+iv)}{2\sqrt{L}}\exp\left(iK_{gz}H\right)$$
$$\exp\left(-ik_{gz}^{(2)}H\right)\left[1-\exp\left\{i\left(k_{gz}^{(2)}-k_{gz}^{(1)}\right)H\right\}\right] \tag{7.102}$$

and

$$\left|\frac{E_g^a}{E_0^a}\right|^2 = \left(\frac{\cos\theta_1}{\cos\theta_2}\right)\frac{\overline{\chi}_g^2}{\left|\chi_{-g}\right|^2}|u+iv|^2\frac{\left|\exp\left(-ik_{gz}^{(2)}H\right)\left[1-\exp\left\{i\left(k_{gz}^{(2)}-k_{gz}^{(1)}\right)H\right\}\right]\right|^2}{4|L|}. \tag{7.103}$$

In (7.103), we put

$$\begin{aligned}\frac{\overline{\chi}_g^2}{\left|\chi_{-g}\right|^2} &= \frac{\overline{\chi}_g^2}{\left|\left|\chi_{gr}\right|\exp(-i\alpha_{gr})+i\left|\chi_{gi}\right|\exp(-i\alpha_{gi})\right|^2}\\ &= \frac{\overline{\chi}_g^2}{\left|\chi_{gr}\right|^2+\left|\chi_{gi}\right|^2+2\left|\chi_{gr}\right|\left|\chi_{gi}\right|\sin\delta}\\ &= \frac{1}{1+\left(1-u^2\right)^{1/2}\sin\delta},\end{aligned} \tag{7.104}$$

and $|u+iv|^2 = u^2+v^2$. The product of these two terms is expressed as

$$k_p(\delta) = \frac{\overline{\chi}_g^2}{\left|\chi_{-g}\right|^2}\left(u^2+v^2\right) = \frac{u^2+\left(1-u^2\right)^{1/2}\cos^2\delta}{1+\left(1-u^2\right)^{1/2}\sin\delta} = 1-\left(1-u^2\right)^{\frac{1}{2}}\sin\delta. \tag{7.105}$$

$k_p(\delta)$ is called a polarity factor representing the polarity of a crystal, which will be described later.

Next, we rewrite the term $\left|\exp\left(-ik_{gz}^{(2)}H\right)\left[1-\exp\left\{i\left(k_{gz}^{(2)}-k_{gz}^{(1)}\right)H\right\}\right]\right|^2$ in (7.103). At first, we rewrite $\exp\left(-ik_{gz}^{(2)}H\right)$. In Fig. 7.3, as we have $\overline{\mathrm{A_2G}} = k_{gr}^{(2)}$ and can write

$$k_{gz}^{(2)} = k_{grz}^{(2)} + ik_{giz}^{(2)} = \kappa_{0r}\cos\theta_2 + Y_0^{(2)}, \tag{7.106}$$

we get,

$$k_{gz}^{(2)} - \kappa_{0r}\cos\theta_2 + \frac{\kappa_{0r}P\overline{\chi}_g}{2|\cos\theta_1\cos\theta_2|^{1/2}}$$
$$\left[\frac{\sin2\beta}{\sin2\theta_B'}W + ig\frac{\cos\theta_B'\sin\beta}{|\cos\theta_1\cos\theta_2|^{1/2}} + \mathrm{Re}\left(\sqrt{L}\right) + i\mathrm{Im}\left(\sqrt{L}\right)\right], \tag{7.107}$$

with

$$\sqrt{L} = \mathrm{Re}\left(\sqrt{L}\right) + i\mathrm{Im}\left(\sqrt{L}\right). \tag{7.108}$$

The product of the imaginary part of $k_{gz}^{(2)}$ in (7.107) with thickness H can be written as

$$-ik_{giz}^{(2)}H = -i\frac{\kappa_{0r}P\overline{\chi}_g H}{2|\cos\theta_1\cos\theta_2|^{\frac{1}{2}}}\left[g\frac{\cos\theta_B'\sin\beta}{|\cos\theta_1\cos\theta_2|^{\frac{1}{2}}} + \mathrm{Im}\left(\sqrt{L}\right)\right]. \tag{7.109}$$

By using a relation

$$2\cos\theta_B'\sin\beta = \cos\theta_1 + \cos\theta_2,$$

Equation (7.109) can be rewritten as

$$-ik_{giz}^{(2)}H = i\left[-\frac{\mu_0\overline{H}}{2} + sH\mathrm{Im}\left(\sqrt{L}\right)\right]. \tag{7.110}$$

Here,

$$\overline{H} = \frac{1}{2}\left(\frac{1}{\cos\theta_1} + \frac{1}{\cos\theta_2}\right)H, \quad s = \frac{\kappa_{0r}P\overline{\chi}_g}{2|\cos\theta_1\cos\theta_2|^{1/2}}. \tag{7.111}$$

Equation (7.109) is then rewritten as

$$\exp\left(-ik_{giz}^{(2)}H\right) = \exp\left(-\frac{\mu_0\overline{H}}{2}\right)\exp\left[sH\mathrm{Im}\left(\sqrt{L}\right)\right]. \tag{7.112}$$

Next, we make a product of $\exp\left[sH\mathrm{Im}\left(\sqrt{L}\right)\right]$ in (7.112) and $\left[1 - \exp\left\{i\left(k_{gz}^{(2)} - k_{gz}^{(1)}\right)H\right\}\right]$ in (7.103):

$$\exp\left[sH\mathrm{Im}\left(\sqrt{L}\right)\right]\left[1 - \exp\left\{i\left(k_{gz}^{(2)} - k_{gz}^{(1)}\right)H\right\}\right]. \tag{7.113}$$

Since we have

$$\begin{aligned} k_{gz}^{(2)} - k_{gz}^{(1)} &= Y_0^{(2)} - Y_0^{(1)} \\ &= s\left[\frac{\sin 2\beta}{\sin 2\theta_B'}W + ig\frac{\cos\theta_B'\sin\beta}{|\cos\theta_1\cos\theta_2|^{1/2}} + \sqrt{L}\right] \\ &\quad - s\left[\frac{\sin 2\beta}{\sin 2\theta_B'}W + ig\frac{\cos\theta_B'\sin\beta}{|\cos\theta_1\cos\theta_2|^{1/2}} - \sqrt{L}\right] \\ &= 2s\sqrt{L} = 2s\left\{\mathrm{Re}\left(\sqrt{L}\right) + i\mathrm{Im}\left(\sqrt{L}\right)\right\}, \end{aligned} \tag{7.114}$$

we can write

$$1-\exp\left\{i\left(k_{gz}^{(2)}-k_{gz}^{(1)}\right)H\right\}=1-\exp\left[i2sH\left\{\mathrm{Re}\left(\sqrt{L}\right)+i\mathrm{Im}\left(\sqrt{L}\right)\right\}\right]. \quad (7.115)$$

Then, (7.113) becomes

$$\begin{aligned}&\exp\left[sH\mathrm{Im}\left(\sqrt{L}\right)\right]\left[1-\exp\left[i2sH\left\{\mathrm{Re}\left(\sqrt{L}\right)+i\mathrm{Im}\left(\sqrt{L}\right)\right\}\right]\right]\\&\quad=-\exp\left[isH\left\{\mathrm{Re}\left(\sqrt{L}\right)\right\}\right]\left[\exp\left(isH\sqrt{L}\right)-\exp\left(-isH\sqrt{L}\right)\right].\end{aligned} \quad (7.116)$$

From the above results, we get

$$\frac{P_g}{P_0}=\left|\frac{E_g^a}{E_0^a}\right|^2\frac{\cos\theta_2}{\cos\theta_1}=k_p(\delta)\exp\left(-\mu_a\overline{H}\right)\frac{|\exp(i\gamma_L)-\exp(-i\gamma_L)|^2}{4|L|} \quad (7.117)$$

with

$$\gamma_L=sH\sqrt{L}. \quad (7.118)$$

The transmittance P_t/P_0 can be calculated as

$$\begin{aligned}\frac{P_t}{P_0}=\left|\frac{E_t^a}{E_0^a}\right|^2&=\exp\left(-\mu_a\overline{H}\right)\\&\frac{\left|\left(W+ig'+\sqrt{L}\right)\exp(i\gamma_L)-\left(W+ig'-\sqrt{L}\right)\exp(-i\gamma_L)\right|^2}{4|L|}\end{aligned} \quad (7.119)$$

by using

$$\left|E_t^a\right|=\left|D_0^{(1)}\exp\left(-ik_{0z}^{(1)}H\right)+D_0^{(2)}\exp\left(-ik_{0z}^{(2)}H\right)\right|$$

obtained by the boundary condition and using the relation between $D_0^{(1)}$ and $D_0^{(2)}$ in (7.99).

7.5 Diffraction Intensity in the Bragg Case

7.5.1 *For a Finite Thickness Crystal*

We will calculate the diffraction intensity in the Bragg case from a crystal with a thickness of H. Once the deviation parameter W is given, ξ_0 is calculated from (7.81) as

$$\xi_0 - \frac{\kappa_{or} P \overline{\chi}_g}{2} \left(\frac{\cos\theta_1}{|\cos\theta_2|} \right)^{\frac{1}{2}} \left[\left(W + ig' \right) \pm \sqrt{B} \right]. \tag{7.120}$$

From (7.82), $\sqrt{B}$. is

$$\sqrt{B} = \left[\left(W + ig' \right)^2 - (u + iv) \right]^{1/2}. \tag{7.121}$$

Since the reflection coefficient is given by

$$r^{(j)} = \frac{D_g^{(j)}}{D_0^{(j)}} = \frac{2\xi_0^{(j)}}{\kappa_{0r}} \frac{1}{P\chi_{-g}},$$

we get

$$r^{(1)} = \frac{\overline{\chi}_g}{\chi_{-g}} \left(\frac{\cos\theta_1}{|\cos\theta_2|} \right)^{1/2} \left[-\left(W + ig' \right) \mp \sqrt{B} \right], \tag{7.122}$$

$$r^{(2)} = \frac{\overline{\chi}_g}{\chi_{-g}} \left(\frac{\cos\theta_1}{|\cos\theta_2|} \right)^{1/2} \left[-\left(W + ig' \right) \pm \sqrt{B} \right]. \tag{7.123}$$

As described in 6.2, we label branch (1) of the dispersion surface where the pseudo-refraction wave propagating downward in the crystal is excited. The boundary conditions for a plane parallel crystal are the same as those used in the Ewald-Laue method in Chap. 6. Then we get

$$D_0^{(1)} = \frac{-\left(W + ig' - \sqrt{B} \right) \exp(-i\gamma_B)}{\left(W + ig' + \sqrt{B} \right) \exp(i\gamma_B) - \left(W + ig' - \sqrt{B} \right) \exp(-i\gamma_B)} E_0^a, \tag{7.124}$$

$$D_0^{(2)} = \frac{\left(W + ig' + \sqrt{B} \right) \exp(i\gamma_B)}{\left(W + ig' + \sqrt{B} \right) \exp(is\gamma_B) - \left(W + ig' - \sqrt{B} \right) \exp(-i\gamma_B)} E_0^a, \tag{7.125}$$

where we put

$$\gamma_B = sH\sqrt{B}. \tag{7.126}$$

According to the boundary condition at the incident plane of $D_g^{(1)} + D_g^{(2)} = E_g^a$, we get

$$E_g^a = r^{(1)} D_0^{(1)} + r^{(2)} D_0^{(2)}. \tag{7.127}$$

From the boundary condition at the bottom plane, we get

$$E_t^a = \exp(iK_{0z}H)\Big[D_0^{(1)}\exp\Big(-ik_{0z}^{(1)}H\Big) + D_0^{(2)}\exp\Big(-ik_{0z}^{(2)}H\Big)\Big]. \quad (7.128)$$

Using these relations, we can obtain the diffracted and transmitted intensities

$$\frac{P_g}{P_0} = \left|\frac{E_g^a}{E_0^a}\right|^2 \frac{|\cos\theta_2|}{\cos\theta_1} = k_p(\delta)\left|\frac{\exp(i\gamma_B) - \exp(-i\gamma_B)}{\left(W + ig' + \sqrt{B}\right)\exp(i\gamma_B) - \left(W + ig' - \sqrt{B}\right)\exp(-i\gamma_B)}\right|^2, \quad (7.129)$$

$$\frac{P_t}{P_0} = \left|\frac{E_t^a}{E_0^a}\right|^2 = \exp(-\mu_a H_\mathrm{B})\left|\frac{2\sqrt{B}}{\left(W + ig' + \sqrt{B}\right)\exp(i\gamma_B) - \left(W + ig' - \sqrt{B}\right)\exp(-i\gamma_B)}\right|^2. \quad (7.130)$$

Here H_B is given by

$$H_\mathrm{B} = \frac{1}{2}\left(\frac{1}{\cos\theta_1} + \frac{1}{\cos\theta_2}\right)H. \quad (7.131)$$

It is noted that the meaning of H_B in the Bragg case is different from that of $\overline{H}$ in the Laue case in (7.111). In the Laue case, $\overline{H}$ expresses the average propagating distance through the crystal. In the Bragg case, $\cos\theta_2$ is negative for $\theta_2 > \pi/2$, H_B does not express the average distance and depends on the asymmetric factor b given in (6.137). In the case of symmetric reflection, for example, $\cos\theta_1 = -\cos\theta_2$ and $H_\mathrm{B} = 0$, then $\exp(-\mu_a H_\mathrm{B}) = 1$ meaning that no absorption occurs.

7.5.2 For a Semi-infinite Crystal

We describe the diffraction intensity from a semi-infinite crystal. In this case, it can be assumed that the X-rays that penetrate deep into the crystal are attenuated and no reflected waves are generated from the bottom surface. Then the pseudo-refracted wave is excited at one tie-point of branch (1). The boundary conditions at the crystal surface gives

$$E_0^a = D_0^{(1)}, \; E_g^a = D_g^{(1)}. \quad (7.132)$$

Since $r^{(1)} = D_g^{(1)}/D_0^{(1)}$, we get

$$\frac{E_g^a}{E_0^a} = r^{(1)} = \frac{\chi_g}{\chi_{-g}}\left(\frac{\cos\theta_1}{|\cos\theta_2|}\right)^{1/2}\Big[-(W + ig') \mp \sqrt{B}\Big]. \quad (7.133)$$

From (7.133), we obtain

$$\frac{P_g}{P_0} = \left|\frac{E_g^a}{E_0^a}\right|^2 \frac{|\cos\theta_2|}{\cos\theta_1} = \frac{k_p(\delta)}{u^2+v^2}\left|-\left(W+ig'\right)\mp\sqrt{B}\right|^2 = k_p(\delta)\left|\frac{1}{(W+ig')\pm\sqrt{B}}\right|^2, \tag{7.134}$$

by using the relation

$$\frac{\overline{\chi}_g^2}{\left|\chi_{-g}\right|^2} = \frac{k_p(\delta)}{u^2+v^2},\quad u^2+v^2 = -\left(W+ig'\right)^2 + B.$$

We can obtain the same result by eliminating the component of branch (2) in (7.129). It is noted that the sign of $\sqrt{B}$ is chosen so that the direction of the pseudo-refracted wave is downward.

7.6 Integrated Reflection Intensity

When resonant scattering is involved, the relation between W and α_0 is given by

$$W = (\alpha_0 - \alpha_{0L})\left(\frac{\cos\theta_1}{|\cos\theta_2|}\right)^{1/2}\frac{\sin 2\theta_B'}{n_0 P\overline{\chi}_g}$$

from (7.83) and we get

$$\frac{dW}{d\alpha_0} = \left(\frac{\cos\theta_1}{|\cos\theta_2|}\right)^{1/2}\frac{\sin 2\theta_B'}{n_0 P\overline{\chi}_g}. \tag{7.135}$$

The integrated reflection intensity $R_g^{\alpha_0}$ is written as

$$R_g^{\alpha_0} = \left(\frac{|\cos\theta_2|}{\cos\theta_1}\right)^{1/2}\frac{n_0 P\overline{\chi}_g}{\sin 2\theta_B'}R_g^W. \tag{7.136}$$

Here, R_g^W is the integrated reflection intensity in terms of W given by

$$R_g^W = \int_{-\infty}^{\infty}\frac{P_g(W)}{P_0}dW. \tag{7.137}$$

Figure 7.7 shows R_g^W as a function of sH for some pairs of (u, v) in the symmetric Laue case. When $u = 1$ and $g = 0$, the diffraction is only due to the normal scattering. R_g^W shows Pendellösung fringes as a function of sH and converges to $\pi/2$ as sH increases. When $u = 1$ and $g \neq 0$, i.e., resonant scattering is involved, R_g^W becomes very small as sH increases. When $u = 0.8$, R_g^W shows the maximum at $sH \approx 1$ and Pendellösung fringes up to the second period. When $u = -1$, R_g^W shows

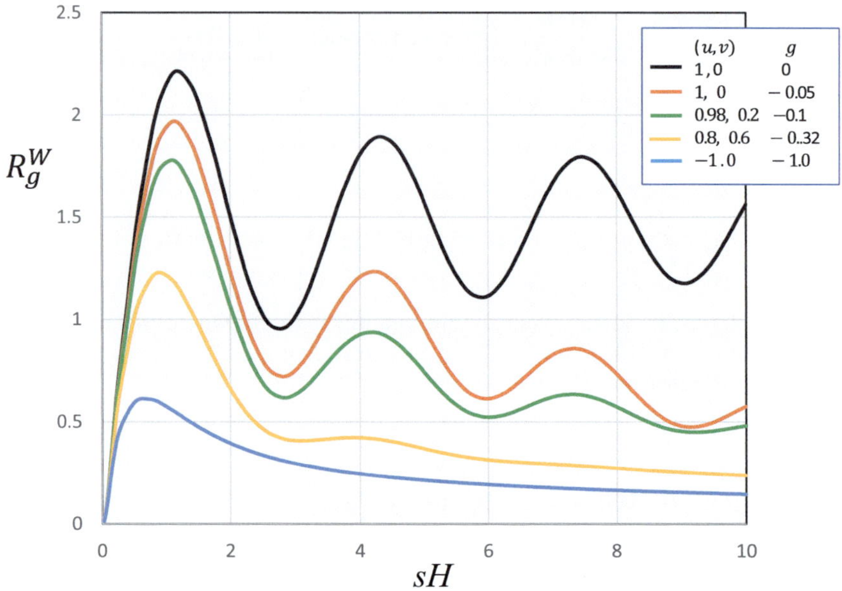

Fig. 7.7 Integrated reflection intensities as a function of sH, when u changes. The numbers in the square bracket shows values of (u, v)

the maximum at $sH = 0.7$, then monotonously decreases as sH increases without showing Pendellösung fringes. Pendellösung fringes are observed in R_g^W in a narrow range of u from 1 to 0.8.

7.7 Complex Dispersion Surface and Diffraction Intensity Due to Change of Parameter *u*

We describe some characteristics of the complex dispersion surface, and diffraction and transmission intensities for typical values of u. We take the symmetric Laue case and the symmetric Bragg case as two examples. We ignore the temperature effects.

7.7.1 Symmetric Laue Case

In the symmetric Laue case ($\beta = \pi/2,\ \cos\theta_1 = \cos\theta_2 = \cos\theta_B'$), the real part (7.68) and imaginary part (7.69) of the dispersion surface are respectively given by

$$Y_{0r}^2\cos^2\theta_B' - X_0^2\sin^2\theta_B' - \left(Y_{0i}\cos\theta_B' - \kappa_{oi}\right)^2 = \frac{\kappa_{0r}^2 P^2 \overline{\chi}_g^2}{4}u, \tag{7.138}$$

$$Y_{0r}\cos\theta_B'\left(Y_{0i}\cos\theta_B' - \kappa_{oi}\right) = \frac{\kappa_{0r}^2 P^2 \overline{\chi}_g^2}{8} v. \tag{7.139}$$

Figure 7.8 shows the dispersion surfaces for some values of (u, v). The calculated diffracted and transmitted intensities are shown for $(u, v) = (1, 0)$ in Fig. 7.9a and for $(u, v) = (-1, 0)$ in Fig. 7.9b. G_{Br} and G_{Bi} are the real and imaginary parts of the Bragg gap given by

$$G_{Br} = \frac{\kappa_{0r} P \left|\chi_{gr}\right|}{\cos\theta_B}, \quad G_{Bi} = \frac{\kappa_{0r} P \left|\chi_{gi}\right|}{\cos\theta_B}. \tag{7.140}$$

G_{Br} is the same as G_B mentioned in Chap. 5 representing the distance between the vertices of the hyperbola. When $u = 1$ in Fig. 7.8a, $\chi_{gi} = 0$, and $G_{Bi} = 0$. When $u = -1$ in Fig. 7.8b, $\chi_{gr} = 0$, then $G_{Br} = 0$ and the Bragg gap disappears. Instead, a gap of G_{Bi} appears between two branches of the imaginary part at $W = 0$. The real part of the dispersion surface (Fig. 7.8a) for $u = 1$ is the north–south opening

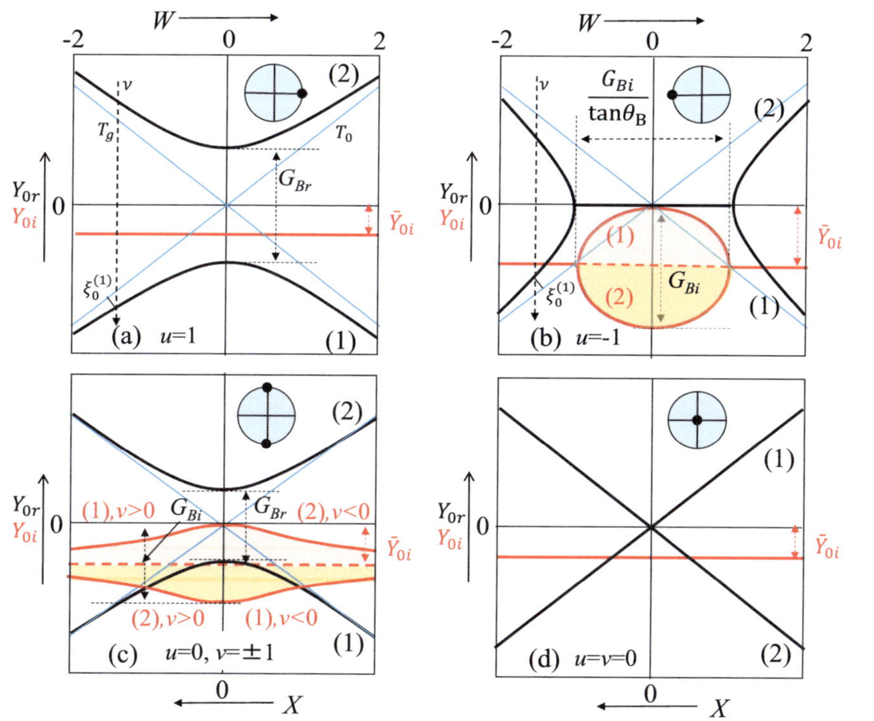

Fig. 7.8 Dispersion surfaces in the symmetric Laue case for **a** $(u, v) = (1, 0)$, **b** $(u, v) = (-1, 0)$, **c** $(u, v) = (0, \pm 1)$, **d** $(u, v) = (0, 0)$. The black lines show the real parts and the red lines show the imaginary parts. The cerulean lines show T_0 and T_g

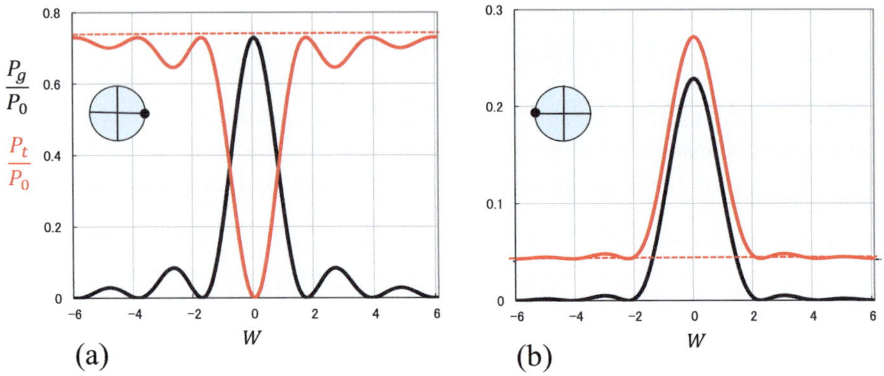

Fig. 7.9 Diffraction and transmission intensities for $u = \pm 1$ and $sH = \pi/2$. The black solid lines show the diffraction intensities and the red one the transmission intensities. The red dashed lines show the transmission intensities due to the mean absorption $(\exp(-\mu_0 \overline{H}))$. **a** $(u, v) = (1, 0)$ and **b** $(u, v) = (-1, 0)$

hyperbola, which is quite different from the east–west opening hyperbola for $u = -1$ (Fig. 7.8b). The sign of $\xi_0^{(1)}$ is reversed for $u = 1$ and for $u = -1$.

7.7.1.1 In Case of $(u, v) = (1, 0)$

The complex dispersion surface in this case is shown in Fig. 7.8a. Since $v = 0$, the right-hand side of (7.139) is zero, and the value of $Y_{0i}\cos\theta_B' - \kappa_{0i}$ on the left-hand side of (7.138) is zero. $Y_{0r}(X_0)$ shows a north–south opening hyperbola. $Y_{0i} = \kappa_{0i}/\cos\theta_B'$ $(\kappa_{0i} = \chi_{0i}K_0/2)$ is a negative constant, and $\mu = \mu_a$. Then $\mu_d = 0$, and neither anomalous transmission nor absorption occurs. In conventional theory, it is known that the Borrmann effect always occurs in absorbing crystals. However, even in absorbing crystals, the Borrmann effect does not occur when $u = 1$. This has been confirmed experimentally by Negishi et al. [10, 11], which will be described later.

Figure 7.9a shows the rocking curves of P_g/P_0 and P_t/P_0 for $u = 1$, which are obtained from (7.117) and (7.119) as

$$\frac{P_g}{P_0} = \exp\left(-\mu_0 \overline{H}\right) \frac{\sin^2\left[sH\left(W^2 + 1\right)^{1/2}\right]}{W^2 + 1}, \tag{7.141}$$

$$\frac{P_t}{P_0} = \exp\left(-\mu_0 \overline{H}\right) - \frac{P_g}{P_0}. \tag{7.142}$$

Here we use $g' = 0$ in the symmetrical reflection. H is the thickness of the plane parallel crystal, and $\overline{H} = H/\cos\theta_B'$. The interference fringes that appear in (7.141) and (7.142) are the same as those due to the normal scattering in (6.20) and (6.21), and are Pendellösung fringes.

7.7.1.2 In Case of $(u, v) = (-1, 0)$

Since $\chi_{gr} = 0$, diffraction is only due to resonant scattering. The complex dispersion surface is shown in Fig. 7.8b. As $v = 0$, the value of $Y_{0i}\cos\theta'_{\mathrm{B}} - \kappa_{oi}$ on the left-hand side of (7.138) is zero. Since $u = -1$, $Y_{0r}(X_0)$ shows an east–west opening hyperbola despite of the Laue case. According to (7.140), the distance between the vertices of the hyperbola is $G_{\mathrm{B}i}/\tan\theta_{\mathrm{B}}$, which is proportional to $\left|\chi_{gi}\right|$. In the region of $Y_{0r} \neq 0$, $Y_{0i} = \overline{Y}_{0i} = \kappa_{0i}/\cos\theta'_{\mathrm{B}}$ which is constant because of $v = 0$, and the Borrmann effect does not occur. In the region of $Y_{0r} = 0$, $Y_{0i}(X_0)$ becomes an ellipse, which is the same as in the Bragg case, but with $Y_{0i} < 0$. The wave excited at a tie-point of the branch above $\overline{Y}_{0i}$ shows anomalous transmission (lightly filled area), and that excited at a tie-point of the branch below it shows anomalous absorption (darkly filled area). The elliptical dispersion surface for $u = -1$ is not related to the extinction effect.

The diffracted intensity and transmitted intensities are given by

$$\frac{P_g}{P_0} = \exp\left(-\mu_0\overline{H}\right)\frac{\sin^2\left[sH\left(W^2-1\right)^{1/2}\right]}{W^2-1}, \quad (|W| \geq 1) \tag{7.143}$$

$$= \exp\left(-\mu_0\overline{H}\right)\frac{\sinh^2\left[sH\left(1-W^2\right)^{1/2}\right]}{1-W^2}, \quad (|W| < 1) \tag{7.144}$$

$$\frac{P_t}{P_0} = \exp\left(-\mu_0\overline{H}\right) + \frac{P_g}{P_0}. \tag{7.145}$$

Figure 7.9b shows the rocking curves of diffracted intensity P_g/P_0 and transmitted intensity P_t/P_0 for $(u, v) = (-1, 0)$. The both curves show the highest peak around $W = 0$, which is due to the anomalous transmission of the branch (2) wave

In the region of $|W| > 1$, interference fringes are observed in the rocking curves of both the diffracted wave and the transmitted wave. These interference fringes show the same phase. This indicates that these interference fringes are not those of "loosely coupled two-coupled pendulum" proposed by Ewald. The in-phase interference fringes have also been observed in experiments by Negishi et al. [11–13], as described later (Fig. 7.23c). In the transmitted intensity of (7.145), the diffracted intensity is added to the intensity attenuated by average absorption, while the latter is subtracted from the diffracted intensity in (7.142) when $u = 1$.

When u changes from 1 to -1, the phase difference between the interference fringes of P_g and P_t changes continuously from π to zero. The phase of the crystal structure factor can be determined by utilizing the measurement of this change of phase difference.

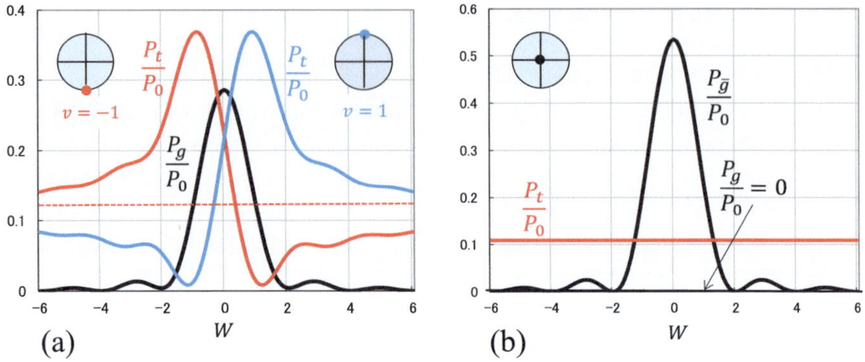

Fig. 7.10 Calculated diffraction and transmission rocking curves for $(u, v) = (0, \pm 1)$ **a**, and $(u, v) = (0, 0)$ **b** by assuming $sH = \pi/2$. The black lines show the diffraction curves. **a** The red and blue lines show transmission curves for $v = -1$ and 1, respectively. The red dotted line shows transmission intensities due to the mean absorption. **b** The red line shows the transmission rocking curves

7.7.1.3 In Case of $(u, v) = (0, \pm 1)$

When $u = 0$, $|\chi_{gr}|$ must be equal to $|\chi_{gi}|$. In addition, when $\delta = 0$, v is equal to 1, and when $\delta = \pi$, v is equal to -1. Under the condition of $(u, v) = (0, \pm 1)$, the dispersion surfaces are shown in Figs. 7.4 and 7.8c, and the rocking curves of the diffracted wave and transmitted wave are shown in Fig. 7.10a. The real part of the dispersion surface shows a hyperbola opening north and south, and the imaginary part always shows $Y_{0i}^{(j)} \neq \overline{Y}_{0i}$ and shows the maximum amplitude in the center ($X_0 = 0$). Anomalous transmission and absorption are seen in the entire region of W, and are the largest at $W = 0$. Y_{0r} does not depend on δ, but Y_{0i} depends on δ. When $\delta = 0$, the lower branch of Y_{0i} corresponds to $j = 1$ and the upper one $j = 2$. When $\delta = \pi$, the branch numbers are swapped. This δ dependence does not appear in the symmetric rocking curve of diffracted wave P_g/P_0, but it appears in the asymmetric rocking curve of transmitted wave P_t/P_0, as shown in Fig. 7.10a. For $\delta = 0$, the peak of the P_t/P_0 rocking curve appears on the right, and for $\delta = \pi$ it appears on the left. This δ dependence has also been experimentally confirmed as described later.

7.7.1.4 In Case of $(u, v) = (0, 0)$

When $(u, v) = (0, 0)$, $\delta = \pm\pi/2$ and $\chi_g \chi_{-g} = 0$, either χ_g or χ_{-g} is zero. The dispersion surface in this case is shown in Fig. 7.8d. As can be seen from (7.138) and (7.139),

$$Y_{0r}\cos\theta'_{\mathrm{B}} = X\sin\theta'_{\mathrm{B}}, \quad Y_{0i}\cos\theta'_{\mathrm{B}} = \kappa_{0i}. \tag{7.146}$$

The real part of the dispersion surface becomes a straight line and coincides with the asymptote. The imaginary part of the dispersion surface is constant, indicating that the X-rays undergo average absorption. When $\delta = \pi/2$, $\chi_g = 0$ and polarity factor $k_p(\pi/2) = 0$, then we get

$$P_g = 0, \tag{7.147}$$

$$\frac{P_{-g}}{P_0} = \exp\left(-\mu_0 \overline{H}\right) \frac{\sin^2(sHW)}{W^2}, \tag{7.148}$$

because of $k_p(-\pi/2) = 2$. P_t is independent of either g or $-g$ and expressed by

$$\frac{P_t}{P_0} = \exp\left(-\mu_0 \overline{H}\right). \tag{7.149}$$

P_{-g} shows a form of the Laue function, and P_t undergoes average absorption. When $(u, v) = (0, 0)$, multiple scattering within the crystal does not occur and the rocking curves are given by the kinematical diffraction theory as shown in Fig. 7.10b.

7.7.2 *Symmetric Bragg Case (for a Thin Crystal)*

The real and imaginary parts of the dispersion surface in the symmetric Bragg case ($\beta = 0$, $\sin\theta_1 = \sin\theta_2 = \cos\theta'_{\mathrm{B}}$, $\cos\theta_1 = -\cos\theta_2 = \sin\theta'_{\mathrm{B}}$) are obtained from (7.68) and (7.69) as

$$X_0^2 \cos^2\theta'_{\mathrm{B}} - Y_{0r}^2 \sin^2\theta'_{\mathrm{B}} + Y_{0i}^2 \sin^2\theta'_{\mathrm{B}} - \kappa_{0i}^2 = \frac{\kappa_{0r}^2 P^2 \overline{\chi}_g^2}{4} u, \tag{7.150}$$

$$Y_{0r} Y_{0i} \sin^2\theta'_{\mathrm{B}} + \kappa_{0i} X_0 \cos\theta'_{\mathrm{B}} = -\frac{\kappa_{0r}^2 P^2 \overline{\chi}_g^2}{8} v. \tag{7.151}$$

7.7.2.1 In Case of $(u, v) = (1, 0)$

The complex dispersion surface for $(u, v) = (1, 0)$ is shown in Fig. 7.11a. In this case, $\chi_{gi} = 0$ and $g = \chi_{0i}/\left|\chi_{gr}\right| = -0.02$. When a tie-point of the imaginary part is in the shaded area in orange, the wave excited at the tie-point shows anomalous absorption. The transmitted wave is excited at a tie-point on the branch (1), and the reflected wave at a tie-point on the branch (2). This dispersion surface should be compared with that in Fig. 6.15c where resonant scattering is not included. In Fig. 6.15c, $Y_{0r} = 0$ for $-1 \le W \le 1$, which corresponds to the Bragg gap. Since

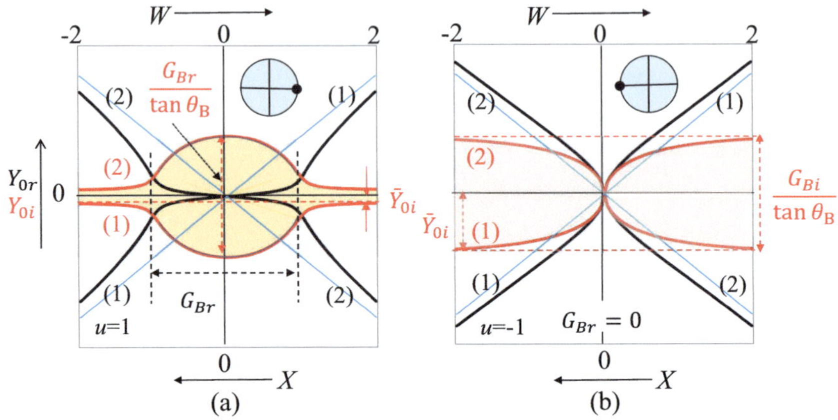

Fig. 7.11 Complex dispersion surfaces for $(u, v) = (1, 0)$ **a** and $(u, v) = (-1, 0)$ **b** in the symmetric Bragg case. The black solid lines represent the real part and the red one the imaginary part of the dispersion surfaces. The red dashed lines show $\overline{Y}_{0i}$. **a** $g = -0.02$, **b** $g = -1$

resonant scattering is included in Fig. 7.11a, $|Y_{0r}|$ becomes significantly small in the region of $-1 \leq W \leq 1$, but it becomes zero only at one point $W = 0$.

When $u = 1$, Y_{0i} has a non-zero value in all regions of X (W). Within the region of $|W| < 1$, $|Y_{0i}|$ becomes significantly large due to the extinction effect. Since $|Y_{0i}| > \left|\overline{Y}_{0i}\right|$, $\mu^{(j)} > \mu_a$ and only anomalous absorption occurs.

Figure 7.12a shows the rocking curves of the diffracted wave (P_g) and transmitted wave (P_t) from a thin crystal. The diffracted rocking curve is symmetrical with respect to W and shows high reflectance in the range of $-1 \leq W \leq 1$, while transmittance (P_t) is significantly small. Interference fringes are seen in the region of $|W| > 1$. These interference fringes in the diffracted waves and the transmitted waves are anti-phase with each other, then these are Pendellösung fringes. The transmitted intensity is always smaller than the attenuation due to average absorption shown by the red broken line because of anomalous absorption.

7.7.2.2 In Case of $(u, v) = (-1, 0)$

The complex dispersion surface in this case is shown in Fig. 7.11b, where $\chi_{gr} = 0$ and $g = -1$. In the real part of the dispersion surface, the Bragg gap completely disappears, and the two branches (black solid lines) are in contact at $X = 0$. The imaginary part is also zero at $X = 0$. Since $|Y_{0i}| < \left|\overline{Y}_{0i}\right|$, only the anomalous transmission of the Borrmann effect occurs. As shown in Fig. 7.12b, the rocking curves of the diffracted wave and the transmitted wave are bilaterally symmetrical and both show sharp peaks at $W = 0$ due to the anomalous transmission. Due to anomalous transmission, the transmitted intensity is always larger than that due to average absorption. When $u = -1$, there is one X (or W) where $Y_{0r} = Y_{0i} = 0$.

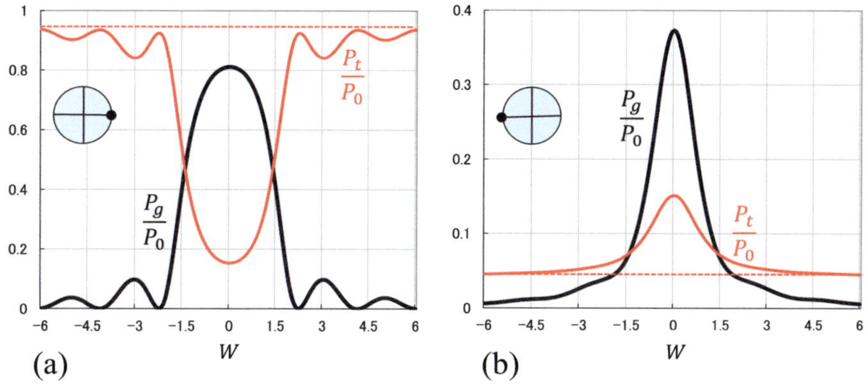

Fig. 7.12 The black solid line shows the diffracted rocking curve and the red one the transmitted rocking curves for $(u, v) = (1, 0)$ **a** and $(-1, 0)$ **b** when $sH = \pi/2$. The red dashed line shows the transmitted intensities due to mean absorption $\exp(-\mu_0 \overline{H})$

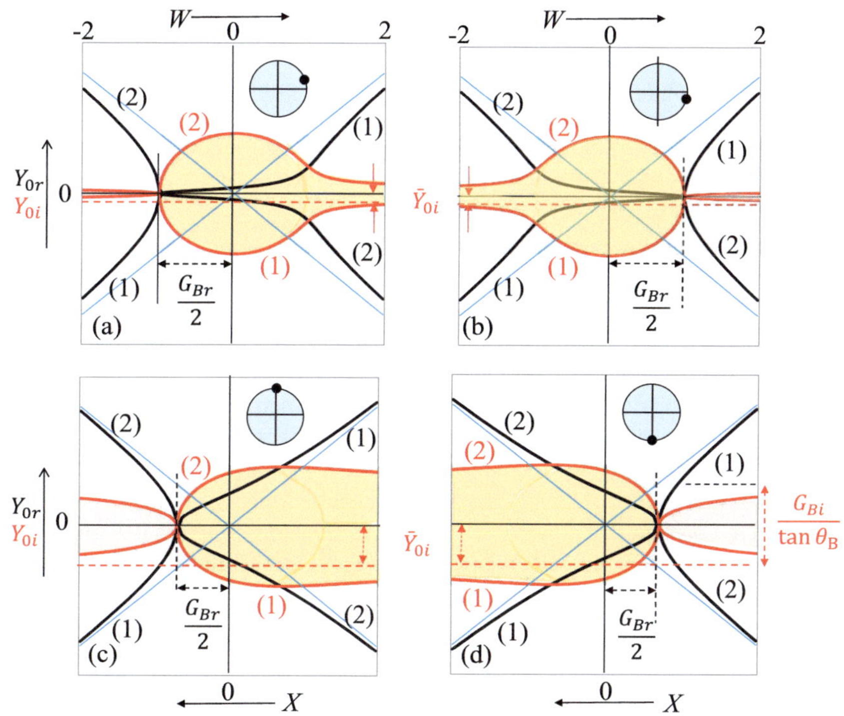

Fig. 7.13 Dispersion surfaces for $(u, v) = (0.98, \pm 0.2)$ **a** and **b**, and $(u, v) = (0, \pm 1)$ **c** and **d**. The black line shows the real part and the red one the imaginary part. The red dashed line indicates $\overline{Y}_{0i}$. **a**, **c** for $v > 0$ and **b**, **d** for $v < 0$

This can be seen when $u < 1$. However, this does not occur when $(u, v) = (1, 0)$, as shown in Fig. 7.11a.

7.7.2.3 In Case of $(u, v) = (0.98, \pm 0.2)$ and $(0, \pm 1)$

We consider the complex dispersion surface and the diffracted intensity when $-1 < u < 1$ with $v > 0$ or $v < 0$. The dispersion surfaces are shown in Fig. 7.13a and c for $v > 0$ and (b and d) for $v < 0$. As we can see from the dispersion surfaces, the left and right sides are reversed when $v > 0$ and $v < 0$. We will focus on the case when $v > 0$.

Figure 7.13a for $u = 0.98$ has many similarities to Fig. 7.11a for $u = 1$. The crucial difference is that for $u = 1$, $Y_{0r} = 0$ and $|Y_{0i}|$ is maximum at $X = 0$, while for $u = 0.98$, $Y_{0r} = Y_{0i} = 0$ when $X = G_{Br}/2$ $(W = -0.995)$. For $u = 0.98$, $|Y_{0i}| < |\overline{Y}_{0i}|$ and anomalous transmission occurs in the region of $X \geq G_{Br}/2$. At $X = G_{Br}/2$, $Y_{0i} = 0$ and the anomalous transmission is most noticeable. In the region of $-G_{Br}/2 < X < G_{Br}/2$, $|Y_{0r}|$ becomes small and $|Y_{0i}|$ becomes extremely large, indicating anomalous absorption due to the extinction effect. In the region of $X < -G_{Br}/2$, Y_{0r} approaches the asymptote and $|Y_{0i}|$ gradually becomes small, but anomalous absorption occurs because of $|Y_{0i}| > \left|\overline{Y}_{0i}\right|$. The absorption coefficient $\mu^{(j)}$ is already shown in Fig. 7.6. This figure shows $\mu^{(j)}$ for $v > 0$ and for $v < 0$, the curve of $\mu^{(j)}$ is reversed left and right, and $\mu^{(j)} = 0$ when $W = -0.995$ for $v > 0$ and $W = 0.995$ for $v < 0$.

When $u = 0$ and $v = \pm 1$, since $\left|\chi_{gi}\right|$ increases, the extinction effect is weaker than that for $u = 0.98$, as shown in Fig. 7.13c and d.

Figure 7.14 shows the rocking curves of the diffracted wave and transmitted wave when $u = 0.98$ and $u = 0$. The two rocking curves for $u = 0.98$ show similar variations to the rocking curves for $u = 0$. For both $u = 0.98$ and $u = 0$, the peak near the center of the diffracted rocking curve is asymmetric. The peak is on the left of the center for $v > 0$, and it is on the right for $v < 0$. As u becomes smaller, the peak becomes sharper.

The larger transmitted intensity than the transmitted intensity due to the average absorption (red broken line) indicates anomalous transmission, and the smaller one indicates anomalous absorption. When $u = 0.98$, the transmitted rocking curve in Fig. 7.14a and b shows a valley almost in the center. When $v > 0$ (a), the left side of the curve $(W < 0)$ shows anomalous transmission, and the right side $(W > 0)$) shows anomalous absorption. When $v < 0$ (b), the left and right are reversed. When $u = 0$ and $v > 0$, the central valley of the rocking curve becomes shallower, and a clear peak due to the anomalous transmission can be seen on the left side.

7.7.2.4 In Case of $(u, v) = (0, 0)$

The complex dispersion surface is shown in Fig. 7.15a, and the corresponding rocking curves are shown in Fig. 7.15b. As $\chi_g \chi_{-g} = 0$, no multiple scattering occurs and

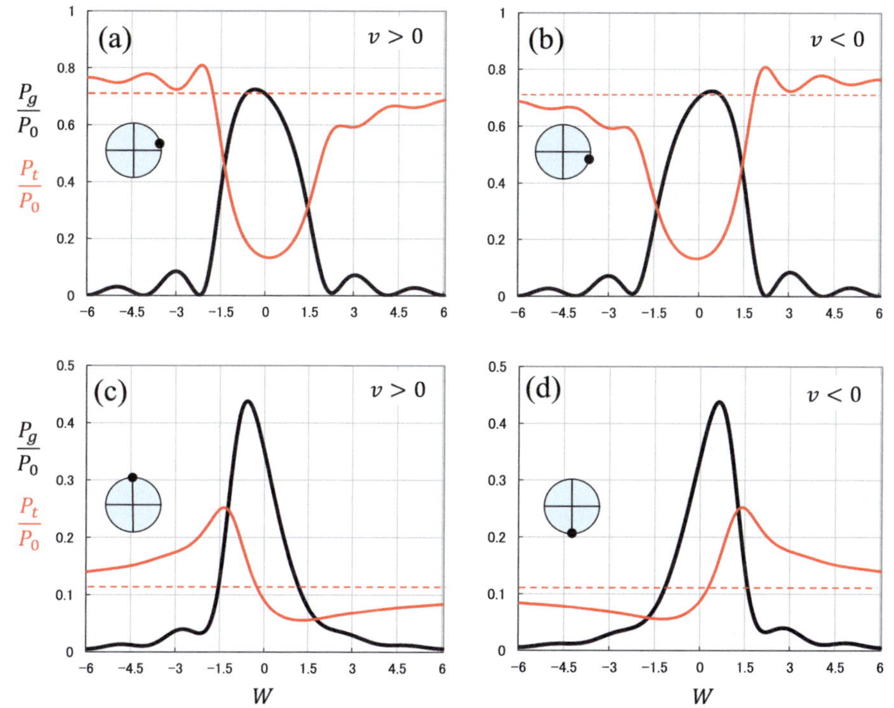

Fig. 7.14 The black solid line shows the diffracted rocking curve and the red one the transmitted rocking curve for $(u, v) = (0.98, 0.2)$ **a**, $(0.98, -0.2)$ **b**, $(0, 1)$ **c** and $(0, -1)$ **d**, when $sH = \pi/2$. The red dashed line shows the transmitted intensities due to the average absorption $\exp\left(-\mu_0\overline{H}\right)$

the dispersion surface coincides with its asymptote. When $\chi_g = 0$ and $\chi_{-g} \neq 0$, the polarity factor k_p is zero for P_g and 2 for $P_{\overline{g}} = P_{-g}$, and the diffracted rocking curves are as shown in Fig. 7.15b. The transmitted intensity becomes constant, and are equal to the intensity due to the average absorption.

7.7.3 *Symmetric Bragg Case (for a Semi-Infinite Crystal)*

We describe the rocking curve of the diffracted wave from a semi-infinite crystal. Figure 7.16 shows the calculated rocking curves of the diffracted wave when u and v are changed. The thin black line is the rocking curve when $u = 1$ and $\chi_{0i} = 0$, and the green line is the rocking curve when $u = 1$ and $\chi_{0i} \neq 0$.

These two rocking curves are symmetric with respect to W, and the smaller $|g|$ becomes, the higher the peak becomes. The brown line in Fig. 7.16a shows the rocking curve for $u = 0.98$ and $v > 0$, showing high reflectance in the region of $|W| \leq 1$ and the highest value of 1 for $W = -0.995$, where $Y_{0r} = Y_{0i} = 0$ on

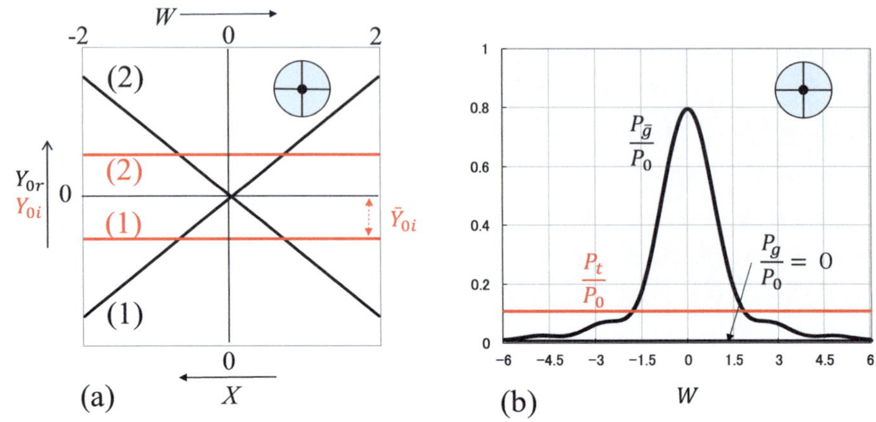

Fig. 7.15 **a** The black solid line shows the real part and the red one the imaginary part of the dispersion surface for $(u, v) = (0, 0)$. **b** The black solid lines show the diffracted and the red line the transmitted rocking curves for $(u, v) = (0, 0)$. $sH = \pi/2$

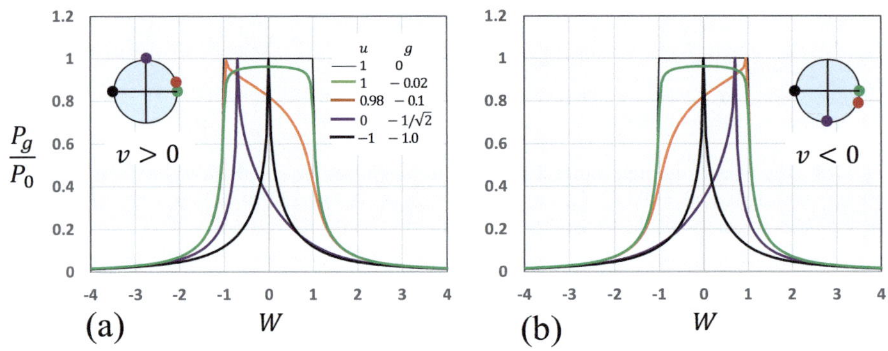

Fig. 7.16 The diffracted rocking curves from a semi-infinite crystal for several (u, v) with $v \geq 0$ (**a**) and $v \leq 0$ (**b**)

the dispersion surface as shown in Fig. 7.13a. This results in an asymmetric rocking curve. In the case of $v < 0$ and $u = 0.98$ in Fig. 7.16b, the rocking curve (brown line) is asymmetric with a reflectance of 1 only for $W = 0.995$.

The purple line in Fig. 7.16a shows the rocking curve when $u = 0$ and $v > 1$. It is asymmetric with a sharp peak at $W = -1/\sqrt{2}$, where $Y_{0r} = Y_{0i} = 0$. When $v < 1$ in Fig. 7.16b, the rocking curve is asymmetric with a sharp peak at $W = +1/\sqrt{2}$.

The thick black lines in Fig. 7.16a and b show the rocking curves for $u = -1$, showing left–right symmetry and a sharp peak at $W = 0$.

7.8 Relations Between the Real Part of the Dispersion Surface and the Poynting Vector

Even when only the normal scattering is involved, the dispersion surface in the symmetric Bragg case becomes complex in the region $-1 \leq W \leq 1$. Then the direction of the Poynting vector is parallel to the tangent of the real part of the dispersion surface; and the dispersion surface and the Poynting vector are not orthogonal. When resonant scattering is involved, the dispersion surface is always complex. We will consider to what extent the orthogonal relation between the tangent of the real part of the dispersion surface and the direction of the Poynting vector holds. We will take the symmetric Laue case as an example to find out how this orthogonal relationship changes as u changes. As shown in Fig. 6.10, let γ_P be the angle between the direction of the Poynting vector and the tangent of the real part of the complex dispersion surface. Differentiating both sides of (7.138) with respect to X_0, we get

$$-X_0\sin^2\theta_B' + \left(\frac{dY_{0r}}{dX_0}Y_{0r} - \frac{dY_{0i}}{dX_0}Y_{0i}\right)\cos^2\theta_B' - \frac{dY_{0i}}{dX_0}\kappa_{0i}\cos\theta_B' = 0. \tag{7.152}$$

Similarly, differentiating both sides of (7.139) with respect to X_0, we get

$$\left(\frac{dY_{0r}}{dX_0}Y_{0i} + \frac{dY_{0i}}{dX_0}Y_{0r}\right)\cos^2\theta_B' - \frac{dY_{0r}}{dX_0}\kappa_{0i}\cos\theta_B' = 0. \tag{7.153}$$

Finding dY_{0i}/dX_0 from (7.153) and substitute it into (7.152), we obtain

$$\tan\sigma = \frac{dY_{0r}}{dX_0} = \frac{X_0 Y_{0r}\sin^2\theta_B'}{Y_{0r}^2\cos^2\theta_B' + \left(\kappa_{0i} - Y_{0i}\cos\theta_B'\right)^2}. \tag{7.154}$$

$\tan\sigma$ is the slope of the tangent of the real part of the dispersion surface. Then we get

$$\gamma_P = \frac{\pi}{2} - |\varphi - \sigma|,$$

where φ is the angle between the Y_r axis and the Poynting vector. When only the normal scattering is involved, $\varphi = \sigma$ and γ_P is 90°, and the Poynting vector is orthogonal to the dispersion surface. Figure 7.17a shows γ_P as a function of W for various u and g in the Laue case. γ_P is obtained in the same way in the Bragg case, and the result is shown in Fig. 7.17b [14].

In the Laue case, for $u = 1$, the direction of the Poynting vector is always orthogonal to the real part of the complex dispersion surface even when $\kappa_{0i} \neq 0$. On the other hand, for $u = -1$, the orthogonality is broken in the region of $|W| > 1$. As $|W|$ approaches 1, γ_P becomes small and is 0° at $|W| = 1$; the direction of the Poynting vector becomes parallel to the tangent of the dispersion surface. When $|W|$ becomes

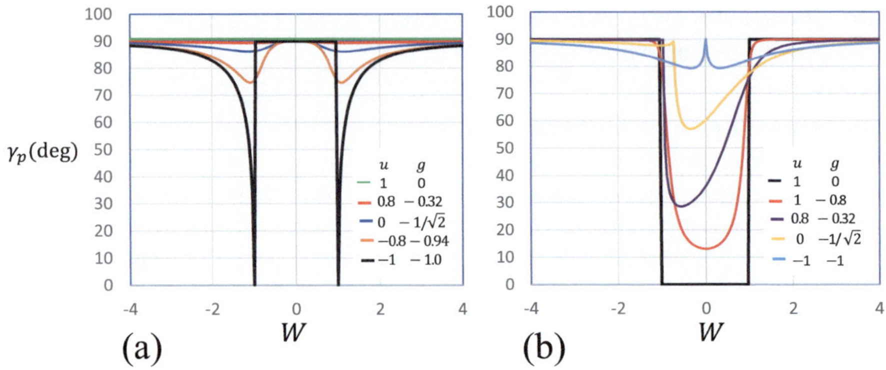

Fig. 7.17 Change of γ_P as a function of W for several values of u in the Laue case (**a**) and in the Bragg case (**b**)

large, γ_P becomes 90° and the orthogonal relation almost holds true. For $0 \le u \le 1$, the orthogonal relation almost holds.

In the Bragg case, the orthogonality relation is significantly broken in the region $-1 \le W \le 1$ even for $u \approx 1$. In the region $|W| > 1$, the orthogonal relation generally holds true.

7.9 Polarity and Phase Determination

7.9.1 Polarity Factor and Polarity Determination

As shown in Chap. 3, by the kinematic theory of diffraction, the polarity of a polar crystal is determined from the measurement of the pair of diffracted intensities in which Friedel's law is violated due to resonant scattering. In the dynamical theory of diffraction, the polarity comes in through χ_g. Figure 7.18 shows the relation between χ_g and χ_{-g}. For a Bijvoet pair, $\left|\chi_g\right| \ne \left|\chi_{-g}\right|$ as shown in Fig. 7.18c. It should be noted here that the dispersion surfaces for $\boldsymbol{g}$ and $-\boldsymbol{g}$ for a Bijvoet pair are the same, because W is given by $\overline{\chi}_g$ which does not depend on δ in (7.46). As a result, there is no change in the full-width at half maximum (FWHM) of the rocking curve, as shown in Fig. 7.19b. However, the diffracted intensity depends on a polarity factor,

$$k_p(\delta) = 1 - \left(1 - u^2\right)^{\frac{1}{2}} \sin\delta,$$

where the sign of δ is reversed for $\boldsymbol{g}$ and $-\boldsymbol{g}$. The intensity ratio of $r_p = P_g/P_{-g}$ is expressed by

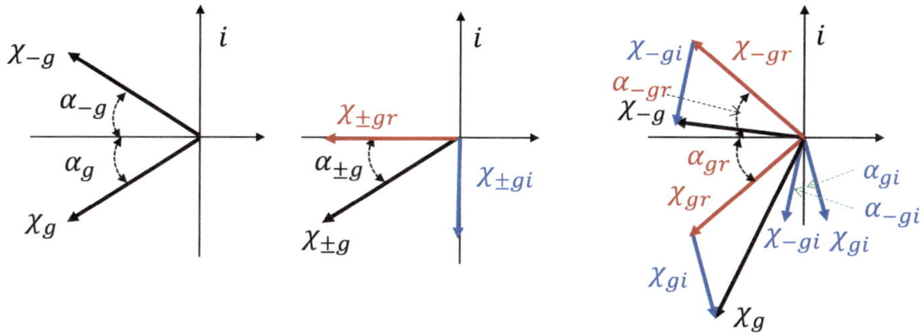

Fig. 7.18 Geometrical representation of relation between χ_g and χ_{-g} **a** for the case Friedel's law holds true, **b** for a Friedel pair and **c** for a Bijvoet pair

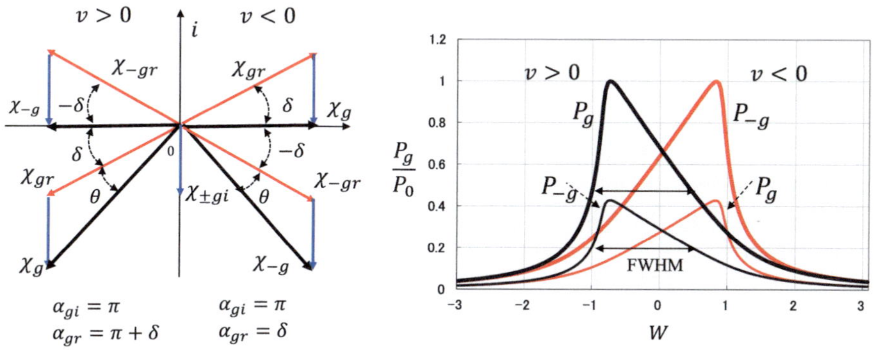

Fig. 7.19 **a** χ_g in a complex plane and the relations between χ_{gr}, χ_{gi} and δ for $v > 0$ and $v < 0$. **b** Rocking curves of the Bijvoet pairs in the Bragg case corresponding to χ_g and χ_{-g} in (**a**)

$$r_p = \frac{P_g}{P_{-g}} = \frac{1 - \left(1 - u^2\right)^{1/2}\sin\delta}{1 + \left(1 - u^2\right)^{1/2}\sin\delta} = \frac{|\chi_g|^2}{|\chi_{-g}|^2} = \frac{|F_g|^2}{|F_{-g}|^2}. \tag{7.155}$$

When $v > 0$, P_g and P_{-g} are shown by thick and thin black lines in Fig. 7.19b, respectively, which have a similar form to each other. P_g and P_{-g} for $v < 0$ are shown by thick and thin red lines. When $|v|$ is the same, r_p is reversed for $v > 0$ and $v < 0$ since the sign of δ changes. The result of $r_p = |F_g|^2/|F_{-g}|^2$ in (7.155) agrees with that obtained by the kinematical theory in Chap. 3.

For a crystal with a center of symmetry, (u, v) takes values on the circumference of the resonance circle. In a polar crystal, (u, v) takes all values within the resonance circle. Among these, in an extreme case, (u, v) takes $(0, 0)$, that is, the origin of the resonance circle, then the diffracted intensity of either $\boldsymbol{g}$ or $-\boldsymbol{g}$ becomes zero since

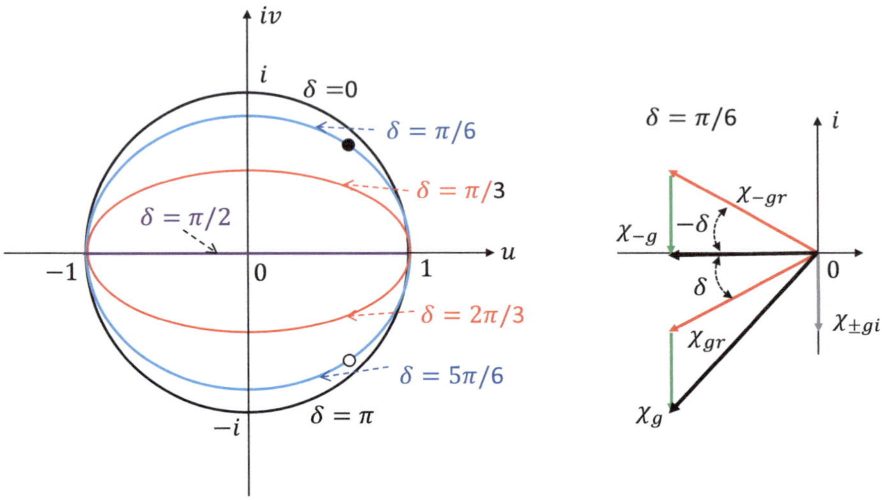

Fig. 7.20 Plots of (u, v) for some values of δ of Bijvoet pairs. The filled circle denotes $(u, v) = (0.6, 0.692)$ and the open circle $(u, v) = (0.6, -0.692)$ of the Bijvoet pair for $\delta = \pi/6$

$\delta = \pi/2$. The rocking curve is expressed as a Laue function, which agrees with the result of the kinematical theory.

Figure 7.20 plots the values of (u, v) for some values of δ. For $0 < \delta < \pi/2$, $v > 0$ and for $\pi/2 < \delta < \pi$, $v < 0$. When $\delta = \pi/2$, the values are on the line of $v = 0$ and u from -1 to $+1$.

Figure 7.19b shows the rocking curve when $\delta = \pi/6$ and $(u = 0.6, v = \pm 0.692)$ and Fig. 7.19a shows the relation between χ_g and χ_{-g}. The shape of the rocking curve is reversed horizontally when the sign of v changes. r_p becomes larger as both u and v approach zero. The sign of v can be determined from the left–right reversal of the peak positions of the transmitted rocking curves in the Laue case, and from that of the diffracted wave as well as the transmitted wave in the Bragg case.

7.9.2 Phase Determination

In Chap. 3, by assuming that the anomalous scattering factor is known, we describe how to determine the phase of the crystal structure factor from intensity measurements at three different wavelengths (multi-wavelength anomalous dispersion method). When we use the dynamical theory of diffraction, we can determine the phase by measuring the rocking curve for one wavelength. Here, instead of the crystal structure factor F_g, the electric susceptibility χ_g is used to consider the phase determination. As already mentioned, χ_g is related to F_g as

$$\chi_g = \chi_{gr} + i\chi_{gi} = -r_e \frac{\lambda^2}{\pi V_c} F_g.$$

We can determine the phase of F_g from χ_g. We write χ_g as

$$\begin{aligned}\chi_g &= \chi_{gr} + i\chi_{gr} = |\chi_{gr}|\exp(i\alpha_{gr}) + i|\chi_{gi}|\exp(i\alpha_{gi}) \\ &= |\chi_{gr}|\left(1 - 2k_a\sin\delta + k_a^2\right)^{1/2}\exp[i(\alpha_{gr} + \theta)],\end{aligned}$$

with

$$k_a = \frac{|\chi_{gi}|}{|\chi_{gr}|}, \quad \theta = \tan^{-1}\frac{k_a\cos\delta}{1 - k_a\sin\delta}.$$

To simplify the explanation of phase determination, we assume that there is one atom that causes resonant scattering in a unit cell. We take the position of the atom as the origin and assume that its anomalous scattering factor is known.

7.9.2.1 For Centrosymmetric Crystal

For a crystal with a center of symmetry, δ takes the value of either 0 or π. It is only necessary to determine the sign of $\cos\delta$, that is, the sign of v. In the Laue case, the sign is positive ($v > 0$) if the peak of the transmitted wave is on the higher angle side of the peak of the diffracted wave, and it is negative ($v < 0$) if it is on the lower angle side. In the Bragg case, if the peaks of the rocking curves of the transmitted wave and the diffracted wave are on the low angle side, the sign of v is positive ($v > 0$), and if they are on the high angle side, the sign is negative ($v < 0$).

7.9.2.2 For Polar Crystal

For a polar crystal, if the ratio r_p of the diffracted intensities of a Bijvoet pair is found and k_a of the ratio of $|\chi_{gr}|$ and $|\chi_{gi}|$ is found from the FWHM or the shape of the diffracted rocking curves, then, δ and θ are obtained. As a result, α_{gr} can be determined (see Fig. 7.19a).

As an example, we describe the phase determination from the measurement of the rocking curves in the symmetric Bragg case. Let $\Delta\Theta$ be the measured FWHM of the rocking curve. If the FWHM (ΔW) of the rocking curve as a function of W is obtained to be 2.12 and $k_a < 0.1$, from (7.83) we express $|\chi_{gr}|$ approximately as

$$|\chi_{gr}| \approx \frac{2\sin\theta_B}{2.12}\Delta\Theta.$$

Since χ_{gi} is known once the incident X-ray energy is fixed, $k_a = |\chi_{gi}|/|\chi_{gr}|$ can be obtained. When $k_a < 0.1$, from (7.155), the ratio r_p of the diffracted intensities from the front and back sides of the crystal is obtained by

$$r_p = \frac{P_{+g}}{P_{-g}} = \frac{|\chi_{+g}|^2}{|\chi_{-g}|^2} = \frac{1-(1-u^2)^{1/2}|\sin\delta|}{1+(1-u^2)^{1/2}|\sin\delta|} \approx \frac{1-2k_a|\sin\delta|}{1+2k_a|\sin\delta|}.$$

Here, the relation $u^2 \approx 1 - 4k_a^2$ is used. Then we get

$$|\sin\delta| = \frac{1}{2k_a}\frac{(r_p - 1)}{(r_p + 1)}$$

and δ can be obtained. By using the relation

$$\theta = \tan^{-1}\frac{k_a\cos\delta}{1-k_a\sin\delta},$$

we get χ_g as

$$\chi_{\pm g} = |\chi_{gr}|(1 \mp 2k_a|\sin\delta|)^{1/2}\exp[i(\alpha_{gi} - \delta + \theta)].$$

Since α_{gi} is known and $\delta = \alpha_{gi} - \alpha_{gr}$, we can obtain the phase α_{gr}. Negishi et al. [15] used this approach to determine the phase of GaAs.

7.10 Experimental Results

The history of research on resonant scattering dynamical diffraction is not so long, but there are some experimental works demonstrating phenomena peculiar to it. We describe some experimental results using parallel plate crystals of GaAs and Ge.

7.10.1 Rocking Curve of GaAs 200 (in the Symmetrical Laue Case)

For a monoatomic crystal, it is possible to make the real part of the atomic scattering factor zero and leave only the imaginary part by using the anomalous scattering factor that changes significantly at the absorption edge of the atom. By setting the real part of the Fourier coefficient of X-ray susceptibility χ_{gr} to be zero, diffraction of only due to χ_{gi}, that is, the condition of $u = -1$, can be obtained. As χ_{gi} cannot be made zero, the condition of $u = 1$ cannot be realized. For a polyatomic crystal, any value of u from -1 to 1 can be realized.

Negishi et al. [10–12] measured the 200 diffracted and transmitted intensities in the symmetric Laue case across the K absorption edges of both Ga and As of a GaAs crystal. The crystal structure factor of the GaAs 200 reflection is given as $F = 4\left(f^0_{\mathrm{Ga}} - f^0_{\mathrm{As}}\right)$ if resonant scattering is ignored. Here f^0_{Ga} and f^0_{As} are the normal atomic scattering factor of Ga and As, respectively. This reflection is also called quasi-forbidden reflection because the atomic number of Ga (Z = 31)) and As (Z = 33) are not so much different, $f^0_{\mathrm{Ga}} \approx f^0_{\mathrm{As}}$, and F is nearly zero. The influence of resonant scattering can be largely reflected. Although this crystal does not have a center of symmetry, the 200 reflections do not depend on polarity (Friedel pair) and (u, v) take values on the circumference of the resonance circle.

Figure 7.21 shows the Fourier coefficient of electric susceptibility (see (5.5)) when Ga is taken as the origin of the unit cell. At point (A) before the K absorption edge of Ga, $\chi_{gi} = 0$ and $(u, v) = (1, 0)$. At points (B), (C), and (D) between the K absorption edges of Ga and As, $(u, v) = (0, -1)$ with $\chi_{gr} = -\chi_{gi}$, $(u, v) = (-1, 0)$ with $\chi_{gr} = 0$, and $(u, v) = (0, 1)$ with $\chi_{gr} = \chi_{gi}$, respectively.

In actual measurement of rocking curves, an optical system is required that makes the divergence angle as small as possible. An example of such an optical system is shown in Fig. 7.22. The divergence angle of the incident beam is suppressed to approximately 0.2 s using the asymmetric monochromator described in 6.3.

Figures 7.23a to d show the measured rocking curves at X-ray energies corresponding to points (A) to (D) in Fig. 7.21. In Fig. 7.23a, the rocking curves of I_h and I_t show left–right symmetry. In the interference fringe, I_t shows a valley at the angle $(\Delta\Theta)$ where I_h shows a peak. The phases of I_t and I_h are opposite, indicating that they are Pendellösung fringes. Since the intensity of I_t on the low-angle side

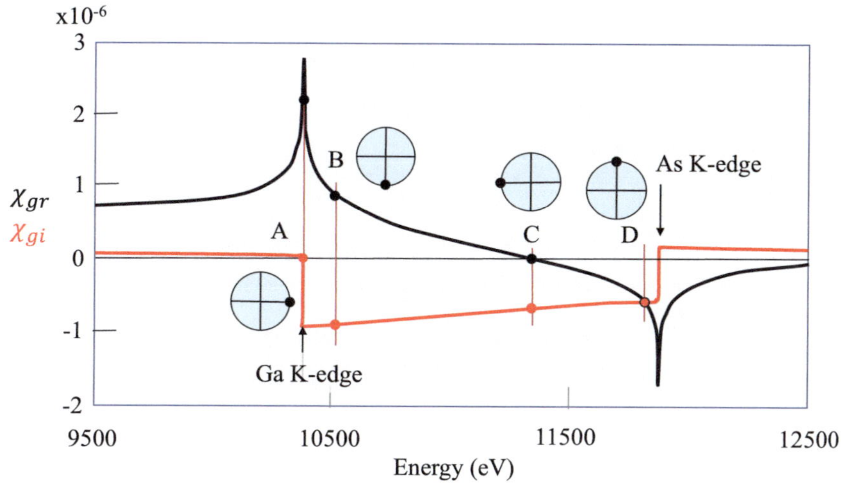

Fig. 7.21 The black and red lines show χ_{gr} and χ_{gi} for GaAs 200 reflection as a function of X-ray energy near the K-absorption edges of Ga (10,368 eV) and As (11,863 eV). A, B, C and D indicate the energies to satisfy the conditions $(u, v) = (1, 0)$, $(0, -1)$, $(-1, 0)$ and $(0, 1)$, respectively

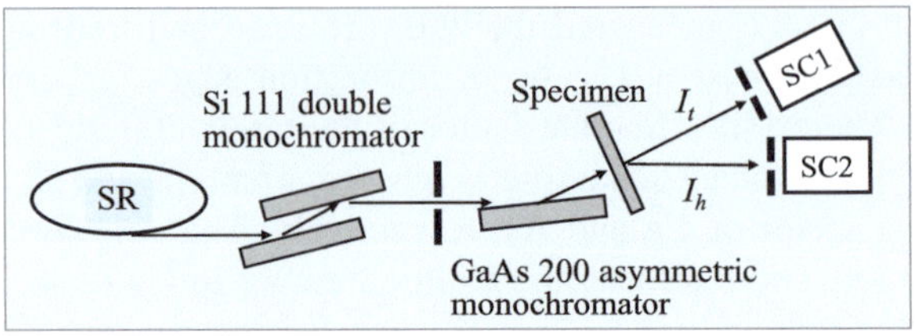

Fig. 7.22 Rocking curve measurement system using GaAs 200 asymmetric monochromator. The asymmetric factor $b = 0.1$ and the divergence angle of the beam to the sample is approximately 0.2 s. I_h and I_t are the diffracted and transmitted intensities

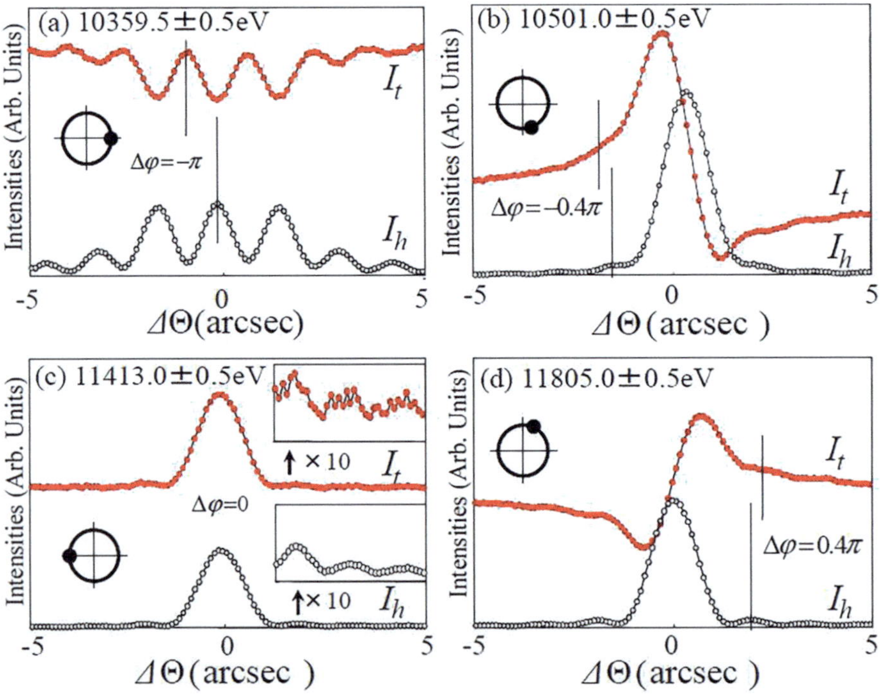

Fig. 7.23 Measured rocking curves of the diffracted wave I_h and transmitted wave I_t of GaAs 200 reflection [12]. The incident X-ray energy is **a** 10,359.5 ±1 eV, **b** 10,501.0 ±1 eV, **c** 11,413.0 ±1 eV and **d** 11,805.0 ±1 eV. The crystal thickness is 45 μm. Adapted from Fig. 7.2 in [12]

of $\Delta\Theta$ is equal to that on the high-angle side, the effect of anomalous transmission has completely disappeared. If any anomalous transmission occurs, the intensity of I_t on the low angle side may not be equal to that on the high angle side, as shown in Fig. 7.10a. This agrees with the calculated result for $u = 1$ shown in Fig. 7.9a.

In Fig. 7.23c, the rocking curves of I_h and I_t show a peak at the center ($\Delta\Theta = 0$) and are symmetrical. The peak in the center is due to anomalous transmission. Although the amplitudes on both sides of the peaks of I_h and I_t are small, interference fringes show the same phase. This agrees with the calculated result for $u = -1$ shown in Fig. 7.9b.

In Fig. 7.23b and d, the rocking curve of I_h is symmetrical, but that of I_t is asymmetrical. In (b), the peak of I_t is on the low angle side, while in (d) it is on the high angle side. This result agrees the tendency of the calculated result of P_t in Fig. 7.10a, and the sign of v can be determined from this result. Negishi et al. [12] measured the phase difference in the interference fringes between I_h and I_t by changing u from 1 to -1 and determined the phase of the crystal structure factor.

7.10.2 Measurement of Anomalous Scattering Factor Near Ge K Absorption Edge

We describe an approach for precisely determining the anomalous scattering factor of Ge by measuring the rocking curves of the diffracted wave and transmitted wave of 844 reflection from a plane parallel single crystal in the symmetric Bragg case.

A Ge crystal shows the diamond cubic structure, and the crystal structure factor of 844 reflection is expressed as

$$F_{844} = F_{844,r} + iF_{844,i} = 8f.$$

f is the atomic scattering factor of Ge and can be written as

$$f = f^0(k_s) + f'(\hbar\omega) + if''(\hbar\omega).$$

The real part and imaginary part of F_{844} are respectively

$$F_{844,r}(\omega) = 8\left[f^0(k_s) + f'(\hbar\omega)\right]\exp(-M), \quad F_{844,i}(\omega) = 8f''(\hbar\omega)\exp(-M).$$

Here, $\exp(-M)$ expresses the correction due to thermal vibration of the atom. The X-ray susceptibility is given by

$$\chi_{844} = \chi_{844,r} + i\chi_{844,i} = -r_e\frac{\lambda^2}{\pi V_c}F_{844}(\omega) = -r_e\frac{\lambda^2}{\pi V_c}\left[F_{844,r}(\omega) + iF_{844,i}(\omega)\right].$$

Figure 7.24 shows the atomic scattering factor of an isolated Ge atom for the 844-reflection, with $f^0(k_s) + f'(\hbar\omega)$ on the horizontal axis and $f''(\hbar\omega)$ on the vertical axis. The normal scattering factor in a crystal is $f^0_{844} = 9.27$, the real part of the anomalous scattering factor at the K absorption edge of an isolated Ge atom is $f'(\hbar\omega_K) = -10.2$. Then $\left|f^0_{844}\right|$ is smaller than $\left|f'(\hbar\omega_K)\right|$ and the absolute value of the real part of the anomalous scattering factor is larger than that of the normal scattering factor. The real part $f^0(s) + f'(\hbar\omega)$ can be made negative by increasing the reflection order to 844.

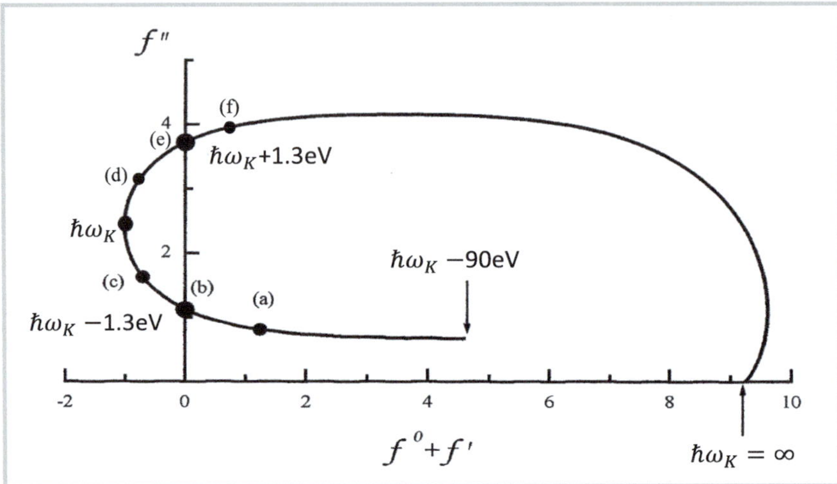

Fig. 7.24 The atomic scattering factor of Ge 844-reflection for X-ray energies at **a** $\hbar\omega_K - 4.3$ eV, **b** $\hbar\omega_K - 1.3$ eV, **c** $\hbar\omega_K - 0.5$ eV, **d** $\hbar\omega_K + 0.5$ eV, **e** $\hbar\omega_K + 1.3$eV and **f** $\hbar\omega_K + 2.1$eV. The energy of Ge K absorption edge $\hbar\omega_K$ is 10,368 eV

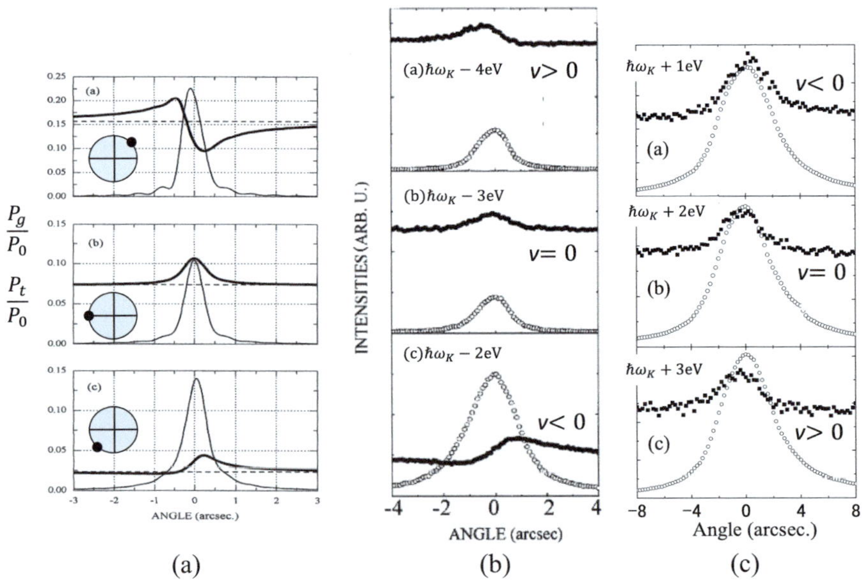

Fig. 7.25 844-reflection rocking curves near Ge K absorption edge. **a** The thin line shows the calculated diffracted curves and the thick line the transmitted curves (adapted from Fig. 7.6 in [2]). **b** and **c** The open circles show measured diffracted intensities and the filled circles the transmitted intensities. The crystal thickness is 82 μm. **b** is adapted from Fig. 7.8 in [2] and **c** is adapted from Fig. 7.5 in [16]

In Fig. 7.24, the real part of the atomic scattering factor $f^0(s) + f'(\hbar\omega)$ is positive, and the imaginary part $f''(\hbar\omega)$ is also positive at point (a). Then, $\chi_{844,r} < 0$, $\chi_{844,i} < 0$, and $v > 0$. At point (b), $f^0(s) + f'(\hbar\omega) = 0$ and $(u, v) = (-1, 0)$, the diffraction only due to resonant scattering occurs. At point (c) $f^0(s) + f'(\hbar\omega) < 0$ and $f''(\hbar\omega > 0)$, then $\chi_{844,r} > 0$, $\chi_{844,i} < 0$ and $v < 0$. At point (d), $\chi_{844,r} > 0$, $\chi_{844,i} < 0$ and $v < 0$. At point (e), $(u, v) = (-1, 0)$, diffraction only due to resonant scattering occurs. At point (f), $\chi_{844,r} < 0$, $\chi_{844,i} < 0$, and $v > 0$.

Figure 7.25a shows the calculated rocking curves of the diffracted wave and transmitted wave under the conditions shown in Fig. 7.24a to c. The thick solid line shows the rocking curve of the transmitted wave, and the thin solid line that of the diffracted wave. In (a) of Fig. 7.25a, the peak of the diffracted wave is almost symmetrical, and the peak position has moved to the lower angle side by about 0.1 s from the center. The rocking curve of the transmitted wave shows larger intensity on the low angle side and is asymmetrical with the peak position at the lower angle by about 0.5 s. In (b) of Fig. 7.25a, the peaks of the rocking curves of the diffracted wave and the transmitted wave are both symmetrical, the peak positions are both at the center, and the background heights on the left and right sides of the transmitted wave are equal. The symmetrical relations in (c) of Fig. 7.25a are opposite to those in (a). In this way, at the X-ray energy where the sign of v is reversed, diffraction only due to resonant scattering occurs.

Figures. 7.25b and c show the corresponding measured results. In Fig. 7.25b, (a) shows the intensity behavior for $v > 0$, (b) shows that for $(u, v) = (-1, 0)$, and (c) shows that for $v < 0$. In the actual Ge crystal used in the experiment, $f^0(s) + f'(\hbar\omega) = 0$ holds and diffraction only due to resonant scattering is obtained at $\omega_K - 3$eV with ω_K being the energy at Ge K absorption edge. From the measured rocking curves of (b) in Fig. 7.25c, the condition of $(u, v) = (-1, 0)$ is satisfied at $\omega_K + 2$eV, and diffraction only due to resonant scattering is obtained.

From these measured rocking curves, the value of the real part of the anomalous scattering factor can be obtained. Since the value of $f^0(k_s)$ is not affected by bonding electrons and is obtained by theoretical calculations with the error within 1% for a high-order reflection, we can obtain $f'(\hbar\omega)$ with the error of about 1% by using the condition of $f^0(k_s) + f'(\hbar\omega) = 0$. It is the advantage of this approach that the measured intensities are not affected either by the accuracy of crystal thickness or by temperature factors.

Yoshizawa et al. [16] changed the reflection indices (844,840,555,660) from a Ge single crystal to find the condition of $f^0(k_s) + f'(\hbar\omega) = 0$, and precisely measured $f'(\hbar\omega)$ near the K absorption edge of Ge. The measured values are shown in Fig. 3.10.

7.10.3 Evaluation of Temperature Factor

We describe the evaluation of the temperature factor of atoms within a crystal using resonant dynamical diffraction. As shown in 2.6, the effect of atomic vibration around the atomic equilibrium position in a crystal appears as the decrease in the diffraction intensities, and is expressed by the Debye–Waller factor $\left[\exp\left(-M_j\right)\right]$ for atom j. Let's focus on the changes in the diffraction intensity near the K absorption edge ($\hbar\omega_K$). When the incident X-ray energy ($\hbar\omega$) is less than $\hbar\omega_K$, the effect of absorption is small, and when it becomes larger than $\hbar\omega_K$, the absorption becomes significantly large. The diffraction intensity of the latter should become significantly weaker than that of the former. (It is noted, however, that the intensity does not necessarily become weaker if the imaginary part of the anomalous scattering factor increases and its influence on $F_g(\hbar\omega)$ is large for $\hbar\omega > \hbar\omega_K$).

The diffracted intensity from a mosaic monatomic crystal I_m is given by

$$I_m \propto \frac{\left|F_g(\hbar\omega)\right|^2 \exp(-2M)}{\mu_a(\hbar\omega)}. \tag{7.156}$$

The effect of thermal vibration does not appear on the intensity ratio of the diffracted waves between below and above the absorption edge. In contrast, in resonant dynamical diffraction, μ_d in the effective absorption coefficient expressed by (7.92) depends on the temperature. Using this fact, the Debye–Waller factor can be obtained by the intensity ratio between below and above the absorption edge [17]. This method has been used to determine the temperature factors of GaAs [18], InSb [19], and CuI [20, 21]. In the case of GaAs and InSb, since their atomic numbers are close to each other, the temperature factor is almost the same. CuI is known to have an abnormally large temperature factor [22]. In addition, Cu ($Z = 29$) is much lighter than I ($Z = 53$), so the temperature factor of Cu is predicted to be larger than that of I, and the thermal vibrational behavior of Cu affects the physical properties. The temperature factors of Cu and I at room temperature are determined independently in [21].

7.10.4 Assessment of Non-stoichiometry

Datsenko and Klad'ko [23] studied atomic concentrations due to lattice defect density using the GaAs 200 reflection. As mentioned above, GaAs 200 reflection is chosen as it is a quasi-forbidden reflection. Since the real part of the crystal structure factor

$$F_r \propto c_{\mathrm{Ga}}\left[f^0_{\mathrm{Ga}} + f'_{\mathrm{Ga}}(\hbar\omega)\right] - c_{\mathrm{As}}\left[f^0_{\mathrm{As}} + f'_{\mathrm{As}}(\hbar\omega)\right] \tag{7.157}$$

can be made zero, it is possible to extract the diffraction due to the crystal structure factor of only the imaginary part of the anomalous scattering factor F_i. In (7.157), c_{Ga} and c_{As} are the atomic concentrations of Ga and As. When $F_r = 0$ is satisfied at one X-ray energy, the diffracted intensity shows its minimum. This critical X-ray energy changes significantly due to the slight change in $c_{\mathrm{Ga}} - c_{\mathrm{As}}$. By measuring this energy change, $c_{\mathrm{Ga}} - c_{\mathrm{As}}$ can be determined, and non-stoichiometry can be assessed.

Machulin et al. [24] observed Pendellösung fringes in the integrated reflection intensity at the energy where $F_i = 0$ is satisfied just below the K absorption edge of Ga in GaAs crystals, and assessed the non-stoichiometry of the crystal.

References

1. B.W. Batterman, H. Cole, Rev. Modern Phys. **36**, 681–717 (1964)
2. T. Fukamachi, R. Negishi, S. Zhou, M. Yoshizawa, T. Kawamura, Acta Cryst. A**58**, 552–558 (2002)
3. A. Authier, *Dynamical Theory of X-ray Diffraction* (Oxford University Press, 2001)
4. S. Miyake, *X-raty Diffraction.* (Askura (In Japanese), 1969)
5. N. Kato, Acta Cryst. A**48**, 829–833 (1921)
6. T. Fukamachi, T. Kawamura, Acta Cryst. A**49**, 384–388 (1993)
7. G. Borrmann, Phys. Z. **42**, 157–162 (1941)
8. W.H. Zachariasen, *Theory of X-Ray Diffraction in Crystal* (John Willy Sonds Inc., New York, 1945)
9. M. von Laue, Acta Cryst. **2**, 106–113 (1949)
10. R. Negishi, T. Fukamachi, M. Yoshizawa, K. Hirano, K. Hirano, T. Kawamura, J. Phys. Soc. Jpn. **77**, 023709 (2008)
11. R. Negishi, T. Fukamachi, M. Yoshizawa, K. Hirano, K. Hirano, T. Kawamura, Phys. Status Solidi A **206**, 1865–1870 (2009)
12. R. Negishi, T. Fukamachi, M. Yoshizawa, K. Hirano, K. Hirano, T. Kawamura, Phys. Status Solidi A **208**, 2567–2570 (2011)
13. R.W. James, Solid State Phys. **15**, 53–220 (1963)
14. R. Negishi, Dissertation (Department of Engineering of The University of Tokyo) (1998)
15. R. Negishi, T. Fukamachi, S. Jongsukswat, K Hirano, K. Hirano, T. Kawamura, J. Appl. Cryst. **46,** 1216–1218 (2013)
16. M. Yoshizawa, S.M. Zhou, R. Negishi, T. Fukamachi, T. Kawamura, Acta Cryst. A**64**, 321–325 (2008)
17. T. Kawamura, T. Fukamachi, Acta Cryst. A**35**, 831–835 (1979)
18. T. Fukamachi, T. Kawamura, K. Hayakawa, Y. Nakano, F. Koh, Acta Cryst. A**38**, 810–813 (1982)
19. M. Yoshizawa, T. Kawamura, T. Fukamachi, K. Hayakawa, Acta Cryst. A**42**, 113–116 (1986)
20. B. Morlon, T. Fukamachi, S. Hosoya, Acta Cryst. A**35**, 714–717 (1979)
21. B. Morlon, T. Fukamachi, S. Hosoya, Acta Cryst. A**37**, 92–94 (1981)
22. S. Miyake, S. Hoshino, Rev. Mod. Phys. **30**, 172–174 (1958)
23. L.I. Datsenko, V.P. Klad'ko, Crystallogr. Rep. **45**, 705–708 (2000)
24. V.F. Machulin, L.I. Datsenko, V.P. Klad'ko, Crystallogr. Rep. **49**, Suppl. 1, S72–S79 (2004)

Chapter 8
Dynamical Diffraction of X-ray Wave-Packet with Finite Spectral Width

So far, we have assumed that a pseudo-refracted wave excited by an ideal plane wave as a plane wave. It is very difficult to actually obtain ideal plane waves of X-rays. In dynamical diffraction, even if the divergence angle of the incident beam is very small and can be regarded as an almost ideal plane wave, it is necessary to treat the pseudo-refracted wave as spherical waves with large angular divergence when the angular amplification effect occurs as described in this chapter [1]. It is appropriate to treat actual X-rays as a spatially localized wave-packets that propagate with a finite spectral width. In this chapter, we discuss X-ray dynamical diffraction of such a wave-packet. We consider non-absorbing crystals, unless otherwise specified.

8.1 Angular Amplification Effect

We will discuss an angular amplification rate defined by the ratio of the divergence angle $\Delta\varphi$ of the pseudo-refracted wave to the divergence angle $\Delta\alpha_0$ of the incident wave given by

$$A = \frac{\Delta\varphi}{\Delta\alpha_0}. \tag{8.1}$$

8.1.1 In the Symmetric Laue Case

When the glancing angle α_0 has a width of $\Delta\alpha_0$ in the symmetric Laue case, the corresponding deviation parameter W also has a width of ΔW given by

T. Fukamachi and T. Kawamura, *Resonant Dynamical X-ray Diffraction*, Springer Series in Solid-State Sciences 208, https://doi.org/10.1007/978-981-96-8406-9_8

$$\left|\frac{\Delta W}{\Delta \alpha_0}\right| = \frac{\sin 2\theta_B}{n_0 P \left|\chi_g\right|} \tag{8.2}$$

using (5.84). Suppose that $n_0 \approx 1$, $\Delta\alpha_0 = 20 \times 10^{-6}$ rad $\left|\chi_g\right| = 5 \times 10^{-6}$, $\sin 2\theta_B = 0.5$ and $P = 1$, for example, we get $\Delta W = 2$. By using (6.141), the angle φ between the reflecting plane and the pseudo-refracted wave is related to the Bragg angle θ_B as

$$\tan\varphi = \frac{W}{\left(W^2 + 1\right)^{1/2}} \tan\theta_B. \tag{8.3}$$

This indicates that when $|W|$ changes from 0 to 2, φ changes from 0 to approximately θ_B. For an incident wave with $\Delta\alpha_0$ approximately 20×10^{-6} rad, which can be regarded as a plane wave (a pseudo-plane wave), the refracted wave is regarded as a spherical wave (a pseudo-spherical wave) as shown in Fig. 8.1b. The incident pseudo-plane wave becomes a pseudo-spherical wave due to dynamical diffraction by a perfect crystal, just as a plane wave becomes a spherical wave when it passes through a narrow slit, as shown in Fig. 8.1a. Kato [2–4] has focused on this effect and developed the dynamical theory of diffraction for spherical wave. He has conducted research on Pendellösung fringes (also called Kato fringes) formed in Borrmann fans and obtained many results related to the fringes. We will not go into the dynamical theory of diffraction for spherical wave, and will treat the pseudo-spherical wave by decomposing it into plane waves propagating to different directions.

According to Authier [1], the angular amplification rate A_L in the symmetric Laue case is expressed by

$$A_L = \left|\frac{\Delta\varphi}{\Delta\alpha_0}\right| = \left|\frac{\Delta\varphi}{\Delta p}\right|\left|\frac{\Delta p}{\Delta W}\right|\left|\frac{\Delta W}{\Delta\alpha_0}\right|. \tag{8.4}$$

By using the relation $p = \tan\varphi/\tan\theta_B$ in (6.142), we get

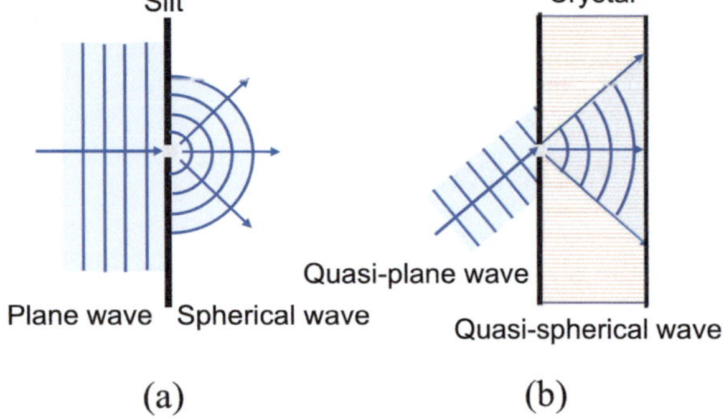

Fig. 8.1 Diffraction by a slit (**a**) and by a plane parallel crystal (**b**)

$$\left|\frac{\Delta\varphi}{\Delta p}\right| = \frac{\tan\theta_B}{1+\tan^2\varphi} = \frac{(W^2+1)\cos^2\theta_B}{W^2+\cos^2\theta_B}\tan\theta_B. \tag{8.5}$$

Since p is related to W as $p = W(W^2+1)^{-1/2}$ using (6.143), we get

$$\left|\frac{\Delta p}{\Delta W}\right| = \frac{1}{(W^2+1)^{\frac{3}{2}}}. \tag{8.6}$$

By substituting (8.5) and (8.6) into (8.4), we obtain

$$A_L = \frac{\sin^2 2\theta_B}{2P|\chi_g|(W^2+\cos^2\theta_B)(W^2+1)^{1/2}}. \tag{8.7}$$

Since $|\chi_g|$ is of the order of 10^{-4} to 10^{-6}, A_L is very large of the order of 10^4 to 10^6. Figure 8.2 shows the plot of A_L for Si 220 as a function of W. The maximum amplification factor is obtained when $W = 0$.

Figure 8.3 shows how $\Delta\varphi$ changes with changes in W and ΔW using the relation in (8.6). Figure 8.3a shows that the pseudo-refracted wave spreads to almost the region of Bormann fan when $W = 0$ and $\Delta W = 1$. When $W = 0$ and $\Delta W = 0.2$, the divergence angle of the pseudo-refracted wave becomes considerably small, as shown in Fig. 8.3b. As $|W|$ increases, the divergence angle of the pseudo-refracted wave becomes even smaller, as shown in Fig. 8.3c.

This angular amplification effect can be used to design a monochromator which makes the divergence angle smaller. The schematic diagram is shown in Fig. 8.4. By extracting a part of the diffracted beam spread by the angular amplification effect in the first crystal (C1, monochromator) through a narrow slit placed just before the

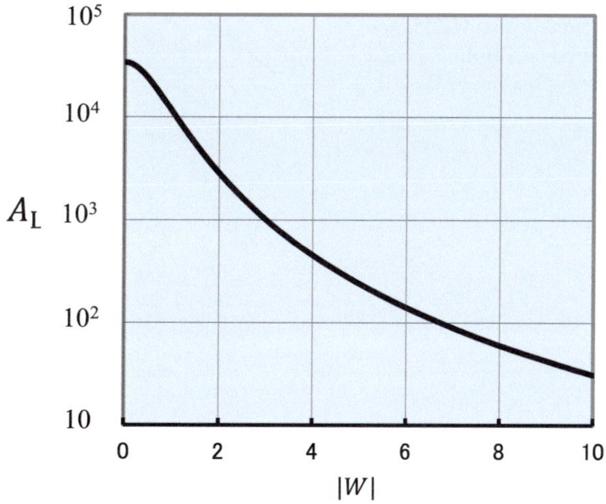

Fig. 8.2 The amplification factor A_L for Si 220 in the symmetric Laue case, when $P = 1$ and $\theta_B = 16.9$. A_L shows the maximum of 3.5×10^4 at $W = 0$

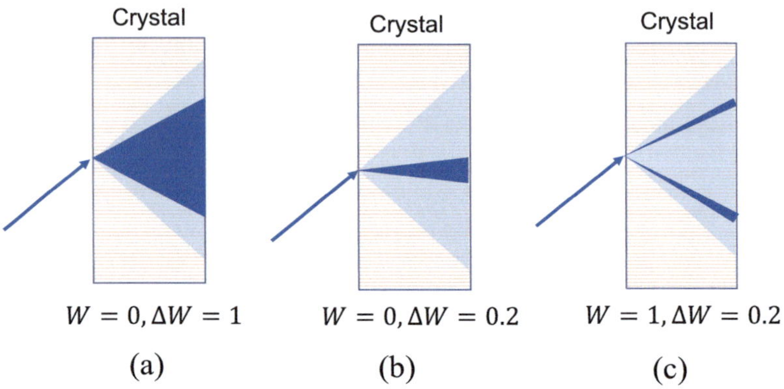

Fig. 8.3 Schematic diagrams to show the relations between the divergent angle of the incident beam and that of the pseudo-refracted wave in a crystal. The light blue filled triangle represents the Borrmann fan and the dark blue filled triangle represents the area where the pseudo-refracted wave spreads

second crystal (C2), the divergent angle of the diffracted beam from the second crystal is suppressed. By slightly shifting the incident angle on C2 from that corresponding to $W = 0$, the paths of the pseudo-refracted waves excited separately by the two tie-points can be observed separately as shown in Fig. 8.3c. (Detailed explanation is found in the book by Author [1]).

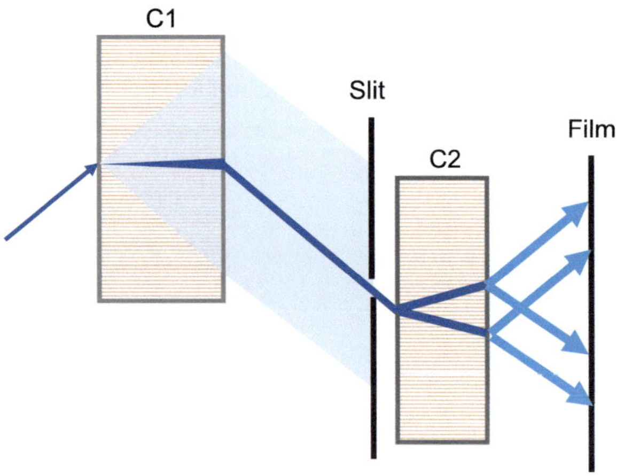

Fig. 8.4 Monochromator utilizing the angular amplification effect (the first crystal, C1). The incident angle on C1 is set to satisfy $W = 0$ and that on the second crystal (C2) is set slightly off from the angle corresponding to $W = 0$

8.1.2 In the Symmetric Bragg Case

In the symmetric Bragg case, we treat the region of $|W| \geq 1$. As shown in (6.146), $p = |\tan\varphi|/\tan\theta_B = \left(W^2 - 1\right)^{1/2}/|W|$, and we get

$$\left|\frac{\Delta\varphi}{\Delta p}\right| = \frac{W^2\cos^2\theta_B}{W^2 - \sin^2\theta_B}\tan\theta_B, \tag{8.8}$$

$$\left|\frac{\Delta p}{\Delta W}\right| = \frac{1}{W^2\left(W^2 - 1\right)^{1/2}}. \tag{8.9}$$

The angular amplification factor A_B in the Bragg case is given by

$$A_B = \frac{\sin^2 2\theta_B}{2P|\chi_g|\left(W^2 - \sin^2\theta_B\right)\left(W^2 - 1\right)^{1/2}}. \tag{8.10}$$

A_B becomes significantly large as $|W|$ approaches 1, and diverges at $|W| = 1$. This indicates that the angular amplification can be made larger in the Bragg case than in the Laue case. Figure 8.5 shows the calculated results of A_B for Si 220 reflection as a function of W. The horizontal axis δW is given by $\delta W = W - 1$.

The monochromator which takes advantage of this high angular amplification in the Bragg case will be described in detail later.

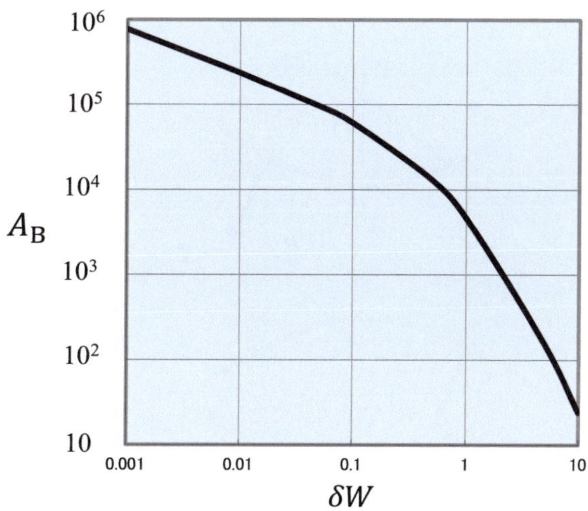

Fig. 8.5 Angular amplification factor A_B for Si 220 in the symmetric Bragg case when $P = 1$ and $\theta_B = 16.9°$

8.2 Interference Fringes Due to Pseudo-Spherical Waves

We discuss interference fringes created by pseudo-spherical waves caused by the angular amplification effect.

8.2.1 In the Laue Case

First, we discuss the interference fringes of pseudo-spherical waves due to the angular amplification effect in the symmetric Laue case. We assume that the incident beam is monochromatic and its divergence angle is $\Delta\alpha_0$ (the corresponding width of W is ΔW). Figure 8.6a shows the dispersion surface and pseudo-refracted waves $\boldsymbol{S}^{(1)}$ and $\boldsymbol{S}^{(2)}$ in the reciprocal space, and Fig. 8.6b shows the diffraction conditions in the real space. When the incident beam has a divergence angle $\Delta\alpha_0$ and a pseudo-refracted wave $\boldsymbol{S}^{(1)}$ is excited at tie-point d_1, another pseudo-refracted wave $\boldsymbol{S}^{(2)}$ may be excited at tie-point d_2 in the direction parallel to $\boldsymbol{S}^{(1)}$. Let the trajectory of the pseudo-refracted wave composed of $\boldsymbol{S}^{(1)}$ and $\boldsymbol{S}^{(2)}$ be $\overline{\mathrm{AB}} = l$. When the incident point A is taken at (0,0), this composed pseudo-refracted wave that reaches the position B(x, H) on the transmission surface is emitted as a diffracted wave $\boldsymbol{E}_g$ and a transmitted wave $\boldsymbol{E}_t$.

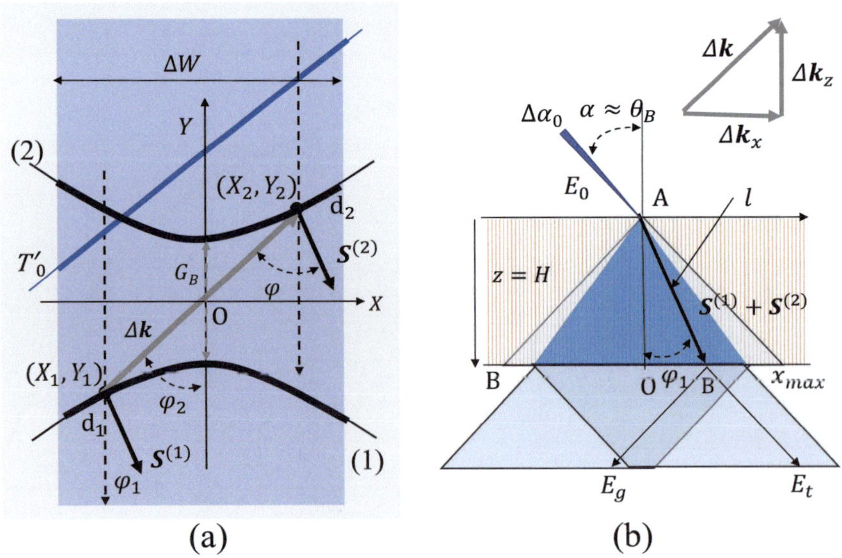

Fig. 8.6 Geometry of pseudo-refracted waves for interference fringes caused by angular amplification effect. (**a**) The waves $\boldsymbol{S}^{(1)}$ and $\boldsymbol{S}^{(2)}$ excited at two tie-points on the dispersion surface in the reciprocal space and (**b**) the wave geometry in the real space

The phase difference $\Delta\theta$ between the two pseudo-refracted waves $\boldsymbol{S}^{(1)}$ and $\boldsymbol{S}^{(2)}$ is obtained from Fig. 8.6, as

$$\Delta\theta = \Delta\boldsymbol{k}\cdot\boldsymbol{l} = \Delta k l \cos\varphi = \Delta k_x x + \Delta k_z H \tag{8.11}$$

by using $\Delta\boldsymbol{k} = \overrightarrow{d_1 d_2}$ and $\overline{\mathrm{OB}} = x$. The angle φ is

$$\varphi = \varphi_1 + \varphi_2 \tag{8.12}$$

and

$$\varphi_1 = \cos^{-1}\left(\frac{H}{l}\right),\ \varphi_2 = \cos^{-1}\left(\frac{2|Y_1|}{\Delta k}\right) \tag{8.13}$$

as shown in Fig. 8.6. Here, Δk is given by

$$\Delta k = |\Delta\boldsymbol{k}| = \left|\boldsymbol{k}_g^{(1)} - \boldsymbol{k}_g^{(2)}\right| = 2\left(X_1^2 + Y_1^2\right)^{1/2} = G_\mathrm{B}\left[W^2\left(1/\tan^2\theta_\mathrm{B} + 1\right) + 1\right]^{1/2}, \tag{8.14}$$

where $X_1 = -WG_\mathrm{B}/(2\tan\theta_\mathrm{B})$, $Y_1 = -G_\mathrm{B}\left(W^2+1\right)^{1/2}/2$, $G_\mathrm{B} = 2\pi/\Lambda_0$ with G_B being the Bragg gap and Λ_0 the spacing of the Pendellösung fringes. The length l of the trajectory of the pseudo-refracted wave is given by

$$l = \left[(px_{max})^2 + H^2\right]^{1/2} = H\left[\frac{W^2\left(\tan^2\theta_\mathrm{B} + 1\right) + 1}{W^2 + 1}\right]^{1/2}. \tag{8.15}$$

Here, $x_{max} = H\tan\theta_\mathrm{B}$ and $p^2 = W^2/\left(W^2+1\right)$. In a similar way to obtaining (6.20), we obtain the diffracted intensity as

$$\frac{P_g}{P_0} = \frac{\sin^2(\Delta\theta/2)}{W^2+1}.$$

Figure 8.7 shows two examples of the calculated diffracted intensities showing interference fringes from the transmission surface due to the angular amplification effect. Here, we assume that the crystal thickness $H = n\Lambda_0$ and $n = 10$ and 10.5.

Figure 8.8a shows the intensity distribution (wave field) inside the crystal due to the angular amplification effect. The darker colored curve represents the stronger wave field. Figure 8.8b shows the intensity distribution appearing on the transmission plane of a plane parallel crystal, and Fig. 8.8c shows that of a wedge-shaped crystal. This interference pattern is called Kato fringe, although Kato derives the fringes using spherical wave theory and the result do not completely agree with the results in Fig. 8.8 using pseudo-plane waves.

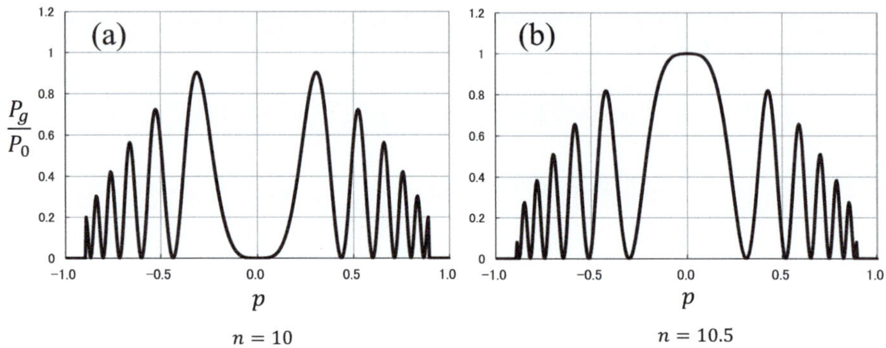

Fig. 8.7 Interference fringes due to angular amplification effect when $\Delta W = 4$. **a** $n = 10$ and **b** $n = 10.5$

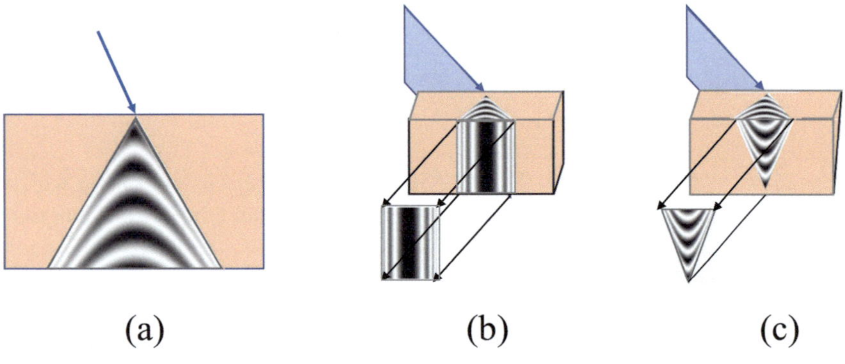

Fig. 8.8 Interference fringes due to angular amplification effect in the Borrmann fan region (**a**), those appeared from a plane parallel crystal (**b**) and from a wedge-shaped crystal (**c**)

8.2.2 *In the Bragg Case*

Kato fringes also appear in the Bragg case when the divergence angle of the incident beam becomes large. Figure 8.9a shows a geometry of pseudo-refracted waves excited at tie-points (d_1 and d'_1) on the dispersion surface, and (b) shows the beam geometry in the real space.

Two pseudo-refracted waves ($\boldsymbol{S}^{(1)}$ and $\boldsymbol{S'}^{(1)}$) propagating in the same direction can be excited at the same time, if $|\Delta W| > 2$ as shown in Fig. 8.9a. Since the phase difference between the beams $\boldsymbol{S}^{(1)}$ and $\boldsymbol{S'}^{(1)}$ at point B is

$$\Delta\theta = \Delta k \overline{\mathrm{AB}} \cos\varphi,$$

interference fringes with period Λ_0 appear on the crystal surface as shown in Fig. 8.9b. Compared to the integrated intensity in the region of $-1 < W < 1$, the intensity of the Kato fringes is very weak, which makes its observation very difficult. If $|\Delta W| < 2$,

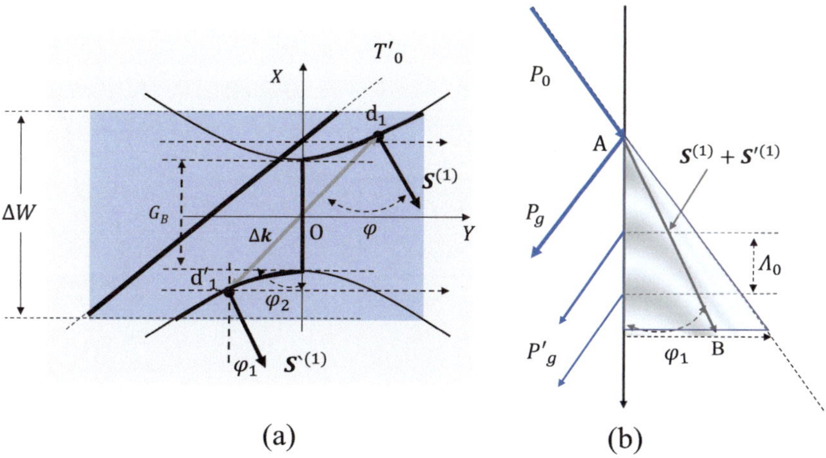

Fig. 8.9 Geometrical conditions of pseudo-refracted waves causing interference fringes due to angular amplification effect in the Bragg case. (**a**) The waves excited at two tie-points on the dispersion surface form the interference fringes. The blue region indicates the expanded area of the pseudo-refracted waves. (**b**) The wave geometry and interference fringes in the real space

two pseudo-refracted waves propagating in the same direction cannot be excited at the same time, and the Kato fringes do not appear.

8.3 Multiple Bragg-Laue Interference Fringes

We discuss Multiple-Bragg-Laue (MBL) interference fringes.

8.3.1 Bragg-Laue Type Diffraction

For discussing Bragg-Bragg-Laue (BbL) type interference fringes, we consider the diffraction intensity in the Bragg-Laue (BL) case, first. The trajectory of the pseudo-refracted wave in the BL case has already been described in Fig. 6.28. Figure 8.10 shows the diffraction conditions when one end of a crystal is cut perpendicular to the incident surface. The incident wave satisfies the boundary conditions in the Bragg case, and the reflected wave also satisfies the Bragg case boundary conditions on the incident surface. When the wave comes out from the lateral surface, the boundary condition in the Laue case is satisfied.

We assume that the incident wave is a plane wave with a finite width. We write the intensity of the incident wave per unit area, i.e., the energy flux as P_0, the diffracted intensity from the incident point as P_g, the intensity of the emitted beam in the direction of diffraction from the lateral surface as P_{lg}, and that in the transmission

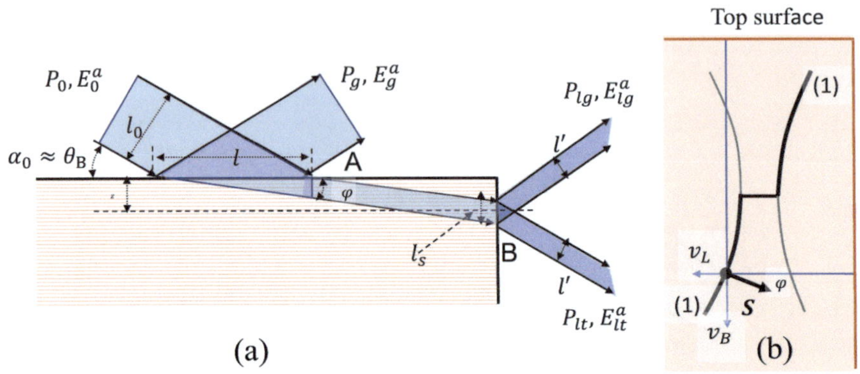

Fig. 8.10 Diffraction conditions in the symmetric Bragg-Laue case. (**a**) Beam geometry in the crystal whose one end is cut perpendicular to the incident surface. (**b**) The pseudo-refracted wave excited at a tie-point on branch (1) of the dispersion surface

direction as P_{lt}. The diffraction intensity in the Bragg case is given by (6.101) as

$$\frac{P_g}{P_0} = \left|\frac{E_g^a}{E_0^a}\right|^2 = \left|\frac{D_g^{1(1)}}{D_0^{1(1)}}\right|^2 = \left|r^{(1)}\right|^2. \tag{8.16}$$

The intensities of the emitted beam P_{lg} and P_{lt} from the lateral surface are respectively given by

$$\frac{P_{lg}}{P_0} = \left|\frac{E_{lg}^a}{E_0}\right|^2 \frac{l'}{l_0},\ \frac{P_{lt}}{P_0} = \left|\frac{E_{lt}^a}{E_0}\right|^2 \frac{l'}{l_0}. \tag{8.17}$$

Since

$$l_0 = l\sin\theta_B,\ l = \frac{l_s}{\tan\varphi},\ l_s = \frac{l'}{\cos\theta_B},$$

we get

$$\frac{l'}{l_0} = \frac{\tan\varphi}{\tan\theta_B} = \frac{1-\left|r^{(1)}\right|^2}{1+\left|r^{(1)}\right|^2}. \tag{8.18}$$

The boundary condition on the lateral surface is expressed by

$$D_g^{1(1)}\exp\left(-ik_{0z}^{(1)}z\right) = E_{\mathrm{lg}}^a\exp(-iK_{0z}z), \tag{8.19}$$

$$D_0^{1(1)}\exp\left(-ik_{0z}^{(1)}z\right) = E_{lt}^a\exp(-iK_{0z}z). \tag{8.20}$$

For an absorbing crystal, from the relation in (7.85), we can write

$$\left|\exp\left(-k_{0zi}^{(1)}z\right)\right|^2 = \exp\left(2Y_i^{(1)}z\right).$$

Then, we get

$$\frac{P_{\mathrm{lg}}}{P_0} = \left|r^{(1)}\right|^2 \frac{1-\left|r^{(1)}\right|^2}{1+\left|r^{(1)}\right|^2} \exp\left(2Y_i^{(1)}z\right), \tag{8.21}$$

$$\frac{P_{lt}}{P_0} = \frac{1-\left|r^{(1)}\right|^2}{1+\left|r^{(1)}\right|^2} \exp\left(2Y_i^{(1)}z\right). \tag{8.22}$$

If we ignore the absorption effect, we have

$$\frac{P_g}{P_0} + \frac{P_{lg}}{P_0} + \frac{P_{lt}}{P_0} = 1$$

and the law of energy conservation is satisfied.

An example of the calculated diffraction intensities for an absorbing crystal is shown in Fig. 8.11 by assuming $k_a = 0.1$. The FWHM (ΔW) of the peak of P_g/P_0 is approximately 2, while that of P_{lt}/P_0 is 0.5 and that of P_{lg}/P_0 is 0.2. Due to the influence of anomalous transmission, the diffraction intensities from the lateral surface become larger for $W < -1$. In contrast, these diffraction intensities become small due to anomalous absorption for $W > 1$. The rocking curves of diffraction from the lateral surface was experimentally observed by Yoshizawa et al. [5] using a Ge crystal.

8.3.2 Multiple-Bragg-Laue (MBL) Interference Fringes

We discuss the interference fringes caused by the angular amplification effect, which are observed in the diffraction from the lateral surface of a parallel plate crystal with a finite size. As shown in Fig. 8.12a, one beam that directly reaches point C on the lateral surface is the pseudo-refracted wave ($\boldsymbol{S}_{\mathrm{BL}}$) in the BL case and another beam that reaches point C after reflected at point E on the bottom surface is a pseudo-refracted wave ($\boldsymbol{S}_{\mathrm{BbL}}$) in the BbL case (satisfying diffraction condition in the Bragg case on the incident surface, and that in the Bragg case on the bottom surface, and that in the Laue case on the lateral surface). We write the distance from the incident point A to the rightmost point B on the incident surface as x_L, and the distance between points B and C as z. In the Kato fringe, the two pseudo-refracted waves pass through the same orbit, and the interference fringes are due to the difference in wave number

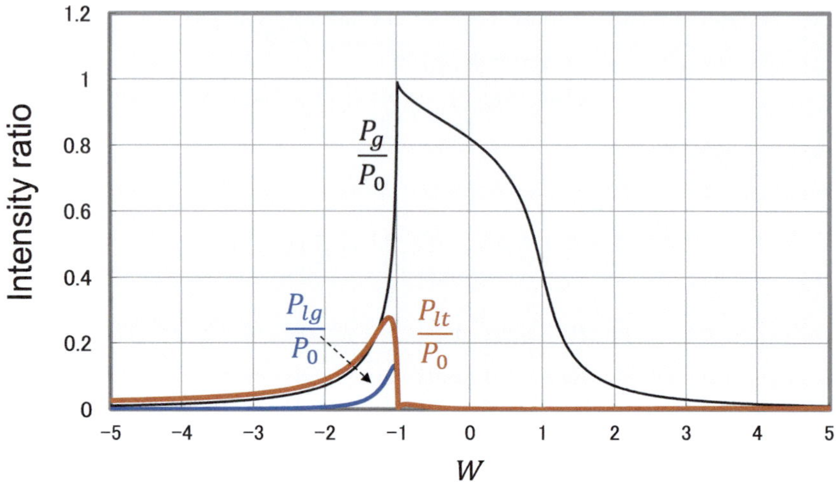

Fig. 8.11 Diffracted intensities in the symmetric Bragg-Laue case for $k_a = 0.1$. The distance from the incident point to the exit point on the lateral surface is 1 mm. The Bragg angle is 16.9°

between the two pseudo-refracted waves. In the MBL interference fringes, the two pseudo-refracted waves have different wave numbers and different trajectories.

Figure 8.12a, c show the relations between the point on the lateral surface where the pseudo-refracted wave reaches and the tie-point where the wave is excited. Let's denote the pseudo-refracted wave in the BL case as $\boldsymbol{S}_{\mathrm{BL}}$ and that in the BbL case as $\boldsymbol{S}_{\mathrm{BbL}}$. When point C where $\boldsymbol{S}_{\mathrm{BL}}$ reaches changes from $z = 0$ to H, the corresponding tie-point moves from d_0 to d_1 on the dispersion surface as shown in Fig. 8.12c. $\boldsymbol{S}_{\mathrm{BL}}$ excited at tie-point d_0 is parallel to the surface. After passing through point E, $\boldsymbol{S}_{\mathrm{Bb}}$ could reach point F if there were the lateral surface. When the point F moves from $z = H$ to $2H$, the corresponding tie-point moves from d_1 to d_2. In the actual situation, $\boldsymbol{S}_{\mathrm{Bb}}$ is reflected at point E and $\boldsymbol{S}_{\mathrm{BbL}}$ reaches point C. When the point C changes from $z = H$ to 0, the corresponding tie-point moves from d_1' to d_2'.

We write the electric field of the wave emitted in the direction of diffracted wave from the lateral surface as E_{lg}^a and that emitted in the direction of transmitted wave as E_{lt}^a. We write the electric field of the wave emitted in the direction of diffraction from the lateral surface in the BL case as $E_{g,\mathrm{BL}}^a$ and that in the BbL case as $E_{g,\mathrm{BbL}}^a$, and express them as

$$E_{g,\mathrm{BL}}^a = \left| E_{g,\mathrm{BL}}^a \right| \exp(-i\theta_1), \tag{8.23}$$

$$E_{g,\mathrm{BbL}}^a = \left| E_{g,\mathrm{BbL}}^a \right| \exp(i\phi) \exp(-i\theta_2). \tag{8.24}$$

Using the boundary condition on the lateral surface, we can write

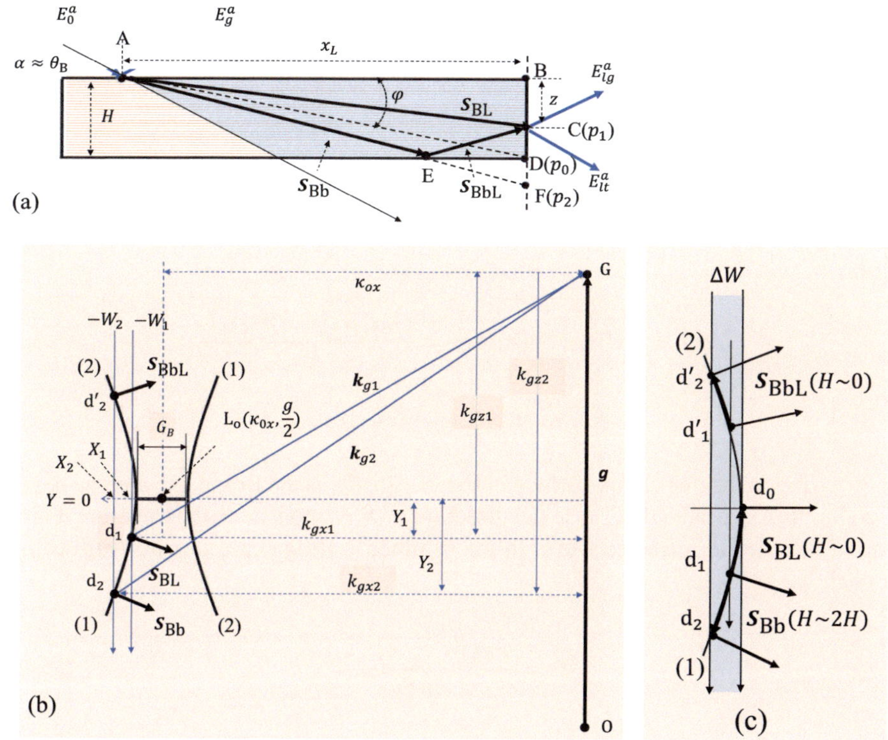

Fig. 8.12 **a** Beam geometry for MBL interference fringes in real space. **b** The pseudo-refracted waves in the BL case ($\boldsymbol{S}_{\text{BL}}$), the Bb case ($\boldsymbol{S}_{\text{Bb}}$) and the BbL case ($\boldsymbol{S}_{\text{BbL}}$), and their corresponding tie-points. (**c**) The range of tie-points for $\boldsymbol{S}_{\text{BL}}$, $\boldsymbol{S}_{\text{Bb}}$ and $\boldsymbol{S}_{\text{BbL}}$

$$\left|E^a_{g,\text{BL}}\right| = \left|\Omega_1 D_g^{1(1)}\right|, \left|E^a_{g,\text{BbL}}\right| = \left|\Omega_2 D_g^{2(1)}\right| \tag{8.25}$$

Here, Ω_j ($j = 1$ or 2) is the correction factor for the cross section that appears in the BL case and is $\left(l'_j/l_0\right)^{1/2}$. In (8.23) and (8.24), the phase ϕ represents the phase change due to the reflection at the bottom surface, and the phases θ_1 and θ_2 are expressed as

$$\theta_1 = \int \boldsymbol{k}_{g1} \cdot d\boldsymbol{r} = k_{gx1}x_L + k_{gz1}z, \tag{8.26}$$

$$\theta_2 = \int \boldsymbol{k}_{g2} \cdot d\boldsymbol{r} = k_{gx2}x_L + k_{gz2}(2H - z). \tag{8.27}$$

Since the amplitude of the electric field of the wave emitted in the direction of diffraction from the lateral surface E^a_{lg} is the sum of those in the BL case and in the BbL case, we get

$$\left|E^a_{lg}\right|^2 = \left|E^a_{g,\mathrm{BL}} + E^a_{g,\mathrm{BbL}}\right|^2 = \left|E^a_{g,\mathrm{BL}}\right|^2 + \left|E^a_{g,\mathrm{BbL}}\right|^2 + 2\left|E^a_{g,\mathrm{BL}}\right|\left|E^a_{g,\mathrm{BbL}}\right|\cos(\Delta\theta - \phi) \tag{8.28}$$

Here, we write

$$\Delta\theta = \Delta\theta_x + \Delta\theta_z, \tag{8.29}$$

with

$$\Delta\theta_x = -(|X_2| - |X_1|)x_L = \frac{G_\mathrm{B}(|W_1| - |W_2|)x_L}{2}, \tag{8.30}$$

$$\Delta\theta_z = |Y_2|(2H - z) - |Y_1|z, \tag{8.31}$$

by referring to the dispersion surface in Fig. 8.12b. G_B is the Bragg gap and the negative sign in $\Delta\theta_x$ is added because the direction of X is opposite to the traveling direction of the pseudo-refracted wave. In the symmetric Bragg case of a non-absorbing crystal, we get

$$Y = \frac{G_\mathrm{B}\left(W^2 - 1\right)^{1/2}}{2\tan\theta_\mathrm{B}}. \tag{8.32}$$

The electric field of the wave emitted in the transmitted wave direction can be obtained by using

$$\left|E^a_{t,\mathrm{BL}}\right| = \left|\Omega_1 D_0^{1(1)}\right|, \left|E^a_{t,\mathrm{BbL}}\right| = \left|\Omega_2 D_0^{2(1)}\right|. \tag{8.33}$$

The electric displacement density $D_0^{i(j)}$ and $D_g^{i(j)}$ here have already been obtained in the partial wave expansion in Chap. 6.

Figure 8.13 shows an example of the calculated results for Si 220 reflection without absorption, showing MBL interference fringes. The X-ray energy is assumed to be 11.1 keV. The solid black line is the intensity of the wave emitted in the diffracted wave direction, and the brown solid line is that in the transmitted wave direction. The contrast of MBL interference fringes in the transmitted wave direction is weaker than that in the diffracted wave direction. We call the interference fringes in the diffracted wave direction as IFLSD (Interference Fringes from Lateral Surface for Diffracted wave direction), and those in the transmission direction as IFLST (IFLS for Transmitted wave direction).

For observing MBL interference fringes, we normally choose the diffraction condition where the anomalous transmission occurs and the path length of the pseudo-refracted wave is long. If we look at the dispersion surface for $u = 0.98$ ($k_a = 0.1$) in Fig. 7.13, anomalous transmission occurs when X is on the positive side (W is on the

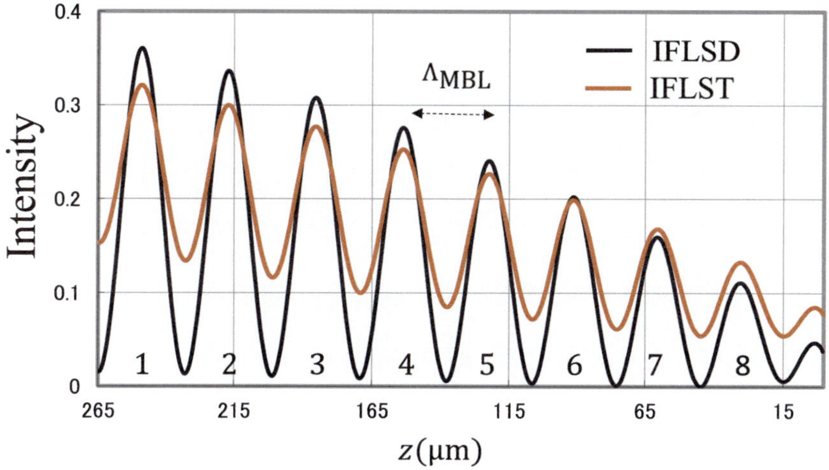

Fig. 8.13 The calculated MBL interference fringes for Si 220. $\theta_B = 16.9°$, $\Lambda_0 = 22$ μm, $x_L = 4.3$ mm and $H = 265$ μm

negative side). When X is positive, the dispersion surface can be approximated by a hyperbola. Then, the interference fringes under anomalous transmission condition for a weakly absorbing crystal can be approximated by those for a non-absorbing crystal. As for the divergence of the incident beam, for observing MBL interference fringes, the condition of $|\Delta W| = |W(z = 2H)| - 1(z = 0)$ is sufficient as shown in Fig. 8.12c. Under the conditions in Fig. 8.13, ΔW is approximately 0.1 and the corresponding divergence angle is about 0.16 s (0.8 μrad).

Figure 8.14a, b show the reflection topography of Si 220 reflection by using synchrotron radiation showing MBL interference fringes. In the experiment, synchrotron radiation X-rays obtained by a deflecting magnet is monochromatized to 11.1 keV by a two-crystal monochromator using Si 111 reflection, and X-rays in the diffracted wave direction and transmitted wave direction from the lateral surface are recorded using a nuclear emulsion plate. Figure 8.14a shows IFLSD and (b) shows IFLST. Both interference fringes agree with the calculated fringes. Figure 8.14c shows the relationship between the distances where the interference fringes are recorded on the nuclear emulsion plate.

Hirano et al. [7] measured rocking curve of the diffracted wave in the direction of diffracted wave from the lateral surface $\left(E^a_{lg}\right)$ with IFLSD as well as those in the direction of transmitted wave $\left(E^a_{lt}\right)$ using a Ge crystal and compared them with the calculated results, showing good agreement between the two. Fukamachi et al. [6] measured MBL interference fringes using a Si crystal and compared them with the calculated results. Yan and Noyan [8] measured diffractions patterns in the BL, BbL and multiple Bragg-Bragg cases, and also measured IFLSD in the BbL case. They point out that these diffractions should be useful for the analysis of strain in a crystal.

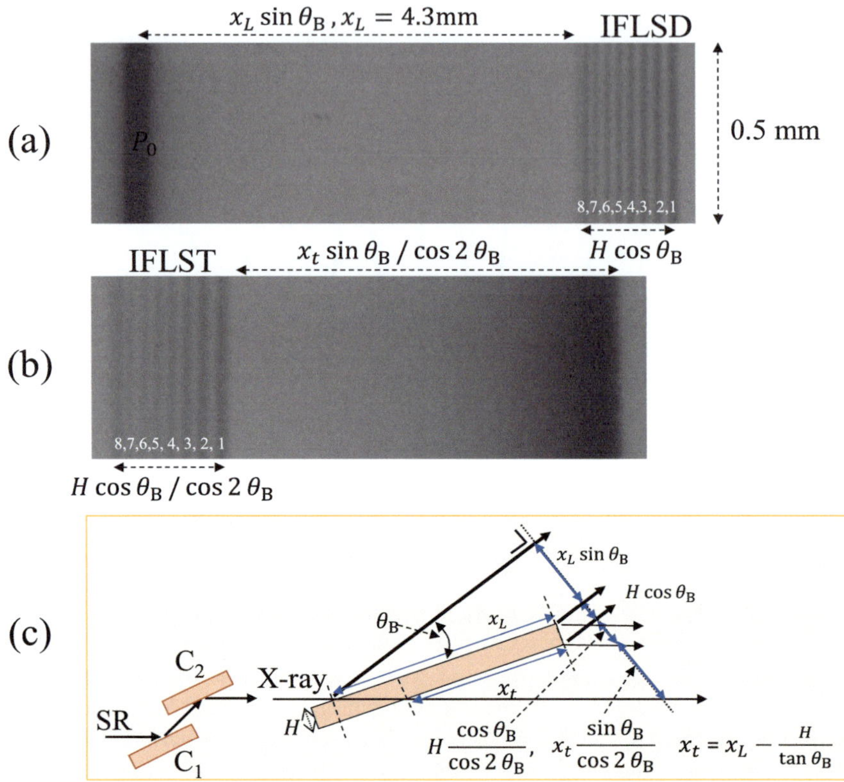

Fig. 8.14 MBL interference fringes of Si 220 recorded on a nuclear emulsion plate and the measuring system. X-ray energy is 11.1 keV, the crystal thickness $H = 280$ μm, and $x_L = 4.3$ mm. Photo recorded in the diffraction direction (**a**) and that in the transmission direction (**b**). (**c**) Layout diagram of the measuring system. The thick line shows the diffraction direction and the thin line the transmission direction. Λ_{MBL} is 32 μm The photographs are taken by the authors (unpublished). The experimental method is described in Fukamachi et al. [6]

8.3.3 Period of MBL Interference Fringes

We describe the period of MBL interference fringes. By putting $p = \tan\varphi/\tan\theta_B$ in Fig. 8.12a and assuming $p^2 < 0.1$, we obtain

$$W^2 = \frac{1}{1-p^2} \approx 1 + p^2, |W| \approx 1 + \frac{p^2}{2}. \tag{8.34}$$

In the figure, we write the value of p when the pseudo-refracted wave reaches the bottom surface at the right end (i.e. point D) as p_0, that when $\boldsymbol{S}_{\rm BL}$ reaches point C as p_1, and that when $\boldsymbol{S}_{\rm BbL}$ reaches point C as p_2. Then we have the following relations

$$p_1 = p_0 - \Delta p, p_2 = p_0 + \Delta p, z = H - \Delta z. \tag{8.35}$$

p_0 and Δp are given by

$$p_0 = \frac{H}{x_L \tan\theta_B}, \Delta p = \frac{\Delta z}{x_L \tan\theta_B}. \tag{8.36}$$

If we write the values of W corresponding to p_1 and p_2 as W_1 and W_2, we obtain the difference in W as

$$\Delta W = |W_1| - |W_2| = -2p_0\Delta p = \frac{-2H\Delta z}{x_L^2 \tan^2\theta_B}. \tag{8.37}$$

From (8.30), we get

$$\Delta\theta_x = \frac{-G_B H \Delta z}{x_L \tan^2\theta_B}. \tag{8.38}$$

Since $W^2 \approx 1 + p^2$, (8.32) becomes

$$Y = \frac{G_B p}{2\tan\theta_B}, \tag{8.39}$$

$|Y_1|$ and $|Y_2|$ in (8.32) are written using (8.39) as

$$|Y_1| = \frac{G_B}{2\tan\theta_B}(p_0 - \Delta p), |Y_2| = \frac{G_B}{2\tan\theta_B}(p_0 + \Delta p). \tag{8.40}$$

If we write $z_1 = H - \Delta z$ and $z_2 = H + \Delta z$, we get

$$\Delta\theta_z = \frac{G_B}{2\tan\theta_B}(p_2 z_2 - p_1 z_1) = \frac{2G_B H \Delta z}{x_L \tan^2\theta_B}, \tag{8.41}$$

$$\Delta\theta = \Delta\theta_x + \Delta\theta_z = \frac{G_B H \Delta z}{x_L \tan^2\theta_B}. \tag{8.42}$$

The period of the MBL interference fringes Λ_{MBL} is given by

$$\Lambda_{MBL} = \frac{\Lambda_0 x_L \tan^2\theta_B}{H}. \tag{8.43}$$

In the case of Fig. 8.13, $\Lambda_{MBL} = 32$ µm.

8.3.4 Borrmann-Lehmann Interference Fringes

We explain the principle of Borrmann-Lehmann interference fringes [9] based on the interference fringes in the Laue-Bragg case. Figure 8.15a shows the dispersion surface and the pseudo-refracted waves excited at tie-points on the dispersion surface in the reciprocal space, and (b) shows the beam trajectory of the pseudo-refracted waves in the real space. The incident wave satisfies the boundary condition in the symmetric Laue case. Let the thickness of the crystal be H, and the distance from the upper left edge of the crystal to the incident point A be z. The range of the divergence (ΔW) of the incident beam is shown in light blue in (a).

There are two pseudo-refracted waves from point A to point C on the exit lateral surface; one is $S_{LL}^{(1)}$ excited at tie-point d_1 of branch (1) and the other is $S_{LL}^{(2)}$ excited at tie tie-point d_2 of branch (2). There are two pseudo-refracted waves reaching to point C after reflected at point B on the top surface: one is $\boldsymbol{S}_{LbL}^{(1)}$ excited at tie-point d_1' of branch (1) and the other is $S_{LbL}^{(2)}$ excited at the tie-point d_2'' of branch (2). The interference between four pseudo-refracted waves results in the so-called Borrmann-Lehmann interference fringes. In the actual experiment, for observing the Borrmann-Lehmann interference fringes, it is better to select only branch (2) so that the Borrmann effect occurs in an absorbing crystal. In the case of Fig. 7. 5, branch (1) shows anomalous absorption and branch (2) anomalous transmission. The pseudo-refracted beams $S_{LL}^{(1)}$ and $S_{LbL}^{(1)}$ cannot propagate over long distance. The pseudo-refracted waves $S_{LL}^{(2)}$ and $S_{LbL}^{(2)}$ reach point C and interference fringes between these two waves appear in the emitted beam from the exit surface. The fringe spacing Λ_{LbL} of the Borrmann-Lehmann interference fringes is given by

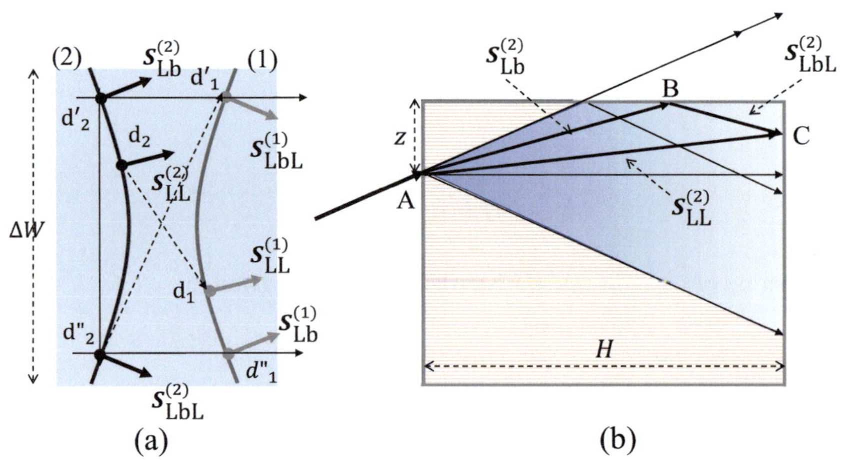

Fig. 8.15 A schematic diagram of the formation of Borrmann-Lehmann interference fringes. (**a**) The directions of pseudo-refracted beams excited at tie-points on the dispersion surface. (**b**) Geometry of pseudo-refracted beams in real space

$$\Lambda_{\mathrm{LbL}} = \frac{\Lambda_0 H \tan^2 \theta_B}{z} \tag{8.44}$$

In (8.43), if the thickness H is replaced with x_L in (8.42) and z is replaced with H, it can be seen that Λ_{LbL} and Λ_{MBL} are given in the same form.

For observing Borrmann-Lehmann interference fringes, the pseudo-refracted wave of branch (1) should be suppressed by the effect of anomalous absorption. For observing MBL interference fringes, one of the two branches is chosen by reducing ΔW of the incident beam. Lang et al. [10] observed Borrmann-Lehmann interference fringes using a diamond crystal. In this case, since the absorption effect is small, anomalous absorption is suppressed, resulting in interference fringes between four waves of branch (1) and branch (2). Authier [1] gives a detailed explanation of Borrmann-Lehmann interference fringes.

8.4 Wave-Packet and Group Velocity

In an actual experiment, we use X-rays from an X-ray tube, or synchrotron radiation X-rays obtained by a deflecting magnet. These X-rays are quasi-monochromatic, and neither ideal plane waves nor ideal spherical waves. These X-rays are generated in pulses and are localized waves that travel in space, as shown in Fig. 8.16a, which are called wave-packets or wave trains. The wave number k of a wave-packet is continuously distributed, and the electric field $E(x, t)$ can be expressed by

$$E(x, t) = \int A(k) \exp\{i(\omega t - kx)\} dk. \tag{8.45}$$

Here, $A(k)$ has values only in a limited range around a certain $\bar{k}$ as shown in Fig. 8.16b.

In a dispersive medium, by assuming $\omega = \omega(k)$, we expand $\omega(k)$ as

$$\omega(k) = \omega\left(\bar{k}\right) + \left(\frac{d\omega}{dk}\right)_{k=\bar{k}} \left(k - \bar{k}\right) + \cdots. \tag{8.46}$$

If the higher-order terms can be ignored, we can write the exponent part as

$$\begin{aligned} \omega t - kx &= \left\{\omega\left(\bar{k}\right) + \left(\frac{d\omega}{dk}\right)_{\bar{k}} \left(k - \bar{k}\right)\right\} t - kx \\ &= \omega\left(\bar{k}\right) t - \bar{k}x + \left(k - \bar{k}\right)\left(v_{gr} t - x\right). \end{aligned} \tag{8.47}$$

In (8.46), we put

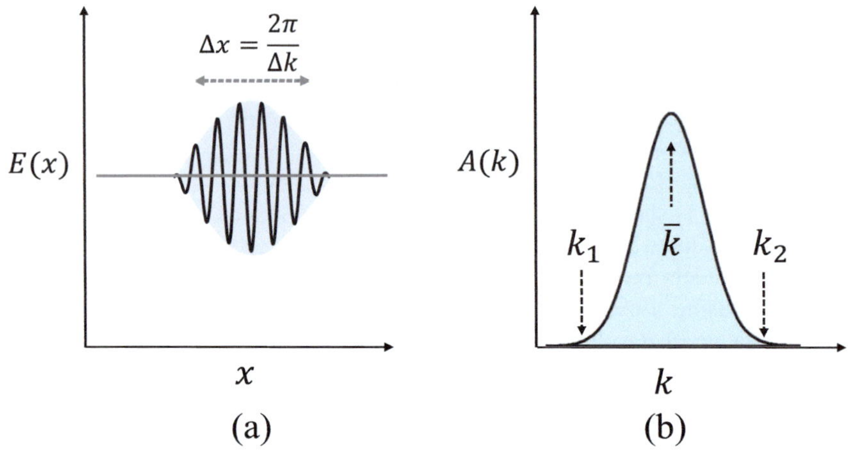

Fig. 8.16 **a** Schematic illustration of a wave-packet. **b** Spectrum of a wave-packet

$$v_{gr} = \left(\frac{d\omega}{dk}\right)_{\bar{k}}. \tag{8.48}$$

Using these equations, we write (8.45) as

$$E(x,t) = \exp\left\{i\left(\omega\left(\bar{k}\right)t - \bar{k}x\right)\right\} \int A(k)\exp\left\{i\left(k-\bar{k}\right)\left(\left(v_{gr}t - x\right)\right\}dk. \tag{8.49}$$

This represents that the wave-packet is regarded as a plane wave having its wave number $\bar{k}$ and angular frequency $\omega\left(\bar{k}\right)$, which shows the same amplitude at $v_{gr}t - x =$ constant. The wave front moves at a velocity of v_{gr} which is called the group velocity. This represents the flow of energy, and when it propagates through three-dimensional space, the velocity is expressed by

$$\boldsymbol{v}_{gr} = \text{grad}\ \ \omega(\boldsymbol{k}) = \nabla\omega(\boldsymbol{k}) \tag{8.50}$$

The direction of $\boldsymbol{v}_{gr}$ is perpendicular to the curved surface with $\omega(\boldsymbol{k}) =$ constant. The dispersion surface we have been dealing with so far is nothing but the locus of $\boldsymbol{k}_0$ or $\boldsymbol{k}_g$ with respect to ω in the reciprocal space. The curved surface with $\omega(k) =$ constant is the dispersion surface itself. That is, the direction of the pseudo-refracted wave of energy flow is perpendicular to the dispersion surface.

The phase velocity v_p can be expressed by

$$v_p = \frac{\omega}{k} = \frac{\omega}{n_0(\omega)K_0} = \frac{c}{n_0(\omega)} \tag{8.51}$$

Since the refractive index $n_0(\omega)$ of X-ray is less than 1, its phase velocity in the medium is larger than the light velocity. The wave number in a crystal can be expressed by

$$k = n_0(\omega)K_0 = \{1 - \delta(\omega)\}K_0 = \frac{1}{c}\left(\omega - \frac{a}{\omega}\right). \tag{8.52}$$

If resonance scattering is ignored, $\delta(\omega)$ is given by (4.19) as

$$\delta(\omega) = \frac{r_e N_e \lambda^2}{2\pi V_c} = \frac{a}{\omega^2}, \text{ with } a = \frac{2\pi r_e c^2 N_e}{V_c}. \tag{8.53}$$

Using (8.53), we have obtained the right-hand side of (8.52). Since $\delta(\omega)$ is of the order of 10^{-6}, the group velocity v_{gr} becomes

$$v_{gr} = \frac{d\omega}{dk} = \frac{c}{1 + \delta(\omega)} = n_0(\omega)c. \tag{8.54}$$

This means that the group velocity is lower than the light velocity. Using (8.51) and (8.54), the following relation is obtained

$$v_{gr}v_p = c^2 \tag{8.55}$$

The X-rays generated in pulses are wave-packets, and the phase relation between individual wave-packets is random. For this reason, there are several points that must be noted when comparing the results of the dynamical theory of diffraction developed under ideal plane waves with the experimental results of X-ray diffraction of wave-packets. One of them is that the glancing angle of the incident X-ray must have a divergence angle of $\Delta\alpha_0$ in order to have a width of $|\Delta W|$ on the dispersion surface. To obtain a divergence angle of $\Delta\alpha_0$ from a crystal monochromator, the corresponding Bragg angle must have an angular width of $\Delta\theta_B$. The wavelength obtained from the monochromator has a wavelength width $\Delta\lambda$ due to $\Delta\theta_B$. That is, an ideal monochromatic plane wave cannot be obtained from a crystal monochromator, nor can a monochromatic spherical wave be obtained. This issue will be discussed in detail in 8.6. Another is that the phase relations between individual wave-packets are random. For discussing the coherence, it is necessary to estimate the degree of overlap of wave-packets. As we know the uncertainty relation

$$\Delta x \Delta k > 2\pi \tag{8.56}$$

between the uncertainty Δx of position in real space and the uncertainty $\Delta k = k_2 - k_1$ of wavenumber in reciprocal space. In order to increase spatial coherence, it is necessary to narrow the spectral width Δk as much as possible. Control of the spectral width can be achieved using a crystal monochromator. We have another uncertainty relation

$$\Delta t \Delta \omega > 2\pi \tag{8.57}$$

between the uncertainty Δt of time and the uncertainty $\Delta \omega$ of angular frequency. This plays an important role in considering the lifetime in resonant scattering.

8.5 Coherence

We consider temporal coherence and spatial coherence.

8.5.1 Temporal Coherence (Longitudinal Coherence)

The measure of the longitudinal interference region between two wave-packets is the longitudinal coherence length (or temporal coherence length) given by

$$l_L = c\Delta t = \frac{\lambda^2}{2\Delta\lambda}. \tag{8.58}$$

This longitudinal coherence length can be derived as follows.

As shown in Fig. 8.17, if two waves with different wavelengths have opposite phases when they travel the distance OA $= l_L$, they will become in-phase when they travel OB $= 2l_L$. When this distance $2l_L$ is equal to N times the wavelength λ as well as $N + 1$ times the wavelength $\lambda - \Delta\lambda$, we get

$$2l_L = N\lambda = (N + 1)(\lambda - \Delta\lambda). \tag{8.59}$$

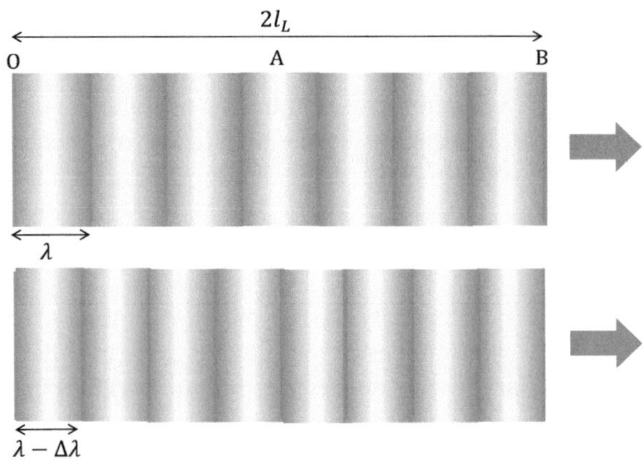

Fig. 8.17 A schematic diagram to show longitudinal coherence. At point A, the wave with its wavelength λ and that with $\lambda - \Delta\lambda$ is anti-phase, while they are in-phase at point B

Since $N \approx \lambda/\Delta\lambda$ is obtained from (8.59), the third equation of (8.58) can be obtained.

8.5.2 *Spatial Coherence (Transvers Coherence)*

As shown in Fig. 8.18, there are two positions P_1 and P_2 which are separated by distance D. We will consider the case where two plane waves of the same wave length passing through positions P_1 and P_2 but the propagation directions are slightly different. The difference in the propagation direction of these two waves at point B is assumed to be $\Delta\theta$. The minimum distance between the maximum and the minimum of the composed wave of these two waves is called the transverse coherence length, and is denoted by l_T. From Fig. 8.18, l_T is given by

$$l_T = \frac{1}{2}\frac{\lambda}{\Delta\theta} = \frac{\lambda}{2}\left(\frac{R}{D}\right). \tag{8.60}$$

For the third-generation synchrotron radiation obtained from a deflection magnet, the transverse coherence length l_T for X-rays with a wavelength of 0.1 nm at a position approximately 20 m away from the light source is 20 μm when the vertical light source size is 50 μm. When X-rays are monochromatized to about $\Delta\lambda/\lambda = 10^{-5}$ using a crystal monochromator, the longitudinal coherence length l_L is about 5 μm. If the optical system satisfies these conditions, interference effects can be observed. It is noted that the condition of $l_T \approx 20$ μm can be easily satisfied, but with a usual two-crystal monochromator, $\Delta\lambda/\lambda$ is approximately 10^{-4} and the condition $\Delta\lambda/\lambda = 10^{-5}$ may not be easily satisfied.

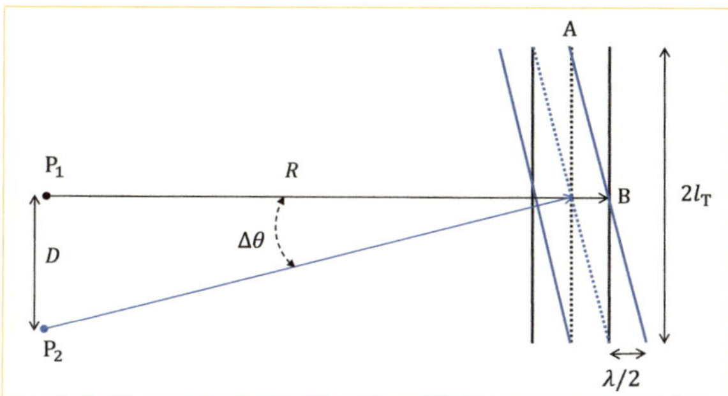

Fig. 8.18 Transverse coherent length. The thick solid lines show the wave fronts of the maximum intensity, and the dashed ones show that of the minimum intensity. At point A, the wave coming from P_1 show the opposite phase to that from P_2, while at point B, these two waves show the same phase

8.6 MBL Interferometer and Its Applications

8.6.1 Correspondence Between Rotating Crystal Method and Wavelength Dispersion Method

In Fig. 8.14, MBL interference fringes can be observed when $|\Delta W|$ is about 0.1, which corresponds to the divergence angle $\Delta\alpha_0$ of the incident glancing angle of about 1 μrad.

In the symmetric Bragg case for σ-polarized X-ray, W is related to α_0 using (6.76) as

$$W = \frac{\sin 2\theta_\mathrm{B}}{|\chi_g|}\left\{\alpha_0 - \left(\theta_\mathrm{B} + \frac{|\chi_0|}{\sin 2\theta_\mathrm{B}}\right)\right\}. \tag{8.61}$$

Then $\Delta\alpha_0$ is related to ΔW as

$$\Delta\alpha_0 = \frac{|\chi_g|}{\sin 2\theta_\mathrm{B}}\Delta W. \tag{8.62}$$

Figure 8.19 shows these relations. T'_0 and T''_0 are dispersion surfaces in vacuum when the incident wave vector is $\boldsymbol{K}_0$ and $\boldsymbol{K}_0 - \Delta\boldsymbol{K}_E$, respectively. When the wave vector of the incident wave is $\boldsymbol{K}_0$, the tie-point of the incident wave is A.

When the crystal is rotated by $\Delta\alpha_0$, $\boldsymbol{K}_0$ changes to $\boldsymbol{K}_0 - \Delta\boldsymbol{K}_\alpha$, and the tie-point moves from A to B. For a monochromatic X-ray, $|K_0| = |K_0 - \Delta K_\alpha|$ should be satisfied. This method of changing the glancing angle by rotating the crystal is called the rotating crystal method. In the experiment of MBL interference fringe, the crystal is fixed and $\Delta\alpha_0$ is zero. One way to change W without changing α_0 is to use white X-rays (X-rays having the continuous range of wavelengths), known as the Laue method. This is called a wavelength-dispersive method or an energy-dispersive method. As shown in Fig. 8.20, if the width of diffraction angle restricted by the slit is $\Delta\theta_B$ and its wavelength width is $\Delta\lambda$, we get

$$\frac{|\Delta\lambda|}{\lambda} = \frac{|\Delta E|}{E} = \left|\frac{\Delta\boldsymbol{K}_E}{\boldsymbol{K}_0}\right| = \frac{|\Delta\theta_\mathrm{B}|}{\tan\theta_\mathrm{B}}, \tag{8.63}$$

by using (1.2) of the Bragg condition. Here, λ is the wavelength and E is the energy of the X-ray. In Fig. 8.19, the parallel component ($\overline{\mathrm{AC}}$) of $\Delta\boldsymbol{K}_\alpha$ to the line connecting $\mathrm{L_o}$ (Lorentz point) and $\mathrm{L_a}$ (Laue point) is given as

$$\overline{\mathrm{AC}} = |\Delta\boldsymbol{K}_\alpha|\sin\theta_\mathrm{B} = \frac{|\Delta\boldsymbol{K}_E|}{\cos\theta_\mathrm{B}} = \frac{G_\mathrm{B}|\Delta W|}{2}. \tag{8.64}$$

This equation gives the connection between the crystal rotation method and the energy-dispersive method. When the energy of the incident X-ray is changed while

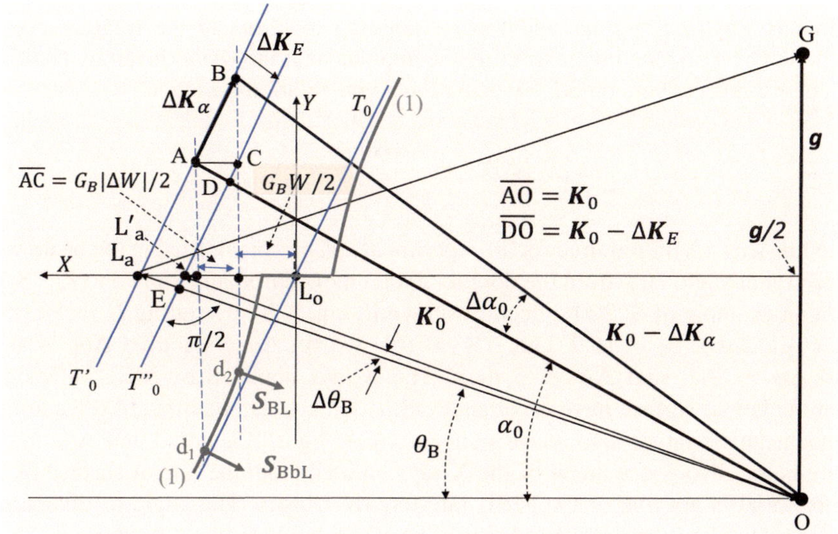

Fig. 8.19 Dispersion surface and beam geometry in reciprocal space. α_0 is the glancing angle of the incident wave and $\alpha_0 \approx \theta_B$. $|K_0| = |K_0 - \Delta K_\alpha|$ and $\angle$DAC $= \angle$CBA $\approx \theta_B$. Δ DAC and Δ $EL_aL'_a$ are similar to each other. d_1 and d_2 are tie-points inside of the crystal. $\overrightarrow{AO}$, $\overrightarrow{BO}$ and $\overrightarrow{L_aO}$ are perpendicular to T'_0

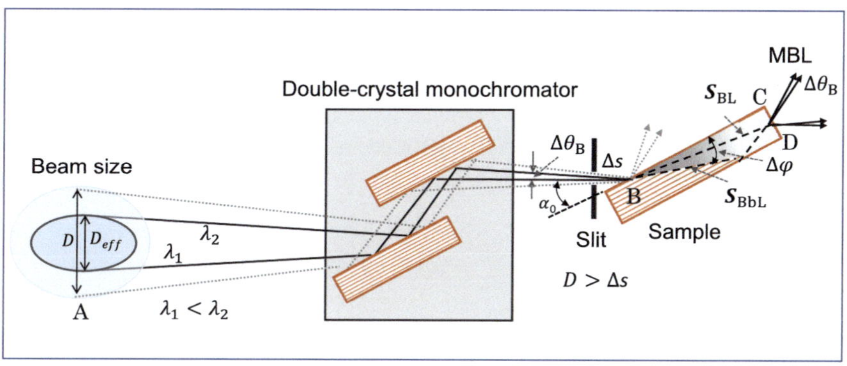

Fig. 8.20 A schematic diagram of MBL diffractometer. The slit width Δs just in front of the sample must be smaller than the beam size D (approximately 60 μm) of the synchrotron radiation source. A typical distance from the light source to the sample is approximately 30 m

keeping α_0 fixed, that is, the wave vector is changed from $\boldsymbol{K}_0$ to $\boldsymbol{K}_0 - \Delta\boldsymbol{K}_E$ (the tie-point moves from point A to point D), the results of the rotating crystal method and those of the energy-dispersive method are the same if $\overline{\mathrm{AC}} = |\Delta\boldsymbol{K}_E|/\cos\theta_\mathrm{B}$. As $|\Delta\boldsymbol{K}_\alpha| = \boldsymbol{K}_0|\Delta\alpha_0|$, by using (8.62) and (8.63), we get

$$|\Delta\theta_\mathrm{B}| = \sin^2\theta_\mathrm{B}|\Delta\alpha_0|. \tag{8.65}$$

In this way, the angular width corresponding to $|\Delta\alpha_0|$ in the rotating crystal method can be obtained by restricting the angular width to $|\Delta\theta_\mathrm{B}|$ given by (8.65) in the energy-dispersive method. By using the above relations, we get

$$|\Delta W| = \frac{2|\Delta\theta_\mathrm{B}|}{|\chi_g|\tan\theta_\mathrm{B}}. \tag{8.66}$$

In this way, $|\Delta W|$ required for an experiment can be obtained using the beam with its divergent angle $\Delta\theta_\mathrm{B}$ from the double-crystal monochromator shown in Fig. 8.20.

In an example of Si 220 reflection shown in Fig. 8.20, by setting $E = 11.1\mathrm{keV}$ ($\lambda = 0.112nm$), $\theta_\mathrm{B} = 16.9^\circ$ and $\Delta W = 0.1$, we get $\Delta\alpha_0 \approx 1$ μrad $\Delta\theta_\mathrm{B} \approx 0.08$ μrad $\Delta\varphi = 7.2^\circ$ and $\Delta E \approx 3$ meV. The X-rays obtained by MBL diffraction from the lateral surface have excellent collimation and can be thought of as almost monochromatic rather than quasi-monochromatic light. It is noted that $\Delta\theta_\mathrm{B}$ in this figure is the divergence angle of the X-rays emitted from the lateral surface of the sample, and is not one of the MBL interference fringes. The angle of divergence from a particular interference fringe will be explained in the next section.

8.6.2 Coherent Length of MBL Diffractometer

We discuss the beam width $\Delta\theta_\mathrm{B}$ and the energy width ΔE when the beam of the first peak of the MBL fringes is taken out from the MBL diffractometer as shown in Fig. 8.21. From (8.37), ΔW is written by

$$\Delta W = |W_2| - |W_1| = \frac{2H\Delta z}{x_L^2\tan^2\theta_\mathrm{B}}. \tag{8.67}$$

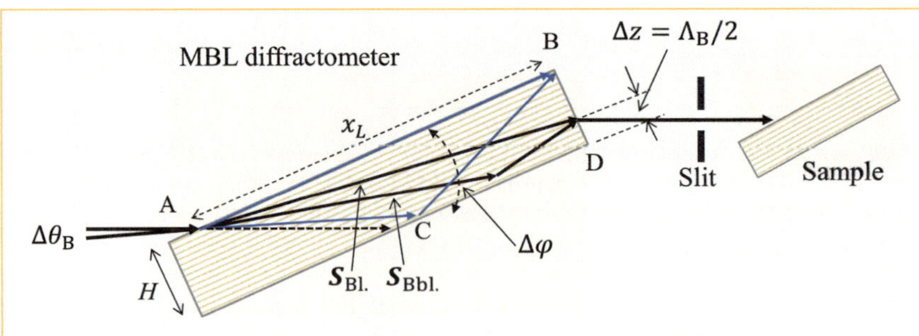

Fig. 8.21 MBL diffractometer. Si 220 reflection is used. The first peak of the MBL interference fringes from the lateral surface (BD) is extracted using a slit. $x_L = 4.3$ mm, $\Lambda_\mathrm{MBL}/2 = 0.016$ mm, $H = 0.28$ mm

The position of the first peak is $\Delta z = \Lambda_{\mathrm{MBL}}/2$, W_1 is given by the pseudo-refracted wave $\boldsymbol{S}_{\mathrm{BL}}$, and W_2 by the pseudo-refracted wave $\boldsymbol{S}_{\mathrm{BbL}}$, Using this ΔW, we can obtain $\Delta\theta_{\mathrm{B}}$ and ΔE of the beam taken from the first peak.

8.6.2.1 Longitudinal Coherent Length

We consider the longitudinal coherent length l_L using the difference in path lengths $\overline{\mathrm{AB}}$ and $\overline{\mathrm{ACB}}$ in Fig. 8.21, which give the maximum difference. By assuming that the thickness of the crystal H is 0.28 mm and the length from the incidence point A to the edge B of the crystal is 4.3 mm, we get $\Delta W \approx 0.1$ using (8.66). Since we get $\Delta\theta_{\mathrm{B}}$ from (8.65) and $\Delta E/E = 2.4 \times 10^{-7}$ using (8.63), we obtain $l_L = 230$ μm. It is noted that the path difference Δl between $\overline{\mathrm{AB}}$ and $\overline{\mathrm{ACB}}$ is 36 μm. As Δl is much smaller than l_L, the condition of the longitudinal coherent length is fully satisfied.

8.6.2.2 Transverse Coherent Length

If we set $\Delta\theta = \Delta\theta_{\mathrm{B}}$ in (8.60), we obtain the transverse coherent length l_T of 750 μm. This value is much larger than the actual effective source size of $D_{eff} = 2.4$ μm seen in the divergent beam of divergent angle $\Delta\theta_{\mathrm{B}}$ shown in Fig. 8.20. The transverse coherent length condition is also fully satisfied in this experimental system.

In Figs. 8.13 and 8.14, Λ_{MBL} is 32 μm at the first peak of the MBL interference fringes from the lateral surface in the transmitted wave direction (IFLST) using Si 220. By inserting $\Delta z = \Lambda_{\mathrm{MBL}}/2$ in Eq. (8.67), we get ΔW is 0.0051. Then, we obtain $\Delta\theta_{\mathrm{B}} = 3.9 \times 10^{-9}$ rad $\Delta E = 1.4 \times 10^{-4}$ eV $\left(\Delta E/E = 1.3 \times 10^{-8}\right)$ from (8.63). If we extract the first peak of the interference fringes and use this crystal as a monochromator, we can extract monochromatic X-rays on the μeV order with the large longitudinal coherent length of $l_L = 4$ mm.

8.6.3 *MBL Interferometer*

Figure 8.22 shows an interferometer using two MBL diffractometers. The first crystal C_1 is used as a monochromator and splitter, and the second crystal C_2 is used as an analyzer. Two beams are selected from the MBL interference fringes in the transmitted wave direction from C_1 using a slit, and the electric field of the lower beam is denoted as $\boldsymbol{E}_1$, and that of the upper beam as $\boldsymbol{E}_2$. We insert a phase plate into the beam path of $\boldsymbol{E}_1$. The electric field in the diffracted wave direction from C_2 denoted by $\boldsymbol{E}_{lg}$ is expressed by

$$\left|\boldsymbol{E}_{lg}\right|^2 \propto |\boldsymbol{E}_1|^2 + |\boldsymbol{E}_2|^2 + 2|\boldsymbol{E}_1||\boldsymbol{E}_2|\cos(\varphi). \tag{8.68}$$

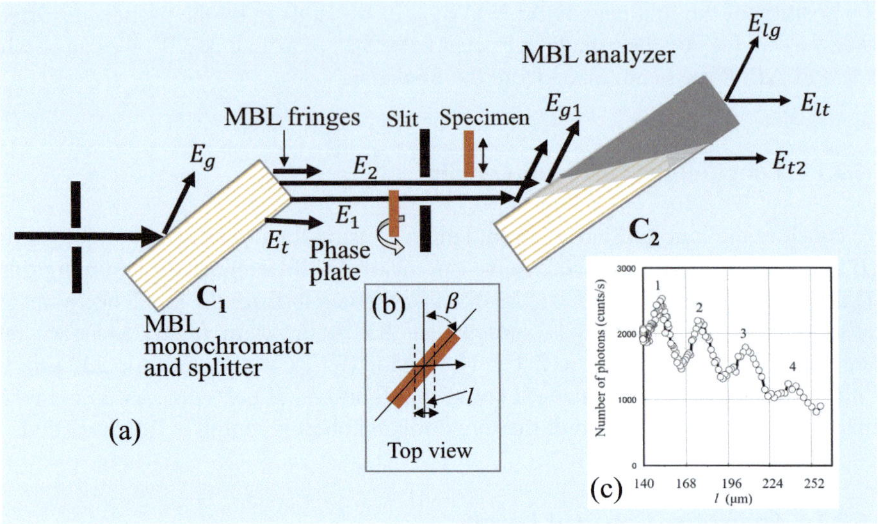

Fig. 8.22 **a** MBL interferometer. **b** Top view of the phase plate. If the thickness of the plate is H, $l = H/\cos\beta$. (**c**) Interference fringes observed by rotating the plate (The specimen is Si and $H = 140\mu m$). The vertical axis is $\left|E_{lg}\right|^2$. (**a**) is adopted from Fig. 8.2 in [11]. (**b**) and (**c**) are adapted from Fig. 8.4a, b in [11], respectively

Here, φ is the phase difference between the phase θ_1 of $|\boldsymbol{E}_1|$ and the phase θ_2 of $|\boldsymbol{E}_2|$ ($\varphi = \theta_2 - \theta_1$). By writing the refractive index of the phase plate as n_0 and the distance that the X-ray passes through the phase plate as l, and by putting $\delta = 1 - n_0$ and $l = l_0 + \Delta l$, we get

$$\theta_1 = 2\pi \frac{l_0 + \Delta l}{\lambda}(1 - \delta) = \theta_0 - \theta_c - \Delta\theta, \theta_2 = 2\pi \frac{l}{\lambda}. \tag{8.69}$$

Here, θ_0, θ_c and $\Delta\theta$ are respectively given by,

$$\theta_0 = 2\pi \frac{l}{\lambda}, \theta_c = 2\pi \frac{l_0\delta}{\lambda}, \Delta\theta = 2\pi \frac{\delta\Delta l}{\lambda}. \tag{8.70}$$

θ_0 is the phase shift when the X-ray passes through the distance l in vacuum, θ_c is the phase shift when it passes through the phase plate, and $\Delta\theta$ is the phase change due to the change in the passing distance through the phase plate by rotating it. As a result, we obtain

$$\varphi = \theta_2 - \theta_1 = \theta_c + \Delta\theta \tag{8.71}$$

To make $\Delta\theta$ between 0 and 2π by rotating the phase plate by β to increase the beam path passing through the phase plate by Δl, we must make Δl between 0 and

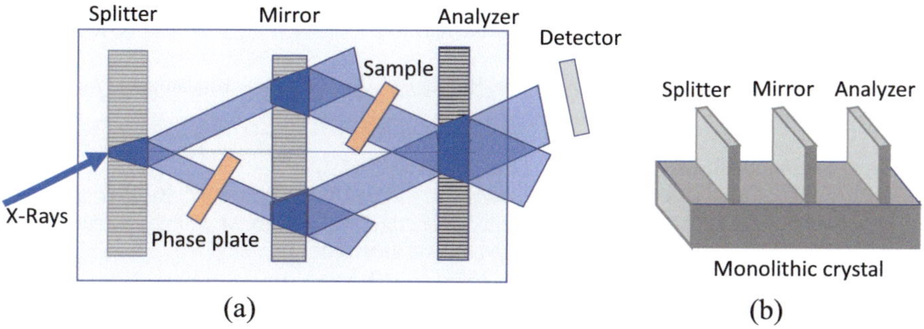

Fig. 8.23 **a** A schematic diagram to explain the principle of the LLL interferometer. **b** LLL interferometer fabricated from a monolithic crystal

λ/δ. For example, in the case of a Si ($Z = 14$) plate, if we set $\delta = 3.91 \times 10^{-6}$ and $\lambda = 0.112$ nm, we get $\Delta l = \lambda/\delta = 29$ μm. Figure 8.22c shows the interference fringes obtained by rotating the phase plate. If a sample is inserted into the path on the $\boldsymbol{E}_2$ side, the refractive index of that sample can be measured. (In the above, we implicitly assumed that the wavelengths of $\boldsymbol{E}_1$ and $\boldsymbol{E}_2$ are the same, but actually they are slightly different. However, the interpretation of the results in Fig. 8.22c still holds true).

A typical X-ray interferometer is one made from a monolithic Si single crystal [12] as shown in Fig. 8.23. As shown in Fig. 8.23b, the splitter, the mirror and the analyzer are cut out of a monolithic crystal and the path of X-rays remains unchanged at atomic sizes for long periods of time. The transmitted waves and diffracted waves from these three plates become coherent waves. Near the exit surface of the crystal, a standing wave with its period equal to the period of the reflecting planes is formed due to interference between the transmitted and the diffracted waves. On the incident plane of analyzer, another standing wave is formed by the two diffracted waves from the mirror. In Fig. 8.23a, if the sample is removed, the diffraction intensity of the analyzer can be changed by rotating the phase plate and changing the path length of X-ray beam. By inserting a sample, the refractive index of the sample can be measured, and the contrast caused by the phase difference due to rotation of the phase plate can be obtained. Several crystal interferometers, such as the so-called LLL, BB and BBB types, are described in detail by Bonze and Graeff [13].

References

1. A. Authier, *Dynamical Theory of X-ray Diffraction* (Oxford University Press, 2001)
2. N. Kato, J. Appl. Phys. **39**, 2225–2230 (1968)
3. N. Kato, J. Appl. Phys. **39**, 2231–2237 (1968)
4. N. Kato, *X-Ray Diffraction*, ed. by L.V. Azarov (McGraw-Hill, New York, 1974)
5. M. Yoshizawa, T. Fukamachi, K. Hirano, T. Oba, R. Negishi, K. Hirano, T. Kawamura, *Acta Cryst.* A**64**, 515–518 (2008)

6. T. Fukamachi, K. Hirano, R. Negishi, Y. Kanematsu, S. Jongsukswat, K. Hirano, T. Kawamura, *Acta Cryst.* A**67**, 154–159 (2011)
7. K. Hirano, T. Fukamachi, M. Yoshizawa, R. Negishi, K. Hirano, T. Kawamura, *Phys. Status Solidi.* A**206**, 1855–1859 (2009)
8. H. Yan, I.C. Noyan, J. Appl. Phys. **98**, 073527 (2005)
9. K. Lehmann, G. Borrmann, Z. Krist. **125**, 234–248 (1967)
10. A.R. Lang, G. Kowalski, A.P.W. Makepeace, M. Moore, Acta Cryst. A**42**, 501–510 (1986)
11. T. Fukamachi, S. Jongsukswat, Y. Kanematsu, K Hirano, R. Negishi, M. Shimojo, J. Dongying, K. Hirano, T. Kawamura, *J. Phys. Soc. Jpn.* **80**, 083001 (2011)
12. U. Bonse, M. Hart, Appl. Phys. Letter. **6**, 155–156 (1965)
13. U. Bonse, W. Graeff, *X-ray Optics*, ed. by H. -J. Queisser, (Springer-Verlang, Berlin Heidelberg New York, 1977), 93–143

Chapter 9
Dynamical Diffraction for Slightly Strained Crystals

The dynamical theory of diffraction described in the previous chapters is basically for discussing diffraction from a perfect crystal. In this chapter, we describe a dynamical theory of diffraction for discussing diffraction from a crystal with a constant strain gradient based on one proposed by Penning and Polder [1], with reference to the treatment of Gronkowski and Melgrenge [2] and that of Authier [3]. We also present some related experimental results.

The lattice constant of a crystal changes uniformly when the temperature changes. In such a case, the reciprocal lattice vector $|\boldsymbol{g}_h| = 2\pi/d$ will also change uniformly. The dynamical theory of diffraction described so far can be applied as is. However, when a plate-shaped crystal is elastically bent (curved crystal), the lattice strain differs depending on the location. The lattice constant can be regarded to be locally uniform in a curved crystal, but it changes slightly depending on the position, such as the depth from the crystal surface, and the strain in the crystal changes accordingly. We consider diffraction from such a crystal.

Authier [3] points out that the paths of pseudo-refracted beam in a crystal with a constant strain gradient are bent in the similar way to rays of light in the region of varying index of refraction, which is called mirage phenomenon. A mirage occurs when light is refracted in the atmosphere. For example, as shown in Fig. 9.1, when the earth is warmed by the heat of the sun and the temperature of the atmosphere decreases continuously as the height from the ground increases. Refracted light bends toward the side with a larger refractive index n_0 (lower temperature and higher density). As a result, light is totally reflected, making distant trees appear upside down. There is a relation between the wave vector of light $\boldsymbol{K}$ and the refractive index n_0 as

$$\delta\boldsymbol{K} \propto \nabla(n_0). \tag{9.1}$$

The penetration depth in the usual X-ray diffraction is about 10–100 μm. In a perfect crystal, the penetration depth can be extended to the order of centimeter even at room temperature due to the anomalous transmission caused by resonance scattering. As

T. Fukamachi and T. Kawamura, *Resonant Dynamical X-ray Diffraction*, Springer Series in Solid-State Sciences 208, https://doi.org/10.1007/978-981-96-8406-9_9

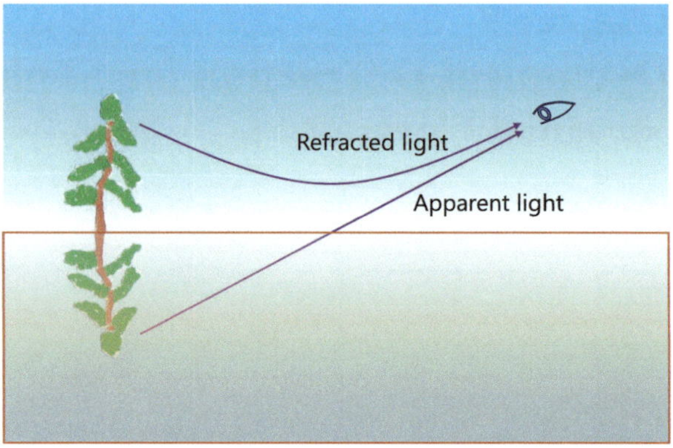

Fig. 9.1 Schematic diagram of the mirage phenomenon

the lattice spacing within the crystal changes continuously due to the strain gradient, the propagating direction of the X-rays changes continuously, causing a mirage phenomenon in which the X-rays are totally reflected. Due to the extremely large angular amplification effect mentioned in Chap. 8, an X-ray mirage diffracted wave with a propagation distance of about a millimeter appears.

In the following, we first describe the theory of X-ray diffraction in crystals with strain gradients. Then, we describe the characteristics of interference fringes caused by X-rays that become pseudo-spherical beams due to the angular amplification effect and their experimental results.

9.1 Local Reciprocal Lattice Vector

Since the crystal has its three-dimensional periodicity, there is a relation between the lattice vector $\boldsymbol{r}$ and the reciprocal lattice vector $\boldsymbol{g}$ as

$$\frac{\boldsymbol{g}\cdot\boldsymbol{r}}{2\pi} = A(\text{constant}). \tag{9.2}$$

If the corresponding minimum spacing of the reflecting planes is $\boldsymbol{d}$, $\boldsymbol{g}\cdot\boldsymbol{d} = 2\pi$. The N_r-th lattice position $\boldsymbol{r}$ can be expressed as $\boldsymbol{r} = (N_x d_x, N_y d_y, N_z d_z)$ when one lattice position is taken as the origin and d_x, d_y and d_z are x-, y- and z-component of $\boldsymbol{d}$. Then we have relations as

$$2\pi N_r = \boldsymbol{g}\cdot\boldsymbol{r} = g_x x + g_y y + g_z z = 2\pi\left(N_x + N_y + N_z\right). \tag{9.3}$$

When a crystal is elastically deformed, the position vector $\boldsymbol{r}$ of the lattice changes, and N_r is thought to deviate from the original integer. That is, N_r becomes a function of $\boldsymbol{r}$, and at the same time, N_x, N_y, and N_z also become functions of $\boldsymbol{r}$ and deviate from integers. Considering this, and using (9.3), we get

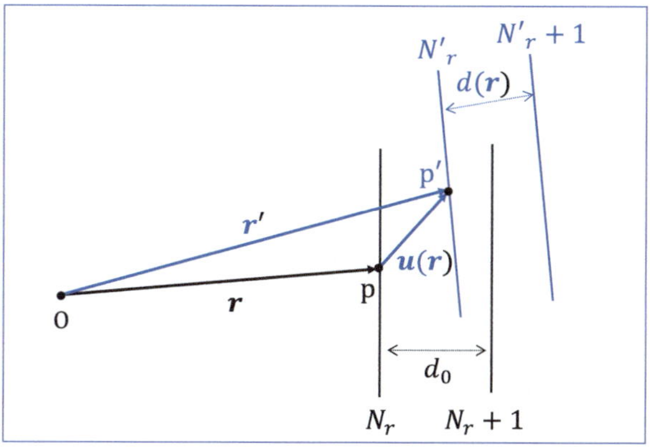

Fig. 9.2 Deformation of lattice planes. Here, $\boldsymbol{u}(\boldsymbol{r})$ is the displacement vector due to deformation, and d_0 and $d(\boldsymbol{r})$ are the lattice spacings before and after deformation

$$\boldsymbol{g} = 2\pi\nabla(N_r) = 2\pi\left(\boldsymbol{i}\frac{\partial N_x}{\partial x} + \boldsymbol{j}\frac{\partial N_y}{\partial y} + \boldsymbol{k}\frac{\partial N_z}{\partial z}\right) = \boldsymbol{i}g_x + \boldsymbol{j}g_y + \boldsymbol{k}g_z. \tag{9.4}$$

If the crystal deforms elastically and $\boldsymbol{r}$ moves to r', as shown in Fig. 9.2, we get

$$\boldsymbol{r} = \boldsymbol{r}' - \boldsymbol{u}(\boldsymbol{r}). \tag{9.5}$$

Here, $\boldsymbol{u}(\boldsymbol{r})$ is a displacement vector due to deformation.

After transformation, N_r' is given by

$$N_r'(\boldsymbol{r}) = \left(\boldsymbol{g}\cdot\boldsymbol{r}' - \boldsymbol{g}\cdot\boldsymbol{u}(\boldsymbol{r})\right)/2\pi = N_{r'}. \tag{9.6}$$

If the lattice spacing after deformation is written as $d(\boldsymbol{r}) = |\boldsymbol{d}(\boldsymbol{r})|$, the reciprocal lattice vector can be written by

$$\boldsymbol{g}'(\boldsymbol{r}) = 2\pi\nabla\left(N_r'(\boldsymbol{r})\right) = \boldsymbol{g} - \nabla(\boldsymbol{g}\cdot\boldsymbol{u}(\boldsymbol{r})). \tag{9.7}$$

$\boldsymbol{g}'(\boldsymbol{r})$ is called a local reciprocal lattice vector, which is perpendicular to the deformed lattice plane and depends on the position vector $\boldsymbol{r}$ with the magnitude being $2\pi/d(\boldsymbol{r})$, The change in the reciprocal lattice vector before and after the deformation $\Delta\boldsymbol{g}$ is given by

$$\Delta\boldsymbol{g} = \boldsymbol{g}'(\boldsymbol{r}) - \boldsymbol{g} = -\nabla(\boldsymbol{g}\cdot\boldsymbol{u}(\boldsymbol{r})). \tag{9.8}$$

Using the strain ε_r in the $\boldsymbol{r}$ direction, the displacement vector $\boldsymbol{u}$ is expressed by

$$\boldsymbol{u} = \varepsilon_r\boldsymbol{r}. \tag{9.9}$$

Then, we get $\Delta \boldsymbol{g} = -\varepsilon_r \boldsymbol{g}$. When the strain ε_r is constant, $\Delta \boldsymbol{g}$ is constant, $\boldsymbol{g}' = \boldsymbol{g} + \Delta \boldsymbol{g}$, that is, the crystal remains a perfect one. On the other hand, when the strain changes linearly with respect to r, i.e., $\varepsilon_r = \varepsilon'_r r$, $\Delta \boldsymbol{g}$ depends on the position and changes depending on the strain gradient ε'_r.

9.2 Local Dispersion Surface

When $d(\boldsymbol{r})$ becomes larger, the Bragg angle becomes smaller. Figure 9.3a shows the dispersion surfaces in vacuum before and after deformation accompanying the change of $\Delta \boldsymbol{g}$. We take the Laue point $\mathrm{L_a}$ as a tie-point. When the crystal is deformed, the magnitude of the incident wavevector $\boldsymbol{K}_0$ does not change and the incident direction changes around the origin O, the Laue point $\mathrm{L_a}$ moves to $\mathrm{L'_a}$ on the dispersion surface T'_0. This shift of Laue point in the vacuum corresponds to the shift of the Lorentz point in the crystal from $\mathrm{L_o}$ to $\mathrm{L'_o}$ on the asymptote of the dispersion surface as shown in Fig. 9.3b. The length of $\overline{\mathrm{L_a L'_a}}$ can be regarded to be equal to that of $\overline{\mathrm{L_o L'_o}}$. Suppose that P is the tie-point on the dispersion surface before deformation. When the crystal is deformed, it moves to P′, which is also on the dispersion surface. If we denote the wavevectors of the incident beam before and after deformation as $\boldsymbol{k}_0$ and $\boldsymbol{k}'_0$, respectively, the change of wavevector $(\Delta \boldsymbol{k}_0)$ due to the deformation can be written as

$$\Delta \boldsymbol{k}_0 = \boldsymbol{k}'_0 - \boldsymbol{k}_0 = \overrightarrow{\mathrm{PP'}}. \tag{9.10}$$

If the wavevector of the diffracted wave changes from $\boldsymbol{k}_g$ to $\boldsymbol{k}'_g$ due to deformation of the crystal, the change of the wavevector $\Delta \boldsymbol{k}_g$ can be written by

$$\Delta \boldsymbol{k}_g = \boldsymbol{k}'_g - \boldsymbol{k}_g = \Delta \boldsymbol{g} - \Delta \boldsymbol{k}_0, \tag{9.11}$$

where $\boldsymbol{k}_g = \overrightarrow{\mathrm{PG}}$, $\boldsymbol{k}'_g = \overrightarrow{\mathrm{P'G'}}$ and $\boldsymbol{k}_g + \Delta \boldsymbol{g} = \Delta \boldsymbol{k}_0 + \boldsymbol{k}'_g$.

Now, we discuss the trajectory of the pseudo-refracted wave in the crystal in the symmetric Laue case as an example. Figure 9.4a shows the trajectory of the pseudo-refracted wave and Fig. 9.4b shows the dispersion surface after deformation. In Fig. 9.4b, P is a tie-point on the dispersion surface before deformation and it is not on the dispersion surface after deformation.

The tie-point after deformation is point P′ which is moved from P by $\Delta \boldsymbol{k}_0$. The tie-point P″ is related to tie-point P as

$$\overrightarrow{\mathrm{P''P}} = -\overrightarrow{\mathrm{L_o L'_o}}, \tag{9.12}$$

where $-\overrightarrow{\mathrm{L_o L'_o}}$ is shown in Fig. 9.3b. From Fig. 9.4b, we get

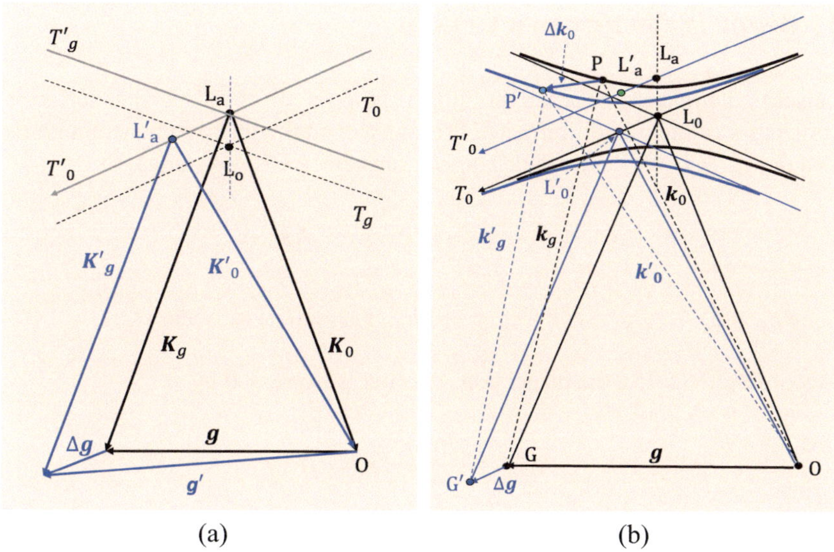

Fig. 9.3 Changes in wave vector and dispersion surface due to deformation of the lattice. In vacuum (**a**) and in a crystal (**b**)

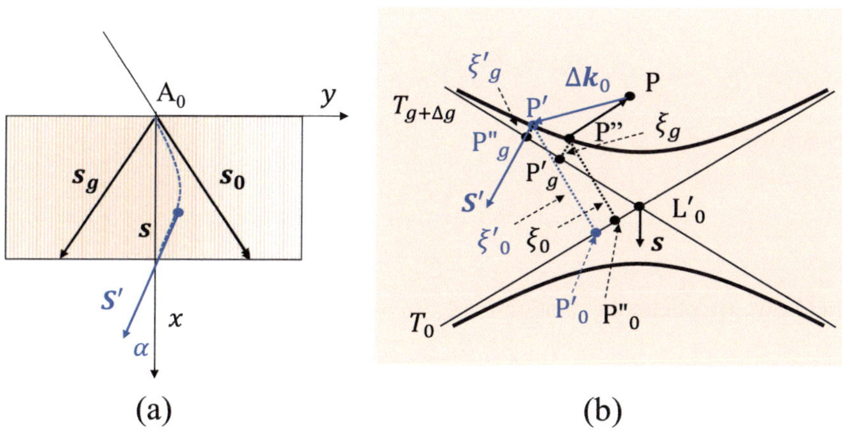

Fig. 9.4 **a** Trajectory of the pseudo-refracted beam in a crystal with constant strain gradient along x, and **b** the corresponding shift of tie-points on the dispersion surface in the symmetric Laue case

$$\overrightarrow{\mathrm{P}''\mathrm{P}'} = \Delta \boldsymbol{k}_0 - \overrightarrow{\mathrm{L}_0\mathrm{L}_0'}. \tag{9.13}$$

The change of $\Delta \boldsymbol{g}$ is reflected to the shift of $\overrightarrow{\mathrm{P}''\mathrm{P}'}$ on the dispersion surface.

9.3 Beam Trajectory in a Crystal

We discuss the relation between the shift of tie-point on the dispersion surface and the trajectory of the pseudo-refracted beam in a crystal with constant strain gradient shown in Fig. 9.4. We define the change in resonance errors ξ_0 and ξ_g as

$$d\xi_0 = \xi_0' - \xi_0 = \overrightarrow{\mathrm{P''P'}} \cdot \boldsymbol{s}_0 = \Delta\boldsymbol{k}_0 \cdot \boldsymbol{s}_0, \tag{9.14}$$

$$d\xi_g = \xi_g' - \xi_g = -\overrightarrow{\mathrm{P''P'}} \cdot \boldsymbol{s}_g = -\Delta\boldsymbol{k}_g \cdot \boldsymbol{s}_g = \Delta\boldsymbol{k}_0 \cdot \boldsymbol{s}_g - \Delta\boldsymbol{g} \cdot \boldsymbol{s}_g. \tag{9.15}$$

According to (5.45), the dispersion equation is expressed by

$$\xi_0\xi_g = \frac{(P\kappa_0)^2}{4}\left(\chi_g\chi_{-g}\right).$$

By differentiate both sides of this equation, we get

$$\xi_0 d\xi_g + \xi_g d\xi_0 = 0. \tag{9.16}$$

Substituting (9.14) and (9.15) into (9.16), we get

$$\Delta\boldsymbol{g} \cdot \boldsymbol{s}_g = \Delta\boldsymbol{k}_0.\left(\frac{\xi_g}{\xi_0}\boldsymbol{s}_0 + \boldsymbol{s}_g\right). \tag{9.17}$$

Since we have

$$\frac{D_0}{D_g} = \frac{P\kappa_0\left(\chi_g\chi_{-g}\right)^{1/2}}{2\xi_0} = \frac{2\xi_g}{P\kappa_0\left(\chi_g\chi_{-g}\right)^{1/2}}$$

from (5.56), we obtain the relation for ξ_g/ξ_0 as

$$\left(\frac{D_0}{D_g}\right)^2 = \frac{\xi_g}{\xi_0} = \frac{1}{r_0^2}, \; r_0 = \frac{D_g}{D_0}. \tag{9.18}$$

Here, r_0 represents the reflection coefficients, $r^{(1)}$ and $r^{(2)}$ given by (7.97) and (7.98). If we write the term in the parentheses of (9.17) as $\boldsymbol{S}'$,

$$\boldsymbol{S}' = \frac{\xi_g}{\xi_0}\boldsymbol{s}_0 + \boldsymbol{s}_g = r_0^{-2}\boldsymbol{s}_0 + \boldsymbol{s}_g. \tag{9.19}$$

$\boldsymbol{S}'$ is proportional to the Poynting vector and parallel to the pseudo-refracted beam.

Next, we examine components of trajectories parallel and perpendicular to the lattice plane. If we write $\boldsymbol{s}$ as a unit vector parallel to the lattice plane (x-axis), we have a relation

$$\boldsymbol{S}' \cdot \boldsymbol{s} = \boldsymbol{S}' \cos\alpha \tag{9.20}$$

or by using (9.19), we have another expression

$$\boldsymbol{S}' \cdot \boldsymbol{s} = \left(r_0^{-2}\boldsymbol{s}_0 + \boldsymbol{s}_g\right) \cdot \boldsymbol{s} = \left(r_0^{-2} + 1\right)\cos\theta_{\mathrm{B}}. \tag{9.21}$$

Using these two relations, we get

$$\boldsymbol{S}' = \left(r_0^{-2} + 1\right)\frac{\cos\theta_{\mathrm{B}}}{\cos\alpha}, \boldsymbol{S}' = \left(r_0^{-2} + 1\right)\frac{\cos\theta_{\mathrm{B}}}{\cos\alpha}\frac{d\boldsymbol{r}}{dr}. \tag{9.22}$$

From (9.17) and (9.19), we get

$$\Delta\boldsymbol{k}_0 \cdot \boldsymbol{S}' = \Delta\boldsymbol{g} \cdot \boldsymbol{s}_g = (d\boldsymbol{r} \cdot \nabla)\boldsymbol{g} \cdot \boldsymbol{s}_g = \nabla\left(\boldsymbol{g} \cdot \boldsymbol{s}_g\right) \cdot d\boldsymbol{r}, \tag{9.23}$$

by using a relation $\Delta\boldsymbol{g} = (d\boldsymbol{r} \cdot \nabla)\boldsymbol{g}$. Using (9.22) and (9.23), we get

$$\left(r_0^{-2} + 1\right)\frac{\cos\theta_{\mathrm{B}}}{\cos\alpha}\Delta\boldsymbol{k}_0 \cdot \frac{d\boldsymbol{r}}{dr} = \nabla\left(\boldsymbol{g} \cdot \boldsymbol{s}_g\right) \cdot d\boldsymbol{r}. \tag{9.24}$$

From this equation, we obtain

$$\Delta\boldsymbol{k}_0 = \left(r_0^{-2} + 1\right)^{-1}\frac{\cos\alpha}{\cos\theta_{\mathrm{B}}}\nabla\left(\boldsymbol{g} \cdot \boldsymbol{s}_g\right) \cdot d\boldsymbol{r}. \tag{9.25}$$

If we use the reciprocal lattice vector $\boldsymbol{g}_0$ without deformation, we have a relation of $\boldsymbol{g} = \boldsymbol{g}_0 - \nabla(\boldsymbol{g} \cdot \boldsymbol{u})$. If x_0 and x_g are variables in the directions of transmitted wave and diffracted wave, respectively, we have relations as

$$\frac{\partial}{\partial x_0} = \boldsymbol{s}_0 \cdot \nabla, \frac{\partial}{\partial x_g} = \boldsymbol{s}_g \cdot \nabla. \tag{9.26}$$

Then we obtain

$$\Delta\boldsymbol{k}_0 \cdot \boldsymbol{s}_0 = -\left(r_0^{-2} + 1\right)^{-1}\frac{\cos\alpha}{\cos\theta_{\mathrm{B}}}\frac{\partial^2(\boldsymbol{g} \cdot \boldsymbol{u})}{\partial x_0 \partial x_g}d\boldsymbol{r}. \tag{9.27}$$

The basic equation (5.39) can be written as

$$k_0^2 - \kappa_0^2 = \kappa_0^2 P\chi_{-g}r_0. \tag{9.28}$$

By differentiating both sides of (9.28), we get

$$dk_0 = d\boldsymbol{k}_0 \cdot \boldsymbol{s}_0 = \frac{\kappa_0 P \chi_{-g}}{2} dr_0, \tag{9.29}$$

using the approximation $k_0 \approx \kappa_0$. In (9.27), if we write $dr \cos\alpha = dx$, we get

$$\left(r_0^{-2} + 1\right) dr_0 = d\left(r_0 - \frac{1}{r_0}\right) = \frac{-2}{\kappa_0 P \chi_{-g} \cos\theta_{\mathrm{B}}} \frac{\partial^2(\boldsymbol{g} \cdot \boldsymbol{u})}{\partial x_0 \partial x_g} dx. \tag{9.30}$$

If we write

$$d\left(r_0 - \frac{1}{r_0}\right) = -2\beta dx, \tag{9.31}$$

β is a parameter that relates the reflection coefficient r_0 to the coordinate x, and is also related to the strain gradient given by (9.111) as will be described later. The reflection coefficient (r_0) changes as a function of depth (x) along the trajectory ($\boldsymbol{S}'$). Comparing (9.30) and (9.31), β is given by

$$\beta = \frac{1}{\kappa_0 P \chi_{-g} \cos\theta_{\mathrm{B}}} \frac{\partial^2(\boldsymbol{g} \cdot \boldsymbol{u})}{\partial x_0 \partial x_g} = \frac{-1}{G_{\mathrm{B}} \cos^2\theta_{\mathrm{B}}} \frac{\partial^2(\boldsymbol{g} \cdot \boldsymbol{u})}{\partial x_0 \partial x_g}, \tag{9.32}$$

where G_{B} is the Bragg gap given by

$$G_{\mathrm{B}} = \frac{P\kappa_0 \left(\chi_g \chi_{-g}\right)^{1/2}}{\cos\theta_{\mathrm{B}}}.$$

As we have relations

$$\frac{\tan\alpha}{\tan\theta_{\mathrm{B}}} = \frac{|D_0|^2 - \left|D_g\right|^2}{|D_0|^2 + \left|D_g\right|^2} = \frac{1 - r_0^2}{1 + r_0^2} = p, \tag{9.33}$$

we get

$$r_0^2 = \frac{1 - p}{1 + p}. \tag{9.34}$$

Since we have a relation given by (6.141)

$$p = \frac{\pm W}{\left(W^2 + 1\right)^{1/2}},$$

we obtain

$$r_0 - \frac{1}{r_0} = \frac{\pm 2p}{\left(1 - p^2\right)^{\frac{1}{2}}} = \pm 2W. \tag{9.35}$$

9.4 Beam Trajectory and Equal Thickness Interference Fringes in the Laue Case

9.4.1 Beam Trajectory in the Symmetric Laue Case

In the symmetric Laue case, the term in the parenthesis on the left side of (9.31) is $r_0 - 1/r_0 = 2W$, we can rewrite (9.31) as

$$\mp dW = \beta dx. \tag{9.36}$$

From (9.33), we have relations

$$p = \frac{\tan\alpha}{\tan\theta_B} = \frac{1}{\tan\theta_B}\frac{dy}{dx} = \frac{\mp W}{\left(W^2 + 1\right)^{\frac{1}{2}}}. \tag{9.37}$$

We also have a relation

$$\mp d\left(W^2 + 1\right)^{\frac{1}{2}} = \frac{\mp W}{\left(W^2 + 1\right)^{\frac{1}{2}}} dW. \tag{9.38}$$

Using (9.36) and (9.37), the left side of (9.38) can be written as

$$\mp d\left(W^2 + 1\right)^{\frac{1}{2}} = \frac{\beta}{\tan\theta_B} dy. \tag{9.39}$$

Integrating (9.36), we get

$$W = \beta x + W_s \tag{9.40}$$

and integrating (9.39), we get

$$\mp\left(W^2 + 1\right)^{\frac{1}{2}} = \frac{\beta}{\tan\theta_B} y \mp \left(W_s^2 + 1\right)^{\frac{1}{2}}. \tag{9.41}$$

Here, W_s is the value of W at the incident point. From (9.40) and (9.41), we obtain

$$\left[\frac{\beta y}{\tan\theta_{\mathrm{B}}} \mp \left(W_s^2+1\right)^{\frac{1}{2}}\right]^2 - (\beta x + W_s)^2 = 1. \tag{9.42}$$

The dispersion Eq. (6.49) in the symmetric Laue case is written as

$$Y^2 - X^2 \tan^2\theta_{\mathrm{B}} = \left(\frac{G_{\mathrm{B}}}{2}\right)^2,$$

with

$$Y^2 = \left(\frac{G_{\mathrm{B}}}{2}\right)^2 (1+W^2),\, X^2\tan^2\theta_{\mathrm{B}} = \left(\frac{G_{\mathrm{B}}}{2}\right)^2 W^2.$$

Subtracting the square of the left-hand side of (9.40) from the square of the left-hand side of (9.41) yields the constant of 1. Then, (9.42) holds and the beam trajectory becomes a hyperbola (hyperbolic trajectory).

Figure 9.5 shows the beam trajectory derived from (9.42) and the corresponding movement of the tie-point on the dispersion surface as the beam propagates from the top to the bottom surface. The movement of the tie-point in the case of $\beta > 0$ is shown in (a). The trajectory of the pseudo-refracted beam in the case of $\beta > 0$ is shown in (b), that in the case of $\beta < 0$ in (c). The darker the background color shows the narrower lattice spacing. If we rotate each dispersion surface in Fig. 9.5a by 90° clockwise, it becomes similar to the trajectory of the hyperbolic refracted beam. The coordinates (x_a, y_a) of the vertices of trajectories of $\boldsymbol{S}_1$ and $\boldsymbol{S}_2$ in Fig. 9.5b, c are given by

$$x_a = -\frac{W_s}{\beta} \text{ and } y_a = \pm\frac{\tan\theta_{\mathrm{B}}}{\beta}\left[\left(W_s^2+1\right)^{\frac{1}{2}} + 1\right]. \tag{9.43}$$

9.4.2 *Local Strain and Angular Amplification Effect*

If we denote the change in the lattice spacing d due to local strain as $\Delta d(\boldsymbol{r})$, and the resulting change in the Bragg angle as $\Delta\theta_{\mathrm{B}}(\boldsymbol{r})$, we have a relation

$$|\Delta\theta_{\mathrm{B}}(\boldsymbol{r})| = \tan\theta_{\mathrm{B}}\left|\frac{\Delta d(\boldsymbol{r})}{d(\boldsymbol{r})}\right|.$$

In Chap. 8, we have described the angular amplification factor A is given by $A = \Delta\varphi/\Delta\theta_{\mathrm{B}}$ with $\Delta\theta_{\mathrm{B}}$ the divergence angle of the incident beam and $\Delta\varphi$ the resulting divergence angle. The value of A can be very large about 10^4, indicating large angular amplification. Using these relations, we get

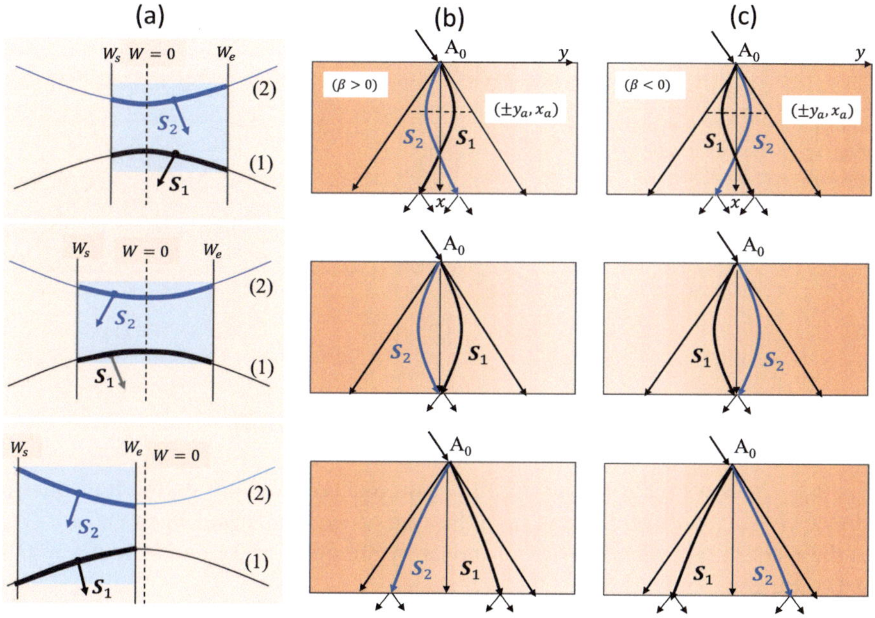

Fig. 9.5 The range of dispersion surface (thick curves) where a tie-point is displaced in the symmetric Laue case for a crystal with constant strain gradient when $\beta > 0$ (**a**). Trajectories of pseudo-refracted beams for $\beta > 0$ (**b**) and $\beta < 0$ (**c**). The darker background color in the crystal shows the narrower lattice spacing in (**b**) and (**c**). The upper row is for $|-W_s| < W_e$, the middle row for $|-W_s| = W_e$ and the lower row for $W_e < 0$. W_s and W_e represent the values of W at the incident point and the emission point, respectively

$$|\Delta\varphi(\boldsymbol{r})| = A|\Delta\theta_{\mathrm{B}}(\boldsymbol{r})| = A\tan\theta_{\mathrm{B}}\left|\frac{\Delta d(\boldsymbol{r})}{d(\boldsymbol{r})}\right|. \tag{9.44}$$

This means that the change in the local lattice spacing $\Delta d(\boldsymbol{r})$ greatly changes the direction of the pseudo-refracted beam due to the angular amplification effect (see Fig. 9.6). For this reason, X-ray mirages can be observed in a region on the order of millimeters.

9.4.3 Pendellösung Fringes

As mentioned in Chap. 6, the Pendellösung fringes in the Laue case appear on the rocking curve. The interference fringes can also be observed by changing the propagation distance, that is, the crystal thickness. The interference fringes are called equal thickness interference fringes. Pendellösung fringes also appear in the Kato fringes caused by pseudo-spherical waves. In this section, we first consider the rocking curve from a crystal with strain gradients when the incident beam is an ideal plane wave.

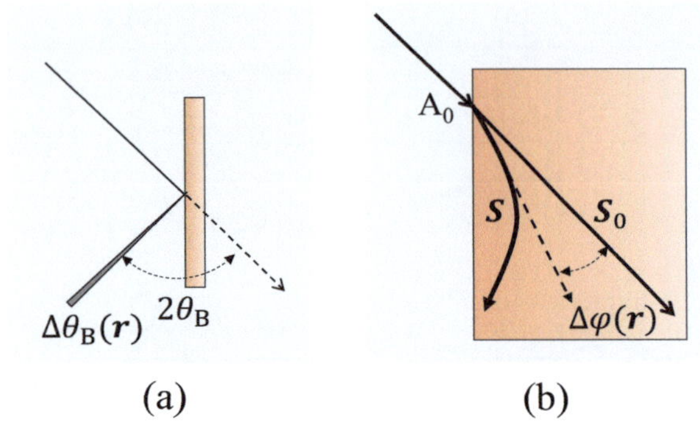

Fig. 9.6 Change in direction of a quasi-refracted beam due to angular amplification effect in a crystal with strain gradients. (**a**) Change in Bragg angle due to strain. (**b**) Change in direction of pseudo-refracted beam due to strain. $\boldsymbol{S}_0$ shows the beam for $\beta = 0$ and $\boldsymbol{S}$ the beam for $\beta > 0$

9.4.3.1 When the Incident Beam is an Ideal Plane Wave

Figure 9.7b shows the trajectory of the beam focused at $y = 0$ on the transmission surface of a crystal of thickness H in the symmetric Laue case, when $|W_s| = |W_e|$. As the x-coordinate of the vertex of trajectory are given as $2x_a = -2W_s/\beta = H$ in (9.43), we get

$$W_s = -W_e = -\frac{\beta H}{2}. \tag{9.45}$$

The change in W from the incident point to the exit point of the crystal is given as

$$\Delta W = W_e - W_s = \beta H.$$

Here, ΔW is constant (βH) regardless of the value of y. Note that in Fig. 9.7 and the following, x-coordinate in the real space in (9.42) is replaced with z, and the X coordinate is replaced with the Z coordinate in the reciprocal space.

From (6.11) and (6.12), the electric displacement $(D_g^{(j)})$ of the pseudo-refracted beam at point A on the incident surface in the symmetric Laue case is given by

$$D_g^{(1)} = \frac{\exp(i\alpha_g)}{2} \frac{-1}{\left(1 + W_e^2\right)^{1/2}} E_0, D_g^{(2)} = -D_g^{(1)}. \tag{9.46}$$

By referring to (6.5), the boundary condition at point B of the transmission surface is given as

$$D_g^{(1)} \exp\left(-i \int \boldsymbol{k}_g^{(1)} \cdot d\boldsymbol{r}\right) + D_g^{(2)} \exp\left(-i \int \boldsymbol{k}_g^{(2)} \cdot d\boldsymbol{r}\right) = E_g \exp(-iK_{gz}H). \tag{9.47}$$

We write the phase $\theta^{(j)}$ as

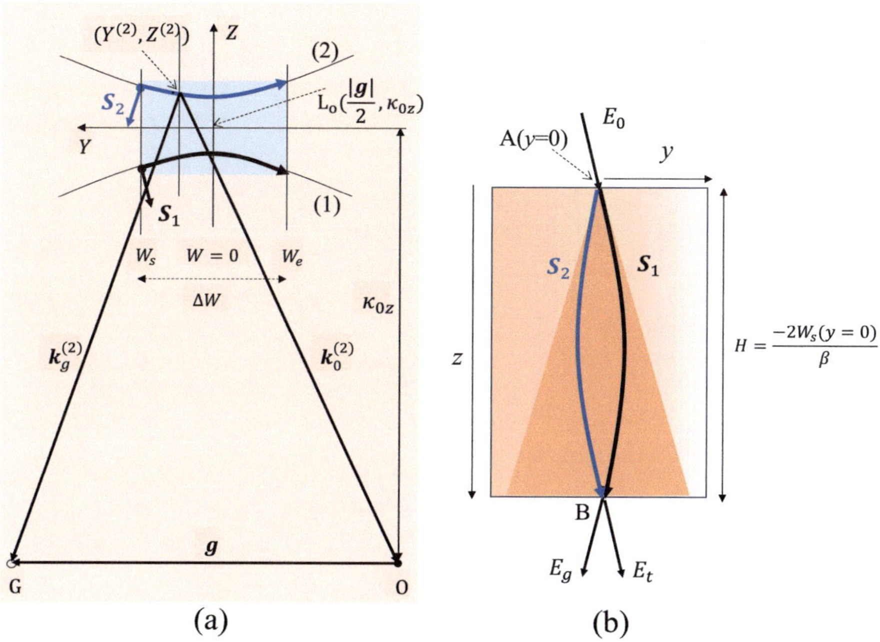

Fig. 9.7 The thick curves show the range of dispersion surface where a tie-point is displaced in the reciprocal space (**a**), when trajectories of the hyperbolic refracted beam passing through the two vertices in the real space (**b**)

$$\theta^{(j)} = \int \boldsymbol{k}_g^{(j)} \cdot d\boldsymbol{r} = \int k_{gy}^{(j)} dy + \int k_{gz}^{(j)} dz, \tag{9.48}$$

and the difference $\Delta\theta = \theta^{(2)} - \theta^{(1)}$ as

$$\Delta\theta = \left(\theta_y^{(2)} - \theta_y^{(1)}\right) + \left(\theta_z^{(2)} - \theta_z^{(1)}\right) = \Delta\theta_y + \Delta\theta_z. \tag{9.49}$$

Using $\Delta\theta$, the diffracted intensity is obtained in the same way as we derived (6.20) as

$$\frac{P_g}{P_0} = \left|\frac{E_g}{E_0}\right|^2 = \frac{\sin^2\left(\frac{\Delta\theta}{2}\right)}{1 + W_e^2}. \tag{9.50}$$

Next, we will find $\Delta\theta$ in (9.49). Referring to Fig. 9.7, we get

$$\theta_y^{(j)} = \int k_{gy}^{(j)} dy = \int \left(\frac{|\boldsymbol{g}|}{2} + Y^{(j)}\right) dy, \tag{9.51}$$

$$\theta_z^{(j)} = \int k_{gz}^{(j)} dz = \int \left(\kappa_{0z} + Z^{(j)}\right) dz. \tag{9.52}$$

By using (9.51), we get

$$\Delta\theta_y = \theta_y^{(2)} - \theta_y^{(1)} = \int \left(Y^{(2)} - Y^{(1)}\right) dy = 0, \tag{9.53}$$

because of

$$Y^{(1)} = Y^{(2)} = \frac{WG_{\mathrm{B}}}{2\tan\theta_{\mathrm{B}}}. \tag{9.54}$$

By using (9.52), we get

$$\Delta\theta_z = \theta_z^{(2)} - \theta_z^{(1)} = \int_0^H \left(Z^{(2)} - Z^{(1)}\right) dz, \tag{9.55}$$

where $Z^{(j)}$ is given as

$$Z^{(j)} = (-1)^j \frac{G_{\mathrm{B}}}{2} \left(W^2 + 1\right)^{1/2}. \tag{9.56}$$

According to (9.36), we have $dW = \beta dz$, then from (9.56) we get

$$\Delta\theta = \Delta\theta_z = \frac{G_{\mathrm{B}}}{\beta} \int_{W_s}^{W_e} \left(W^2 + 1\right)^{1/2} dW = -\frac{G_{\mathrm{B}}}{2\beta} \left[Q(W_e) - Q(W_s)\right]. \tag{9.57}$$

Here, we write

$$Q(W) = W\left(W^2 + 1\right)^{1/2} + \log\left|W + \left(W^2 + 1\right)^{1/2}\right|. \tag{9.58}$$

Figure 9.8 shows an example of the rocking curves calculated by varying the value of β. When $\beta = 0$, the rocking curve is symmetric about $W = 0$. The rocking curve shows the maximum at $W = 0$ when the crystal thickness H is a half-integer multiple of the interval of Pendellösung fringes Λ_0, while it shows the minimum at $W = 0$ when it is an integer multiple of Λ_0. For $\beta \neq 0$, the rocking curve is asymmetric about $W = 0$ and the maximum peak shifts to the left for $\beta < 0$, and to the right for $\beta > 0$.

9.4.3.2 When the Incident Beam is a Quasi-spherical Wave

When the incident beam is a plane wave and the condition $W_s = -\beta H/2$ is satisfied, the two beams excited at the tie-points of branches (1) and (2) are focused at y = 0

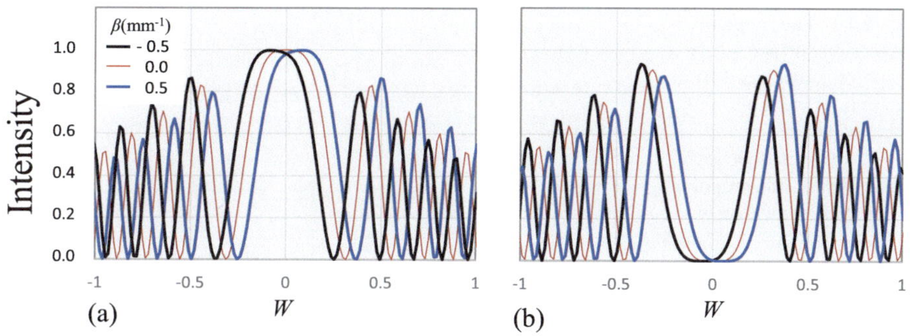

Fig. 9.8 Strain dependence of rocking curves when $H = 10.5\,\Lambda_0$ (**a**) and $H = 10\Lambda_0$ (**b**). Λ_0 is the interval of Pendellösung fringes

Fig. 9.9 Trajectories of pseudo-refracted waves focused on the exit surface

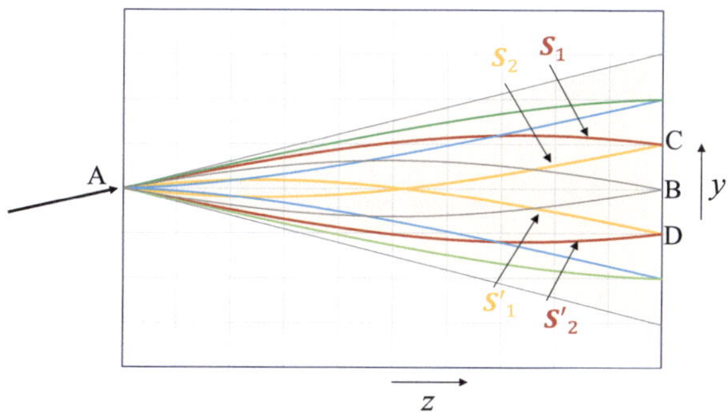

on the transmission surface as shown in Fig. 9.7. But it is not possible to be focused at a point y $\neq$ 0.

If the incident beam has a small divergence angle, a part of the beam can be focused at a point y $\neq$ 0 due to the angular amplification effect. By giving the values of β and H in (9.42), the initial condition of W_s under which the two beams are focused at an arbitrary point of y can be obtained. Figure 9.9 shows an example of trajectories that focus at $y \neq 0$ point. The beams $\boldsymbol{S}_1$ and $\boldsymbol{S}_2$ are focused at point C ($y \neq 0$) and the beams $\boldsymbol{S}'_1$ and $\boldsymbol{S}'_2$ are focused at point D ($y \neq 0$) on the exit surface. Figure 9.10a, b show the ranges of tie-points on the dispersion surface where the beams that focus on points C and D are excited, respectively.

9.4.3.3 Equal Thickness Interference Fringes

For plane waves, if the thickness of the crystal changes, the crystal must be rotated to satisfy condition (9.45) in order to focus the beam to the point $y = 0$. For pseudo-spherical waves, we can always obtain W_s for a wave that focuses at point $y = 0$ without rotating the crystal even if the thickness is changed.

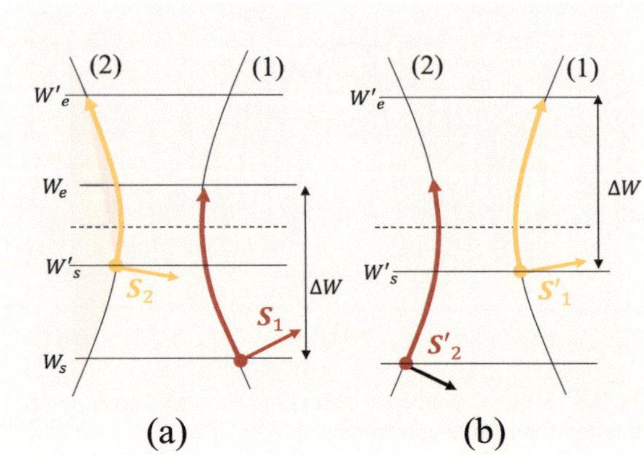

Fig. 9.10 The range of dispersion surface where a tie-point is displaced from each initial value to final value of W for pseudo-refracted waves $\boldsymbol{S}_1$ and $\boldsymbol{S}_2$ that focus on point C in Fig. 9.9, respectively (**a**) and for waves S'_1 and S'_2 that focus on point D (**b**)

In this case, $\Delta\theta$ is given by Hart [4] as

$$\begin{aligned}\Delta\theta = \Delta\theta_z &= \frac{2G_\mathrm{B}}{\beta}\int_{W_s}^{0}\left(W^2+1\right)^{\frac{1}{2}}dW \\ &= -\frac{G_\mathrm{B}}{\beta}\left[Q(W_s)\right] \\ &= \frac{G_\mathrm{B}H}{2\beta}\left[\left(q^2+1\right)^{\frac{1}{2}}+\frac{1}{q}\sinh^{-1}q\right],\end{aligned} \tag{9.59}$$

with $q = W_s = -(\beta H/2)$ and $\sinh^{-1}q = -\log\left\{q+\left(q^2+1\right)^{1/2}\right\}$. Figure 9.11 shows the β dependence of equal thickness interference fringes. The interval of the equal-thickness interference fringes becomes narrower as β increases. This β dependence was observed experimentally by Hart [4].

The relation between the curvature radius and β in a pure bent crystal is given in (C.4) in Appendix C.

9.5 Beam Trajectory and Diffraction Intensity in the Bragg Case

9.5.1 Beam Trajectory in the Symmetric Bragg Case

In the region of total reflection for $|W| < 1$ in the Bragg case, the beam cannot penetrate deep into the crystal due to the extinction effect. Then, we consider the trajectory of the pseudo-refracted beam for $|W| \geq 1$.

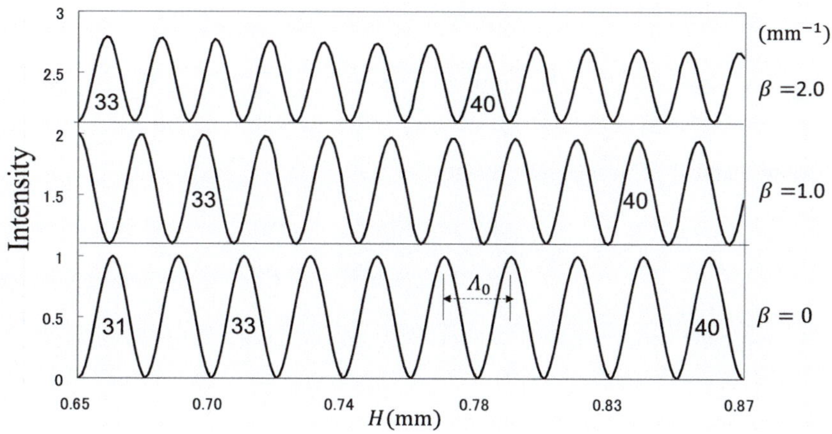

Fig. 9.11 β dependence of equal thickness interference fringes. When $\beta = 0$, the interval of the fringes Λ_0 is 0.022 mm .The number shown at some peaks represents the number of peaks counted from $H = 0$

In the symmetric Bragg case, the beam excited at one of two tie-points can enter the crystal as shown in Fig. 9.12b. We have described how to label the branch in Bragg case in 6.2. In this figure, when the beam is pointing downward, the corresponding tie-point is on the branch (1) of the dispersion surface, and when the beam is pointing upward, it is on the branch (2).

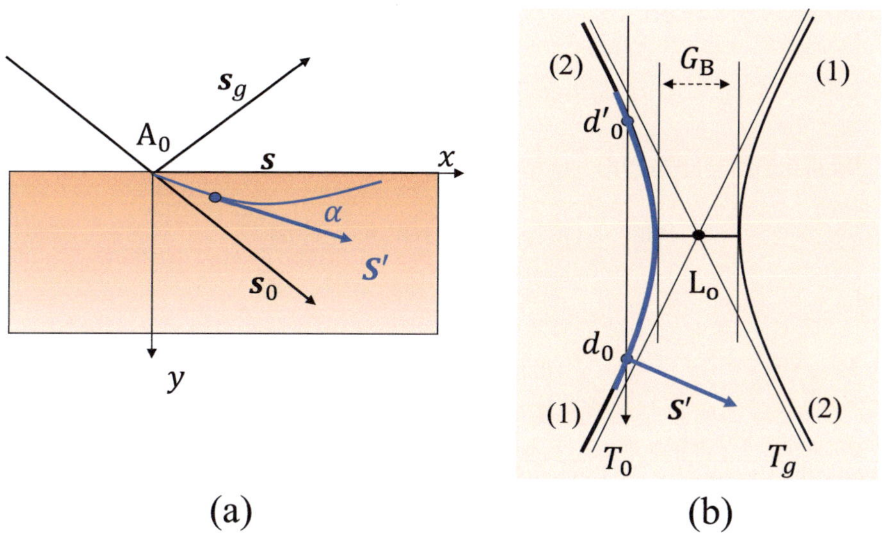

Fig. 9.12 Trajectory of a pseudo-refracted wave in the Bragg case (**a**) and the corresponding tie-points in the symmetric Bragg case (**b**)

In the Bragg case, there is a relation

$$W = \frac{1}{\left(1 - p^2\right)^{\frac{1}{2}}}$$

between p and W. From (9.35), we get

$$r_0 - \frac{1}{r_0} = \frac{2p}{\left(1 - p^2\right)^{\frac{1}{2}}} = \frac{2W}{p}. \tag{9.60}$$

By using the above relation, we get

$$\frac{1}{p} = \frac{W}{\left(W^2 - 1\right)^{1/2}}.$$

Equation (9.31) can be written by

$$\frac{1}{p} dW = d\left(W^2 - 1\right)^{1/2} = \beta dx. \tag{9.61}$$

From (9.37), there is a relation between dx and dy as

$$dx = \frac{1}{p \tan \theta_{\mathrm{B}}} dy,$$

then from (9.61) we get

$$dW = \frac{\beta}{\tan \theta_{\mathrm{B}}} dy. \tag{9.62}$$

By using $dx = dW/(p\beta)$ and integrating (9.61) and (9.62), we obtain

$$\left(W^2 - 1\right)^{1/2} = \beta x + s(W_s)\left(W_s^2 - 1\right)^{1/2}, \tag{9.63}$$

and

$$W = \frac{\beta y}{\tan \theta_{\mathrm{B}}} + W_s. \tag{9.64}$$

Here, $s(W_s)$ is 1 when $W_s > 1$, and -1 when $W_s < 1$. By combining (9.63) and (9.64), the formula for the trajectory of the pseudo-refracted beam is obtained as

$$\left(\frac{\beta y}{\tan \theta_{\mathrm{B}}} + W_s\right)^2 - \left(\beta x + s(W_s)\left(W_s^2 - 1\right)^{\frac{1}{2}}\right)^2 = 1. \tag{9.65}$$

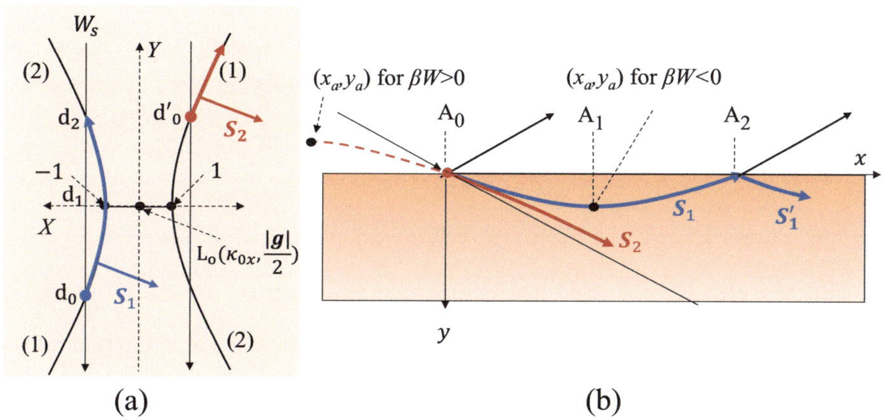

Fig. 9.13 (**a**) The range of tie-point where the pseudo-refracted beam is excited in the symmetric Bragg case in the reciprocal space. (**b**) Trajectory of the beam in the real space

The hyperbolic trajectory of this beam is shown in Fig. 9.13b. When $\beta W < 0$, the trajectory of the beam is downward as shown by $\boldsymbol{S}_1$ and the beam returns to the incident surface after totally reflected within the crystal. When $\beta W > 0$, the trajectory is also downward as shown by $\boldsymbol{S}_2$ and the beam penetrates deep into the crystal. When $|W| = 1$, the direction of the pseudo-refraction beam is parallel to the crystal surface. The coordinate (x_a, y_a) of the vertex of hyperbolic trajectory is expressed by

$$x_a = \frac{-s(W_s)\left(W_s^2 - 1\right)^{1/2}}{\beta}, \; y_a = \frac{-\tan\theta_\mathrm{B}}{\beta}(W_s - s(W_s)). \tag{9.66}$$

The vertex of $\boldsymbol{S}_1$ trajectory is inside the crystal, and that of $\boldsymbol{S}_2$ is outside the crystal.

The movement of tie-point when the beam $\boldsymbol{S}_1$ propagates in the crystal for $\beta > 0$ is shown in Fig. 9.13a. In Fig. 9.13b, the tie-point of the beam $\boldsymbol{S}_1$ at the incident point A_0 in real space is located at d_0 ($W = W_s$) on branch (1). The tie-point of the beam at the vertex A_1 of the hyperbolic trajectory is located at d_1 ($W = -1$) on the branch (1), and the direction of the beam is parallel to the crystal surface. When the pseudo-refracted beam passes through A_1, the corresponding tie-point transfers to the branch (2) (therefore, the pseudo-refracted beam is totally reflected within the crystal). When the beam turns its direction to upward to reach the point A_2 on the incident surface, the tie-point changes its position from d_2 to d_0. A part of $\boldsymbol{S}_1$ is emitted from the surface as a diffracted beam, and the rest is reflected and propagates into the crystal. In the case of $\boldsymbol{S}_2$, the tie-point is higher than point d'_0 on branch (1) of $W \geq 1$ side, and the direction of $\boldsymbol{S}_2$ is always downward.

Figure 9.14 shows the variation of trajectory of pseudo-refracted beam as a function of W_s, when $\beta W < 0$ and $\beta = 1.0\ \mathrm{mm}^{-1}$. The larger $|W_s|$ becomes, the longer the propagation distance in the crystal becomes. Authier [3] calls this phenomenon

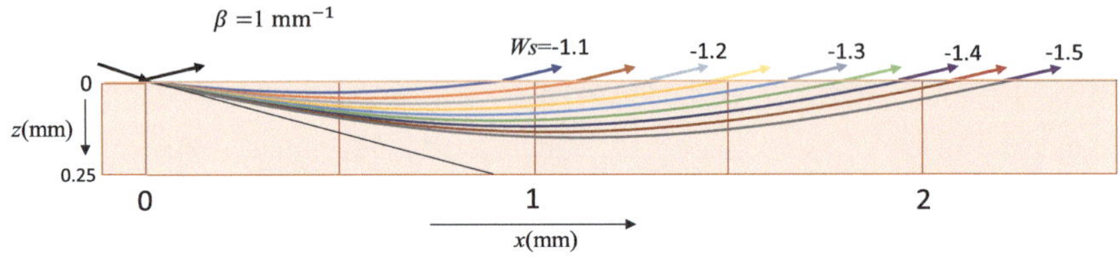

Fig. 9.14 Change in trajectory of pseudo-refracted beam as a function of W_s, when $\beta = 1.0\ \text{mm}^{-1}$ and $\tan\theta_B = 0.3$

an X-ray mirage in which the pseudo-refracted beam undergoes total internal reflection in the crystal and returns to the incident surface. We will call the diffracted beam obtained in this manner the mirage diffracted beam.

9.5.2 Propagation of Pseudo-Refracted Waves and Mirage Diffraction (X-ray Crystal Waveguide)

Figure 9.15 shows a schematic diagram of the propagation of pseudo-refracted waves and mirage diffraction. When an X-ray mirage appears, the pseudo-refracted waves that enter the crystal are totally reflected within the crystal and return to the crystal surface. Some parts of the waves become diffracted waves and are emitted outside the crystal, while the rest are reflected and re-enter the crystal. The latter waves become pseudo-refracted waves and propagate inside the crystal. They return to the crystal surface after totally reflected within the crystal. Some parts of the waves are emitted, and the rest are reflected to re-enter the crystal. In this way, the pseudo-refracted waves are repeatedly reflected on the crystal surface and propagate over long distance within the crystal. This acts as an X-ray waveguide. We will consider the pseudo-refracted waves that are repeatedly reflected from the entrance surface and the diffracted waves emitted from the crystal surface during the process.

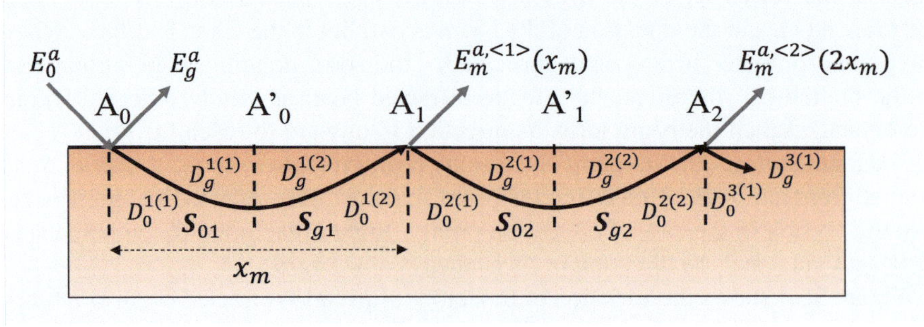

Fig. 9.15 Propagation of pseudo-refracted waves in mirage diffraction

In Fig. 9.15, when the wave of electric filed E_0^a is incident at point A_0 on the surface, a part of the incident wave is reflected at the incident point and emitted as the primary diffracted wave of electric field E_g^a. The rest enters the crystal as the pseudo-refracted wave $\boldsymbol{S}_{01}$ with the electric displacements of $D_0^{1(1)}$ and $D_g^{1(1)}$. The superscript 1 in the electric displacement indicates that the number of reflections from the entrance surface is one, and the number in parentheses indicates the branch index. The boundary condition at A_0 is expressed as

$$E_0^a = D_0^{1(1)}, E_g^a = D_g^{1(1)}.$$

Then we get

$$E_g^a = r_1 D_0^{1(1)}$$

by using a relation $r_1 = D_g^{1(1)}/D_0^{1(1)}$. The number of subscripts of r_1 represents the branch index of 1. At point A'_0, the direction of the pseudo-refracted wave changes from downward to parallel to the crystal surface, and the wave undergoes total internal reflection within the crystal, leading to $\boldsymbol{S}_{g1}$. As can be seen in Fig. 9.13a, the tie-point changes from branch (1) to branch (2). The connection conditions at A'_0 are:

$$\begin{aligned} D_0^{1(1)} \exp\left(-\frac{i\gamma_0}{2}\right) + D_g^{1(2)} \exp\left(-\frac{i\gamma_g}{2}\right) &= 0, \\ D_g^{1(1)} \exp\left(-\frac{i\gamma_g}{2}\right) + D_0^{1(2)} \exp\left(-\frac{i\gamma_0}{2}\right) &= 0. \end{aligned} \tag{9.67}$$

When connecting the 0 wave to the g wave, the 0 wave (such as $D_0^{1(1)}$, $D_0^{2(1)}$) is connected to the g wave (such as $D_g^{1(2)}$), and the g wave (such as $D_g^{1(1)}$) is connected to the 0 wave (such as $D_0^{1(2)}$), because the direction of the pseudo-refracted wave is reversed from downward to upward at point A'_0. γ_0 and γ_g are phase shifts and will be described in detail later.

A part of pseudo-refracted wave $\boldsymbol{S}_{g1}$ is emitted from point A_1 as a mirage diffracted wave $E_m^{a,\langle 1\rangle}(x_m)$. The rest is reflected from the entrance surface as a pseudo-refracted wave $\boldsymbol{S}_{02}$. The boundary condition at A_1 is given by

$$D_0^{1(2)} \exp(-i\gamma_0) + D_0^{2(1)} \exp(-i\gamma_0) = 0, \tag{9.68}$$

$$D_g^{1(2)} \exp(-i\gamma_g) + D_g^{2(1)} \exp(-i\gamma_g) = E_m^{a,\langle 1\rangle} \exp(-iK_{gx}x_m). \tag{9.69}$$

Since we have relations $D_g^{1(2)} = -D_0^{1(1)}$ and $D_g^{2(1)} = r_1 D_0^{2(1)} = r_1^2 D_0^{1(1)}$, (9.69) is rewritten by

$$(r_1^2 - 1) D_0^{1(1)} \exp(-i\gamma_g) = E_m^{a,\langle 1\rangle} \exp(-iK_{gx}x_m). \tag{9.70}$$

Using a relation $D_0^{1(1)} = E_0^a$, (9.70) is expressed by

$$E_m^{a,\langle 1\rangle}\exp\left(-iK_{gx}x_m\right) = (r_1^2 - 1)E_0^a\exp\left(-i\gamma_g\right). \tag{9.71}$$

For a non-absorbing crystal, $\left|E_0^a\exp\left(-i\gamma_g\right)\right| = \left|E_0^a\right|$ holds and the diffracted intensity denoted as $I_m^{\langle 1\rangle}$ from point A_1 is given by

$$I_m^{\langle 1\rangle} = \left|E_m^{a,\langle 1\rangle}(x_m)\right|^2 = (r_1^2 - 1)^2\left|E_0^a\right|^2. \tag{9.72}$$

In a similar way, the electric field $E_m^{\langle 2\rangle}(2x_m)$ of a mirage diffracted wave emitted at A_2 is obtained by using

$$E_m^{a,\langle 2\rangle}\exp\left(-i2K_{gx}x_m\right) = r_1(r_1^2 - 1)E_0^a\exp\left(-i2\gamma_g\right). \tag{9.73}$$

The electric field $E_m^{a,\langle n\rangle}(nx_m)$ of the n-th ($n \geq 1$) mirage diffracted beam that emitted at point A_n can be obtained by using

$$E_m^{a,\langle n\rangle}\exp\left(-inK_{gx}x_m\right) = r_1^{n-1}(r_1^2 - 1)E_0^a\exp\left(-in\gamma_g\right). \tag{9.74}$$

Equation (9.74) is further rewritten as

$$E_m^{a,\langle n\rangle}(nx_m) = E_m^{a,\langle n\rangle}\exp\left(-inK_{gx}x_m\right) = R_nE_0^a\exp\left(-in\gamma_g\right), \tag{9.75}$$

where R_n is the amplitude given as

$$R_n = r_1^{n-1}(r_1^2 - 1), n > 0. \tag{9.76}$$

For an absorption crystal, by writing γ_g as

$$\gamma_g = \gamma_r + i\gamma_i, \tag{9.77}$$

we obtain

$$E_m^{a,\langle n\rangle}\exp\left(-inK_{gx}x_m\right) = E_0^aR_n\exp(-in\gamma_r)\exp(n\gamma_i), \tag{9.78}$$

where $n\gamma_r$ is the phase shift of the pseudo-refraction wave, and $-n|\gamma_i|$ represents the absorption of the wave.

9.5.3 Phase Shift in Mirage Diffraction

The real part γ_r of the phase shift γ_g that appeared in (9.71) can be calculated as

$$\gamma_r = \gamma_{rx} + \gamma_{rz} = \int k_x dx + \int k_z dz, \tag{9.79}$$

by replacing the axis in the depth direction from y to z. The coordinate of Lorentz point L_o is expressed by $(\kappa_{0x}, |\mathbf{g}|/2)$ as shown in Fig. 9.13a. As for γ_{rx}, we can write

$$\gamma_{rx} = \int k_x dx = \kappa_{0x} x_m + 2 \int_o^{x_m/2} X dx. \tag{9.80}$$

The first term $\kappa_{0x} x_m$ does not depend on the movement of tie-point on the dispersion surface. We define the phase shift accompanying the movement as γ'_{rx} which is given by

$$\gamma'_{rx} = 2 \int_o^{x_m/2} X dx = \frac{-G_B}{\beta} \int_{W_s}^{-1} W d\left(W^2 - 1\right)^{1/2} = \frac{-G_B}{\beta} \int_{W_s}^{-1} \frac{W^2}{\left(W^2 - 1\right)^{1/2}} dW, \tag{9.81}$$

by using relations as $X = -\frac{G_B}{2} W$, $dx = \frac{1}{\beta} d\left(W^2 - 1\right)^{\frac{1}{2}}$.

As for γ_{rz}, by referring to Fig. 9.13, we get

$$\gamma_{rz} = \int k_z dz = -\int \left(\frac{|\mathbf{g}|}{2} + Y\right) dz = \frac{-|\mathbf{g}|}{2} \left(\int_0^{za} dz + \int_{za}^{0} dz \right) - 2 \int_0^{za} Y dz. \tag{9.82}$$

The first term on the right side becomes zero and (9.82) can be rewritten as

$$\gamma_{rz} = \frac{-G_B}{\beta} \int_{W_s}^{-1} \left(W^2 - 1\right)^{1/2} dW, \tag{9.83}$$

by using relations as $dz = \frac{\tan\theta_B}{\beta} dW$, $Y = \frac{G_B}{2\tan\theta_B} \left(W^2 - 1\right)^{1/2}$.

Then we obtain the phase shift as

$$\begin{aligned} \gamma'_{rx} + \gamma_{rz} &= \frac{-G_B}{\beta} \int_{W_s}^{-1} \left[\frac{W^2}{\left(W^2 - 1\right)^{\frac{1}{2}}} - \left(W^2 - 1\right)^{\frac{1}{2}} \right] dW \\ &= \frac{-G_B}{\beta} \int_{W_s}^{-1} \frac{1}{\left(W^2 - 1\right)^{1/2}} dW \\ &= \frac{G_B}{\beta} M\left(W_s\right), \end{aligned} \tag{9.84}$$

where $M(W)$ is given by

$$M(W) = \log\left|W + \left(W^2 - 1\right)^{\frac{1}{2}}\right|. \tag{9.85}$$

Then the phase shift γ_r can be written as

$$\gamma_r = \frac{G_{\mathrm{B}}}{\beta} M(W_s) + \kappa_{0x} x_m. \tag{9.86}$$

9.5.4 Absorption Effect in Mirage Diffraction

The mirage diffracted waves propagate over distances of the order of 10 mm under anomalous transmission condition, as will be explained later, so that absorption effect cannot be ignored even when it is small. In the symmetric Bragg case, when absorption is small ($k_a < 0.1$), (7.73) becomes

$$Y = Y_r + iY_i \approx \frac{G_{\mathrm{B}}}{2\tan\theta_{\mathrm{B}}}\left[(W + ig)^2 - (1 + i2k_a)\right]^{\frac{1}{2}}, \tag{9.87}$$

by assuming that the crystal has a center of symmetry. By squaring both sides of (9.87), we get

$$(Y_r + iY_i)^2 \approx Y_r^2 + i2Y_rY_i = \left(\frac{G_{\mathrm{B}}}{2\tan\theta_{\mathrm{B}}}\right)^2 \left[W^2 - 1 + i2g\left(W - g''\right)\right]. \tag{9.88}$$

In (9.88), the term Y_i^2 is ignored as it is small compared to Y_r^2 for $|W| > 1$ and g'' is given by

$$g'' = \frac{k_a}{g} = -\frac{|\chi_{gi}|}{|\chi_{0i}|}. \tag{9.89}$$

From the real part of (9.88), we obtain

$$Y_r = \frac{\pm G_{\mathrm{B}}}{2\tan\theta_{\mathrm{B}}}\left(W^2 - 1\right)^{1/2}, \tag{9.90}$$

and from the imaginary part, we obtain

$$Y_i = \frac{\pm 1}{Y_r}\left(\frac{G_{\mathrm{B}}}{2\tan\theta_{\mathrm{B}}}\right)^2 g\left(W - g''\right) = \frac{\pm g G_{\mathrm{B}}}{2\tan\theta_{\mathrm{B}}}\frac{\left(W - g''\right)}{\left(W^2 - 1\right)^{1/2}}. \tag{9.91}$$

By using the relations

$$\frac{gG_{\rm B}}{2\tan\theta_{\rm B}} = \frac{K_0\chi_{0i}}{2\sin\theta_{\rm B}} = \frac{-\mu_a}{2\sin\theta_{\rm B}}, \tag{9.92}$$

Equation (9.91) is rewritten as

$$Y_i = \frac{\mp\mu_a}{2\sin\theta_{\rm B}}\frac{(W - g'')}{(W^2 - 1)^{1/2}}. \tag{9.93}$$

In the region of anomalous transmission ($W < -1$), we obtain

$$\begin{aligned}\gamma_i &= \int Y_i dz = \frac{\mp\mu_a}{\sin\theta_{\rm B}}\frac{\tan\theta_{\rm B}}{\beta}\int_{W_s}^{-1}\frac{(W - g'')}{(W^2 - 1)^{1/2}}dW \\ &= \frac{\pm\mu_a}{\beta\cos\theta_{\rm B}}\left[\left(W_s^2 - 1\right)^{\frac{1}{2}} - g''M(W_s)\right]. \end{aligned} \tag{9.94}$$

Since the emission point of the mirage diffracted beam is given by $x_m = 2x_a$, and the vertex position (x_a) of hyperbolic trajectory is given in (9.43), we get

$$x_m = \frac{2\left(W_s^2 - 1\right)^{\frac{1}{2}}}{|\beta|}. \tag{9.95}$$

Then we obtain the absorption parameter as

$$\gamma_i = \frac{-\mu_a x_m}{2\cos\theta_{\rm B}}\left[1 + \frac{2|g''|}{x_m|\beta|}M(W_s)\right]. \tag{9.96}$$

A negative sign is added on the right side of the above equation, since $\gamma_i < 0$ in (9.78).

9.5.5 *Mirage Diffraction Intensity and Interference Fringes (Plane Wave Approximation)*

From the above considerations, the diffraction intensity $I_m(nx_m)$ from the emission point nx_m is given by

$$I_m(nx_m) = \left|E_m^{a,(n)}(nx_m)\right|^2 = \left|R_n\exp(-n\gamma_g)\right|^2\left|E_0^a\right|^2. \tag{9.97}$$

By using (9.74), the electric field of mirage diffracted wave E_m^a is expressed as

$$E_m^a = \left[R_1 \exp(-i\gamma_g) + R_2 \exp(-i2\gamma_g) + R_3 \exp(-i3\gamma_g) + \cdots\right] E_0^a \tag{9.98}$$

and is rewritten as

$$\begin{aligned} E_m^a &= \sum_{n=1}^{\infty} E_0^a R_n \exp(-in\gamma_g) \\ &= \left[\left(r_1^2 - 1\right) \exp(-i\gamma_g)[1 + r_1 \exp(-i\gamma_g) + r_1^2 \exp(-i2\gamma_g) + \cdots\right] E_0^a, \end{aligned} \tag{9.99}$$

by using (9.75). From (9.99), we get

$$E_m^a = \frac{\left(r_1^2 - 1\right) \exp(-i\gamma_g)}{1 - r_1 \exp(-i\gamma_g)} E_0^a. \tag{9.100}$$

The intensity ratio of the mirage diffracted wave to the incident wave is written by

$$\frac{P_m}{P_0} = \left|\frac{E_m^a}{E_0^a}\right|^2 = \left|\frac{\left(r_1^2 - 1\right) \exp(-i\gamma_g)}{1 - r_1 \exp(-i\gamma_g)}\right|^2. \tag{9.101}$$

All diffracted beams from the entrance surface are expressed as

$$E_{g,m}^a = E_g^a + E_m^a. \tag{9.102}$$

Since $E_g^a = r_1 E_0$, the intensity ratio of the total diffracted beam to the incident wave is written by

$$\frac{P_g}{P_0} = \left|\frac{E_{g,m}^a}{E_0^a}\right|^2 = \left|\frac{r_1 - \exp(-i\gamma_g)}{1 - r_1 \exp(-i\gamma_g)}\right|^2. \tag{9.103}$$

If absorption is ignored, $P_g/P_0 = 1$ and the law of energy conservation holds. Figure 9.16 shows the calculated results for P_m/P_0 in black line and $\left|E_g^a/E_0^a\right|^2 = |r_1|^2$ in blue line. Very fine interference fringes can be seen. The number l attached to the peak is related with γ_r in (9.79) as $\gamma_r = (2l + 1)\pi$. P_m/P_0 can be observed by setting a slit as shown in Fig. 9.17.

This mirage diffraction can be observed if the incident wave is almost an ideal plane wave, but it is not easy to satisfy the condition. For example, when the difference in the deviation parameter ΔW is 0.015 for two peaks of $l = 29$ and 30, the corresponding difference in glancing angle $\Delta\alpha_0$ is estimated to be 0.14 μrad, by assuming that the incident X-ray energy is 11.1 keV, $\theta_B = 16.9°$ and $\left|\chi_g\right| = 5 \times 10^{-6}$ for Si220 reflection. Then, to observe adjacent peaks in Fig. 9.16 separately, a resolution of approximately 1/10 of this value of $\Delta\alpha_0$ is required.

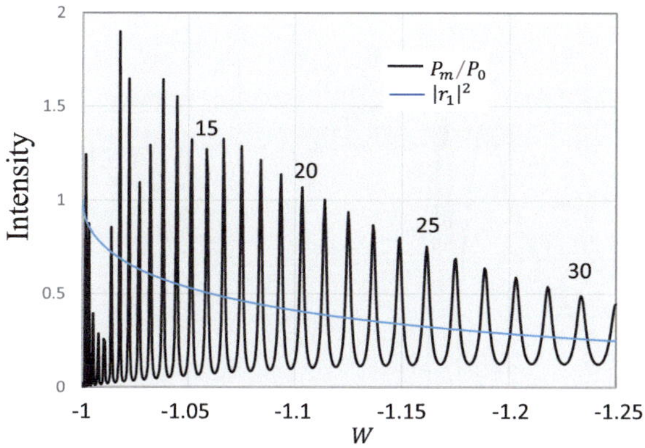

Fig. 9.16 P_m/P_0 and $\left|E_g^a/E_0^a\right|^2 = |r_1|^2$ as a function of W. $\Lambda_0 = 22$ μm, $\theta_B = 16.9°$, $\beta = 1.0\,\text{mm}^{-1}$, $k_a = 0.02$ and $g'' = -0.96$

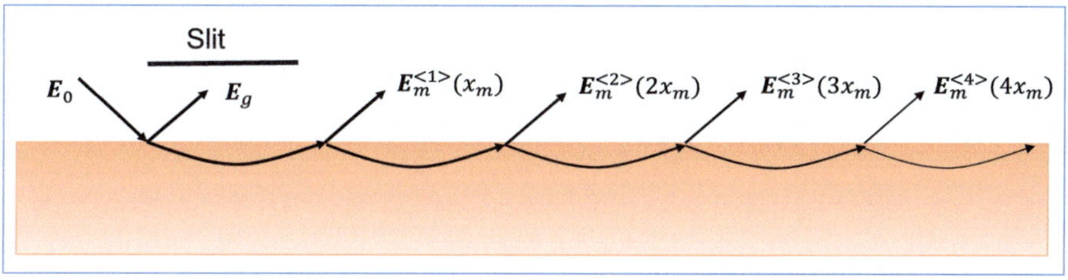

Fig. 9.17 By setting a slit, only the mirage diffracted intensities are observed

9.5.6 *Observation of Mirage Diffraction Intensity (Crystal Rotation Method)*

Figure 9.18 shows an example of a mirage diffraction intensity curves measured by the crystal rotation method [5, 6]. In the three-crystal diffractometer shown in the inset, C_1 is a collimator, C_2 is a monochromator using mirage diffraction and C_3 a sample. All of them are plane-parallel crystals of Si. C_2 and C_3 are slightly bent under cantilever load. As shown in Fig. 8.4, by extracting a part of the mirage diffracted wave from C_2, we can easily extract a beam with a very small divergence angle by using the angular amplification effect, which will be described later. When only the first peak (${}^2P_m^{\langle 1\rangle}$) of the mirage interference fringes from C_2 is extracted through the slit of S_1, its divergence angle is on the order of 0.1 s. Here, the superscript 2 on the left of ${}^2P_m^{\langle 1\rangle}$ represents the reflection from C_2 crystal. The intensity curve (${}^3P_g$) of the primary reflection from C_3 crystal (such as the intensity curve in Fig. 8.11) can be measured. The intensity curve ${}^3P_t + {}^3P_{lt}$ of the sum of the transmitted ${}^3P_t$ and the diffracted intensities ${}^3P_{lt}$ can be observed, which is on the lower angle side than the peak of ${}^3P_g$. The low angle side corresponds to the anomalous transmission

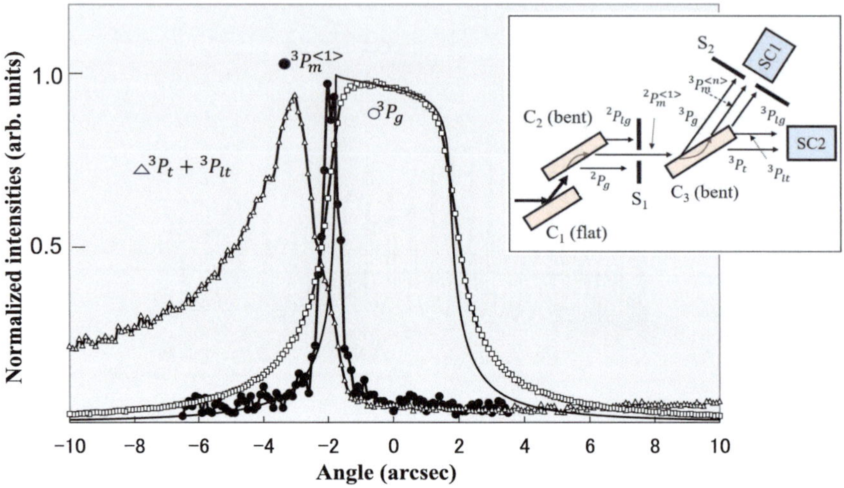

Fig. 9.18 Measured Si 220 rocking curves using mirage diffractometer (adapted from Fig. 9.6 in [5]). The solid curve shows the calculated rocking curve of the diffracted beam from a semi-infinite crystal. The inset shows three-crystal diffractometer. C_1 is a flat plane-parallel crystal, C_2 and C_3 are plane-parallel crystals slightly bent by cantilevers

region. The intensity curve of the first peak of the mirage interference fringes from C_3 ($^3P_m^{\langle 1\rangle}$) appears between the peaks ^3P_g and $^3P_t + ^3P_{lt}$. The width of the peak $^3P_m^{\langle 1\rangle}$ is approximately 0.5 s and 1/8 of ^3P_g.

If $^2P_m^{\langle 1\rangle}$ is an ideal plane wave, the mirage interference fringes shown in Fig. 9.16 should be observed on the intensity curve of $^3P_m^{\langle 1\rangle}$. To observe mirage interference fringes using the crystal rotation method, the divergence angle of the incident beam must be at least 0.01 s or less, which is not the case in Fig. 9.18. As will be described in the next Sects. 9.6.1 and 9.6.2, mirage interference fringes can be observed relatively easily by using a fixed-crystal method that takes advantage of the angular amplification effect.

9.6 Interference Fringes in the Bragg Case Using Pseudo-Plane Waves (Fixed-Crystal Method)

We discuss interference fringes in the symmetric Bragg case caused by quasi-plane waves in a slightly bent plane-parallel crystal. Figure 9.19 shows the geometry of beams that cause the interference fringes when $\beta > 0$ and Fig. 9.20 shows the geometry of beams that cause the interference fringes when $\beta < 0$.

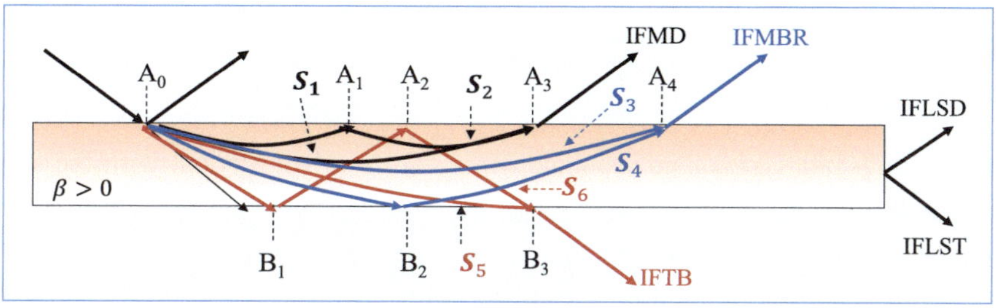

Fig. 9.19 Geometry of pseudo-refracted waves causing interference fringes for $\beta > 0$ in the symmetric Bragg case. The darker contrast in the crystal shows the larger strain

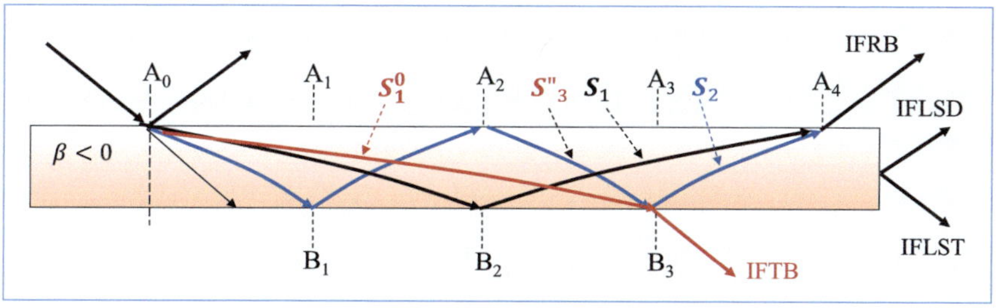

Fig. 9.20 Geometry of pseudo-refracted waves causing interference fringes for $\beta < 0$ in the symmetric Bragg case. The darker contrast in the crystal shows the larger strain

9.6.1 Interference Fringes Observable from the Incident Surface (When $\beta > 0$)

There are two types of interference fringes from a bent crystal with the strain decreasing from the top surface to the bottom surface ($\beta > 0$).

IFMD: As shown in Fig. 9.19, this is a mirage interference fringe caused by the interference between $\boldsymbol{S}_1$ and $\boldsymbol{S}_2$. $\boldsymbol{S}_1$ is a pseudo-refracted wave that directly reaches A_3 from the incident point A_0, that is, a mirage diffracted wave. $\boldsymbol{S}_2$ is another pseudo-refracted wave reflected once at A_1 and reaches A_3, that is, another mirage diffracted wave. Thus, the interference fringes are caused by two mirage diffracted waves and is abbreviated as IFMD. It is possible that there are waves that are reflected multiple times, more than once, from the incident surface and reach A_3.

IFMRB: This is caused by interference between the pseudo-refraction wave $\boldsymbol{S}_3$ which directly reaches A_4 from A_0 and $\boldsymbol{S}_4$ which is reflected once at B_2 on the bottom surface and reaches A_4, and is called IFMRB (Interference Fringes between Mirage diffracted beam and Reflected beam from the Bottom surface).

9.6.2 Interference Fringes Observable from the Incident Surface (When $\beta < 0$)

There is only one type of interference fringes from a bent crystal with the strain decreasing from the bottom surface to the top surface of the crystal ($\beta < 0$).

IFRB: The vertex of the hyperbolic trajectory of the pseudo-refracted wave is outside the crystal, as shown by the trajectory of $\boldsymbol{S}_2$ in Fig. 9.13. As shown in Fig. 9.20, the direction of the downward pseudo-refracted wave gradually becomes downward as it propagates into the crystal, and the slope of the upward pseudo-refracted wave gradually decreases as it approaches the incident plane. These interference fringes are caused by interference between the pseudo-refracted waves $\boldsymbol{S}_1$ and $\boldsymbol{S}_2$. The former is once reflected at B_2 on the bottom surface and reaches A_4 on the incident surface. The latter is reflected three times within the crystal (at B_1, A_2 and B_3) and reaches A_4. We call this IFRB (Interference Fringes between two beams Reflected from Bottom surface).

9.6.3 Interference Fringes Observable from the Transmission Surface

IFTB: In Figs. 9.19 and 9.20, the interference fringes can be observed in the transmitted beam from point B_3. One is caused by the interference between the pseudo-refraction wave $\boldsymbol{S}_5$ and $\boldsymbol{S}_6$ in Fig. 9.19 and another is caused by the interference between $\boldsymbol{S}_1^0$ and $\boldsymbol{S}''_3$ in Fig. 9.20. These are called IFTB (Interference Fringes in the Transmitted Beams).

9.6.4 Mirage Interference Fringes (IFMD) Using Pseudo-Plane Waves

We consider mirage interference fringes (IFMD) where the incident wave is a quasi-plane wave (wave-packet), and describe some experimental results. Figure 9.21a shows the trajectories of two pseudo-refracted beams forming IFMD, and Fig. 9.21b shows the movements of their tie-points on the dispersion surface. As already mentioned, even if the incident wave has a very small divergence angle, the pseudo-refracted wave excited by the angular amplification effect becomes the pseudo-spherical waves. In Fig. 9.21b, the tie-points of the pseudo-refracted waves lie on the dispersion surface within a width ΔW corresponding to the divergence angle of the incident beam.

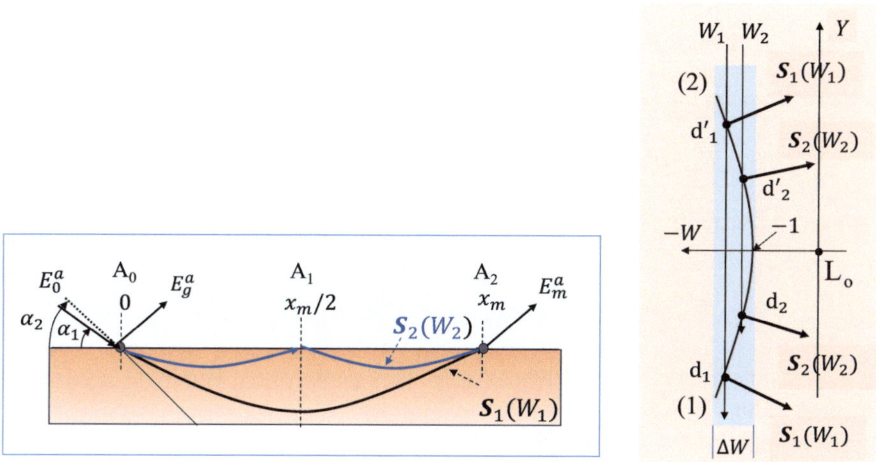

(a) (b)

Fig. 9.21 Principle of mirage interference fringes caused by two pseudo-refracted waves. (**a**) Trajectories of the pseudo-refracted waves. (**b**) Tie-points and the directions of pseudo-refracted waves on the dispersion surface

Two pseudo-refracted waves reach the point A_2 which is x_m away from the incident point A_0. One is $\boldsymbol{S}_1(W_1)$ which directly reaches the point A_2 (the tie-point moves from d_1 to d'_1). The other one is $\boldsymbol{S}_2(W_2)$ which reaches A_2 after being reflected once at the point A_1 on the incident surface (the tie-point moves from d_2 to d'_2). These two waves interfere with each other resulting in IFMD. From (9.73), the electric field of the wave that is reflected n times from the bottom surface and reaches A_2 is expressed as

$$E_m^{a,\langle n\rangle}\exp(-in\gamma_0) = E_0^a R_m^{\langle n\rangle}\exp\big(-in\gamma_m^{\langle n\rangle}\big), \tag{9.104}$$

where γ_0 is the phase shift in vacuum and is a real number. From (9.75), $R_m^{\langle n\rangle}$ is given by

$$R_m^{\langle n\rangle} = r_1^{n-1}(r_1^2 - 1). \tag{9.105}$$

Since $\gamma_m^{\langle n\rangle}$ is the phase shift in the crystal and is a complex number, by writing $\gamma_m^{\langle n\rangle} = \gamma_{mr}^{\langle n\rangle} - i\gamma_{mi}^{\langle n\rangle}$, we get

$$E_0^a R_m^{\langle n\rangle}\exp\big(-in\gamma_m^{\langle n\rangle}\big) = E_0^a R_m^{\langle n\rangle}\exp\big(-n\gamma_{mi}^{\langle n\rangle}\big)\exp\big(-in\gamma_{mr}^{\langle n\rangle}\big) = E_0^a A_m^{\langle n\rangle}\exp\big(-in\gamma_{mr}^{\langle n\rangle}\big). \tag{9.106}$$

$A_m^{\langle n\rangle}$ on the right side of this equation is

$$A_m^{\langle n\rangle} = R_m^{\langle n\rangle} \exp\left(-n\gamma_{mi}^{\langle n\rangle}\right). \tag{9.107}$$

Using this, the total electric field E_m^a reaching point A_2 is

$$E_m^a = E_0^a \exp\left(-i\gamma_{mr}^{\langle 1\rangle}\right)\left[A_m^{\langle 1\rangle} + A_m^{\langle 2\rangle} \exp\left(-i\Delta\gamma_{mr}^{\langle 2\rangle}\right) + \cdots\right]. \tag{9.108}$$

From (9.85) and (9.86), we get

$$\Delta\gamma_{mr}^{\langle n\rangle} = n\gamma_{mr}^{\langle n\rangle} - \gamma_{mr}^{\langle 1\rangle} = \frac{-2\pi}{\Lambda_0\beta}\left[nM\left(W_s^{\langle n\rangle}\right) - M\left(W_s^{\langle 1\rangle}\right)\right]. \tag{9.109}$$

Here, Λ_0 is the interval between two consecutive peaks of the Pendellösung fringes appeared in Sect. 9.4.3.1.

An example of the calculated results of IFMD for Si 220 as a function of the distance x (x_m) is shown in Fig. 9.22. The blue line shows the two-beam interference fringes with $n = 1$ and 2, and the black line shows the sum of interference fringes with n from 1 to 10. In either case, the fringe interval of IFMD becomes narrower as x becomes larger. β-dependence of calculated Si 220 IFMD is shown Fig. 9.23. Interval of IFMD becomes narrower as β becomes larger.

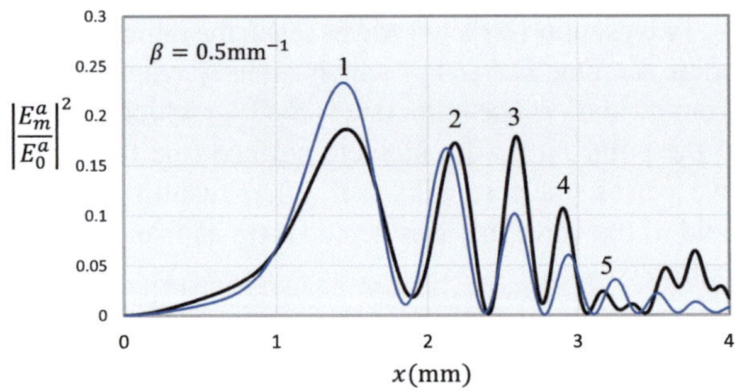

Fig. 9.22 Calculated Si 220 IFMD when $\theta_B = 16.9°$, $\Lambda_0 = 0.022$ mm, $\beta = 0.5$ mm^{-1}, $k_a = 0.02$ and $g'' = -0.96$

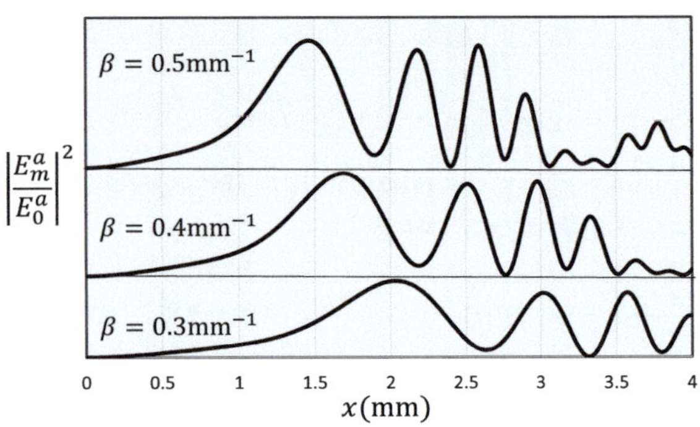

Fig. 9.23 Calculated β dependence of Si 220 IFMD when $\theta_B = 16.9°$, $\Lambda_0 = 0.022$ mm, $k_a = 0.02$ and $g'' = -0.96$

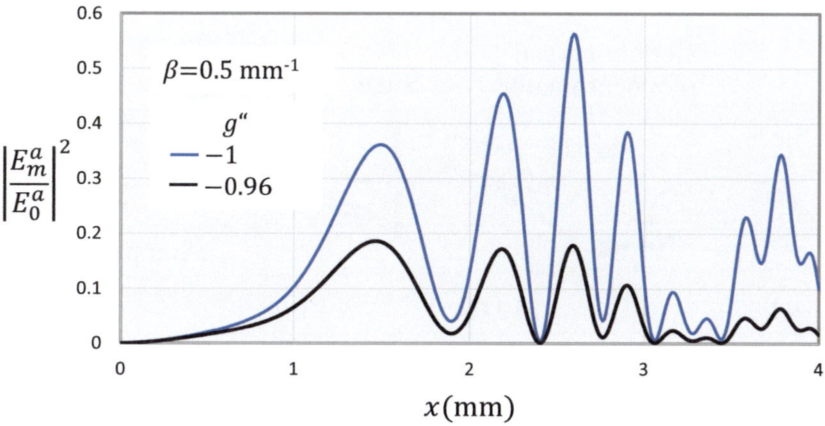

Fig. 9.24 The effect of temperature on IFMD of Si 220. The parameters used are $\theta_B = 16.9°$, $\Lambda_0 = 0.022$ mm, $\beta = 0.5$ mm^{-1} and $k_a = 0.02$

Under anomalous transmission condition, the effect of thermal vibrations appears conspicuously on $\gamma_{mi}^{\langle n\rangle}$ via g''. The parameter g'' is expressed as

$$\left|g''\right| = \frac{|\chi_{hi}|}{|\chi_{0i}|} = \exp(-M) \tag{9.110}$$

and is nothing but the Debye–Waller factor $\exp(-M)$. Figure 9.24 shows the comparison of IFMD for $\left|g''\right| = 1.0$ and 0.96, which correspond to 0 K (without the thermal vibration effect) and room temperature, respectively. The intensity of IFMD becomes smaller as x increases.

9.6.4.1 Observation of Mirage Interference Fringes (IFMD)

The optical system for observing IFMD is shown in Fig. 9.25a. The crystal is bent by fixing one end of the crystal and applying a force to the other end (cantilever), as shown in Fig. 9.25b.

When a force is applied to the end of a cantilever, the strain gradient parameter β is proportional to strain gradient and Dl, where D is the deflection and l is the distance between the load point and the incident point of X-rays, as explained in Appendix C.3. A cantilever bends under gravity even when there is no load. It receives a uniformly distributed load of gravity, and β is proportional to Dl^2 as explained in Appendix C.4.

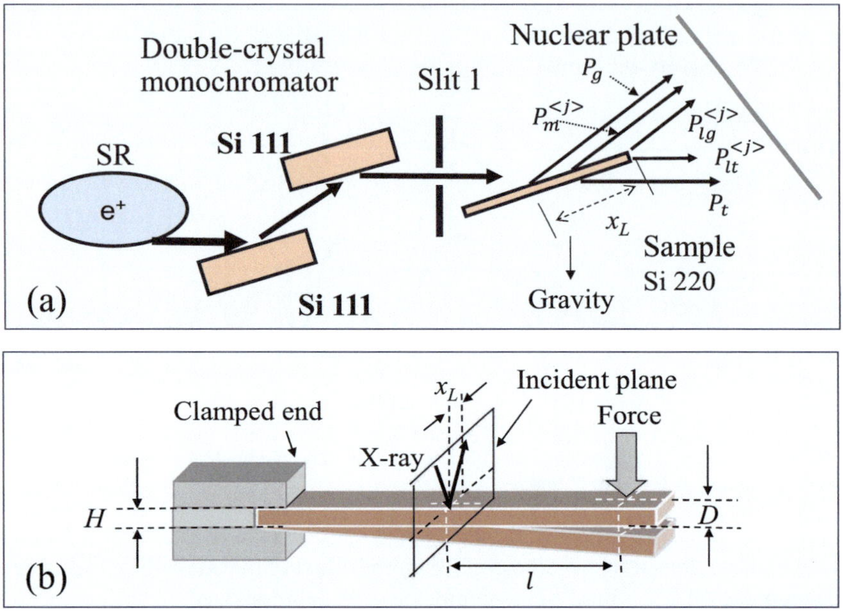

Fig. 9.25 A schematic diagram of IFMD measuring system. **a** The optical system. **b** Sample geometry. One end of a sample crystal is clamped and the force is applied to the other end in the direction of gravity. x_L is the distance from the incident point of X-rays to the crystal edge of the entrance surface

9.6.4.2 Deflection Dependence of IFMD

Figure 9.26a shows IFMD recorded on a nuclear emulsion plate when the deflection D of the end of the sample was changed. When $D = 0$ μm (left photo), no IFMD is observed between the first-order reflection and the interference fringes in the direction of diffraction from the lateral surface (IFLSD). When $D = 20$ μm, four fringes of IFMD are observed, and when $D = 40$ μm, ten fringes of IFMD are observed. The fringe interval of IFMD becomes narrower as x increases, and it also becomes narrower as β becomes large. In Fig. 9.26b, the brown square represents the value of β which is obtained from the third peak position of the observed IFMD as a function of D. The strain gradient β increases linearly as a function of deflection D. As shown in Fig. 9.21a, the traveling direction of the pseudo-spherical wave $\boldsymbol{S}(W)$ differs depending on the value of W. When $|W_2| > |W_1|$, $\boldsymbol{S}(W_2)$ is a wave traveling below $\boldsymbol{S}(W_1)$. As shown in Fig. 9.27, if the trajectory of the wave $\boldsymbol{S}(W_2)$ is a hyperbola with the vertex touches the bottom surface at (x_a, H) and reaches the position Δz from the bottom on the lateral surface, the trajectory is uniquely determined. Using this, the value of β can be obtained. The gray circle represents the value of β obtained from the measurement of $\Delta z\cos\theta_B$, which is the length of the blank section shown in Fig. 9.26a.

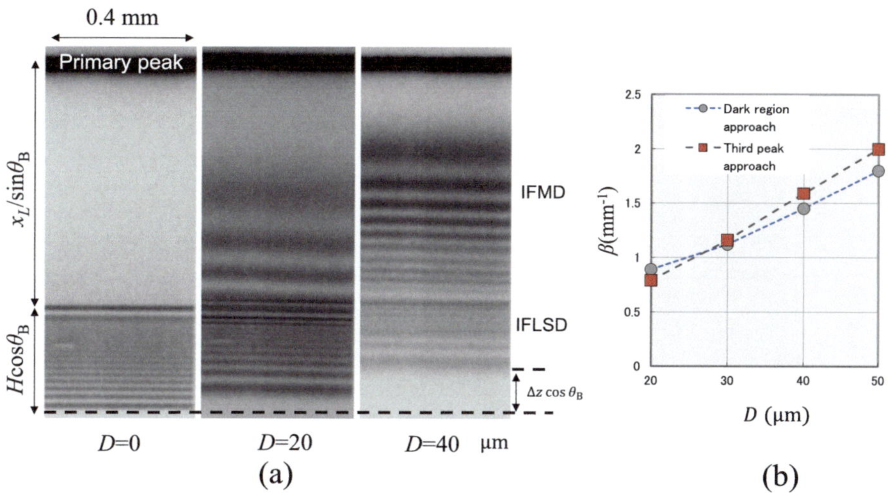

Fig. 9.26 IFMD when D is changed. (**a**) Si 220 reflection topography when $l = 21$mm and $x_L = 2.2$mm in case of $D = 0$ (adapted from Fig. 9.6 in [15]). (**b**) Plots of β for D from 20 μm to 50 μm. ■ represents the value obtained from IFMD and ● that obtained from the length of the blank section Δz (adapted from Fig. 9.8a in [15])

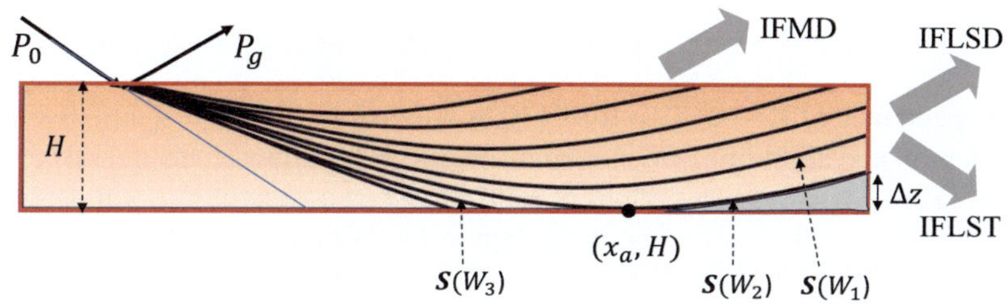

Fig. 9.27 Trajectories of pseudo-spherical waves $\boldsymbol{S}(W_1)$, $\boldsymbol{S}(W_2)$ and $\boldsymbol{S}(W_3)$ when $|W_3| > |W_2| > |W_1|$. The hyperbolic trajectory of $\boldsymbol{S}(W_2)$ touches the bottom surface at its vertex and reaches the position of height Δz on the lateral surface

The beam of $\boldsymbol{S}(W_1)$ which satisfies the condition $|W_1| < |W_2|$ directly reaches the lateral surface. On the other hand, the beam of $\boldsymbol{S}(W_3)$ which satisfies the condition $|W_3| > |W_2|$ cannot directly reach the lateral surface. This causes a blank section of length Δz from the bottom, from which we can obtain the value of W_2.

9.6.4.3 Strain Dependence of IFMD

Figure 9.28a shows the reflection topographies of Si 220 when the deflection D is fixed and the distance l from the free end of the crystal was varied to investigate the

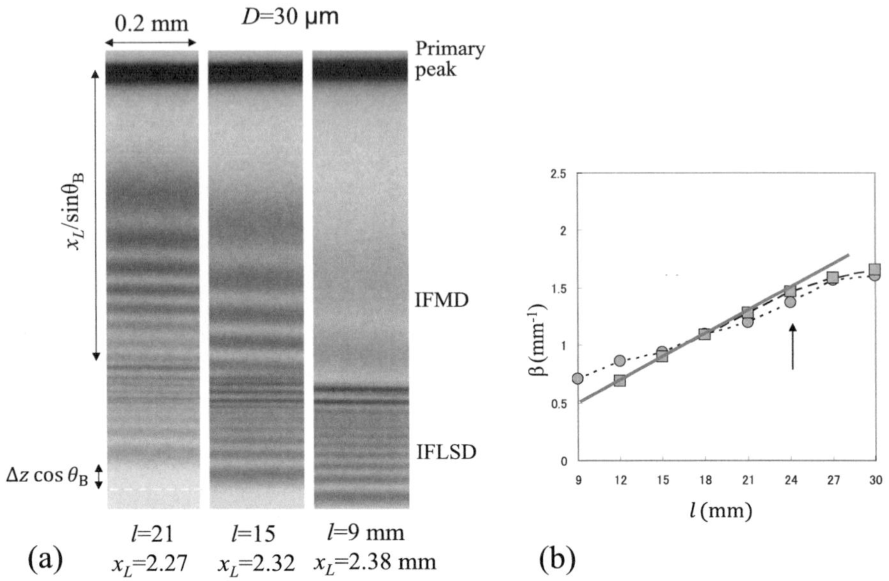

Fig. 9.28 IFMD when varying the distance l from the free end of the crystal to the observation point with deflection $D = 30\mu$m. (**a**) Si 220 reflection topography (adapted from Fig. 9.7 in [15]). (**b**) $l - \beta$ curves. ■ is obtained from IFMD, and ● is obtained from the value of Δz (adapted from Fig. 9.8b in [15])

strain dependence of IFMD. Two IFMD lines are visible when $l = 9$ mm, five lines when 15 mm, and eight lines when 21 mm. Figure 9.28b shows the values of β with respect to l; $l - \beta$ curves. The values of β lie almost on a straight line, although they are slightly below the straight line in the region $l > 24$ mm.

9.6.4.4 β and Strain Gradient

The parameter β is proportional to the strain gradient ε'. The relation between β and ε' in the symmetric Bragg case is shown in (C.21) in Appendix C. According to it,

$$\varepsilon\prime = \frac{d_0}{\Lambda_0 \tan^2\theta_B}\beta. \tag{9.111}$$

Here, d_0 is the lattice spacing in the thickness direction before deformation. The strain ε is expressed by

$$\varepsilon = \frac{d - d_0}{d_0} = \varepsilon'\left(z - \frac{H}{2}\right), \tag{9.112}$$

where d is the lattice spacing after deformation, and H is the thickness of the crystal. From this equation, ε becomes zero when $z = H/2$, and a neutral layer exists there.

We can obtain ε' from β using (9.111). According to the experimental results shown in Fig. 9.28b, the strain gradient is obtained as $\varepsilon\prime = 1.0 \times 10^{-4}\ \text{mm}^{-1}$ by using $\beta = 1.1\text{mm}^{-1}$ as $l = 18\text{mm}$ and $D = 30\mu\text{m}$. The strain ε on the incident surface ($z = 0$) is obtained as $\varepsilon = -1.4 \times 10^{-5}$ from (9.112). This value of ε is one to three orders of magnitude smaller than those of the conventional measurement range where the $\sin^2\psi$ method described by Cullity [7] is applied. It is expected that IFMD should be useful to measure much smaller stress than usual.

9.6.5 *Interference Fringes Between Mirage Diffraction Waves and Bottom Reflected Waves (IFMRB)*

We consider IFMRB, which is the interference fringes between the mirage diffracted wave and the reflected wave from bottom surface. If β is small and the crystal is thin, the pseudo-refracted wave $\boldsymbol{S}_1$ reaches the bottom surface, and reflected from the bottom surface as $\boldsymbol{S}'_1$ which can interfere with the mirage diffracted wave $\boldsymbol{S}_0$ to form IFMRB as shown in Fig. 9.29. The observed IFMRB are shown in Fig. 9.30a. The crystal plate is deformed by gravity with its one end fixed (uniform load distribution). The fringe interval of IFMRB becomes wider as the distance x from the incident point increases, which is opposite to the x dependence of the fringe interval of IFMD. These two interference fringes can be clearly distinguished.

The IFMRB is observed in the range from x_{min} to x_{max} in the example shown in Fig. 9.29. x_{max} is the arrival point of mirage diffracted beam $\boldsymbol{S}_2$ on the incident surface when its vertex of the hyperbolic trajectory coincides with $z_a = H$. x_{min} is the minimum distance of the arrival point of the pseudo-refracted wave $\boldsymbol{S}'_1$ that is reflected from the bottom and returns to the incident surface. We call $x < x_{max}$ the first zone, and $x > x_{max}$ the second zone. The interference fringes in the second zone are the interference between $\boldsymbol{S}'_0$, which has been reflected once on the entrance surface, and the mirage diffracted wave, which has been reflected more than once.

If we write the position A_1 where a part of the mirage diffracted wave $\boldsymbol{S}_0$ is emitted from the surface as x_0, the initial value of deviation parameter W_{s0} for $\boldsymbol{S}_0$ can be written by

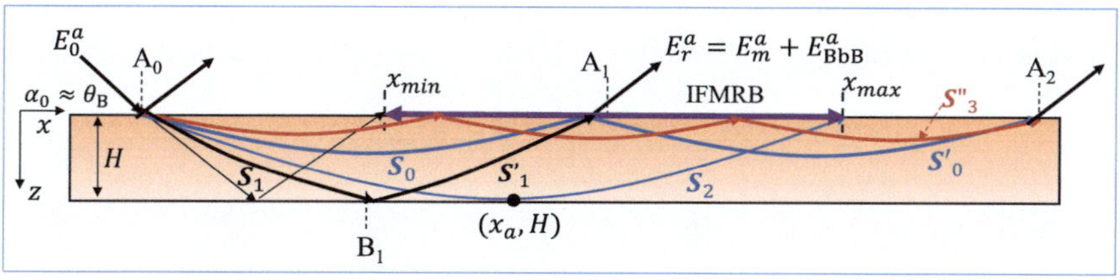

Fig. 9.29 Beam geometry to form IFMRB

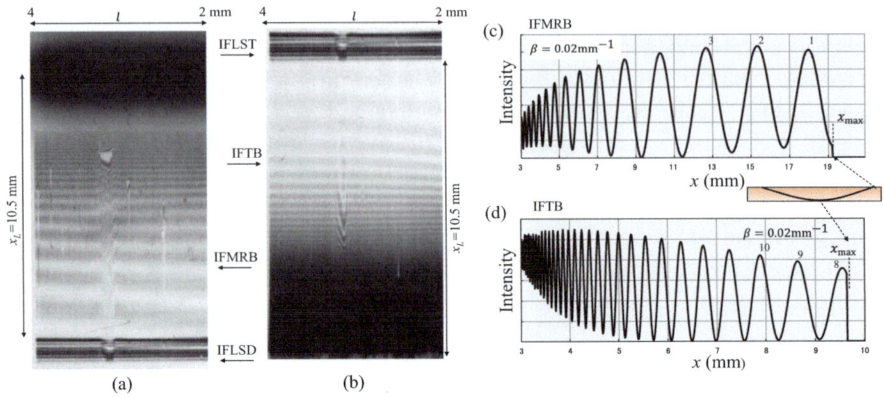

Fig. 9.30 Observed Si 220 reflection and transmission topographies and the corresponding calculated results of IFMRB and IFTB. (**a**) and (**b**) are the topographies. (**c**) and (**d**) are the calculated results of IFMRB and IFTB, respectively, by using $\beta = 0.02\,\text{mm}^{-1}$ and $H = 0.28\,\text{mm}$. The topographies are taken by the authors (unpublished). The experimental conditions are described in [17]

$$W_{s0} = -\left[1 + \left(\frac{\beta x_0}{2}\right)^2\right]^{\frac{1}{2}}. \tag{9.113}$$

As for the wave $\boldsymbol{S}_1$, we write the initial value of deviation parameter at the incident point A_0 as W_{s1}, and that at B_1 on the bottom surface as W_{e1}. W_{s1} is obtained from (9.65) by putting $x = x_0/2$ and $y = H$, and W_{e1} is obtained from (9.64) as

$$W_{e1} = \frac{\beta H}{\tan\theta_B} + W_{s1}. \tag{9.114}$$

Let write the electric field of the diffracted wave at A_1 as E_r^a, that of the mirage diffracted wave $\boldsymbol{S}_0$ as E_m^a, and that of reflected wave from the bottom as E_{BbB}^a. The subscript 'b' of E_{BbB}^a means that the reflection from the bottom surface satisfies the boundary condition in the Bragg case. For a non-absorbing crystal, we get

$$E_m^a = R_1 E_0^a \exp(-i\gamma_{r1}),\ E_{\text{BbB}}^a - R_2 E_0^a \exp(-i\gamma_{r2}). \tag{9.115}$$

Square of the absolute value of the electric field E_r^a of the diffracted wave which is the sum of E_m^a and E_{BbB}^a is given by

$$\left|E_r^a\right|^2 = \left|E_m^a + E_{\text{BbB}}^a\right|^2 = \left[R_1^2 + R_1^2 + 2R_1R_2\cos(\Delta\gamma)\right]\left|E_0^a\right|^2. \tag{9.116}$$

Here, R_1, R_2 and $\Delta\gamma$ are given by

$$R_1 = r_1^2(W_{s0}) - 1, R_2 = r_1(W_{s1})\left(\frac{r_1(W_{s1})}{r_2(W_{e1})} - 1\right) \tag{9.117}$$

and

$$\Delta\gamma = \gamma_{r2} - \gamma_{r1} = \frac{G}{\beta}[\{M(W_{s1}) - M(W_{e1})\} - M(W_{s0})], \tag{9.118}$$

with $M(W)$ being given by (9.85). If the peak number from the x_{max} side is n, then $\Delta\gamma_{rn}$ is given by

$$\Delta\gamma_{rn} = (2n - 1)\pi.$$

The observed results of IFMRB are shown in Fig. 9.30a and the corresponding calculated results are shown in Fig. 9.30c. In this calculation, the value of $\beta = 0.02\text{mm}^{-1}$ is used, which is approximately two orders of magnitude smaller than the value obtained by IFMD.

9.6.6 *Mirage Interference Fringes Modulated by Bottom Reflected Waves (MIFMD)*

As already mentioned, the fringe interval of IFMD becomes narrower as the distance x increases, while that of IFMRB becomes wider as x increases. The two interference fringes can be clearly distinguished. However, IFMD and IFMRB can coexist depending on the value of β. An example of the observed result of this coexistence is shown in Fig. 9.31a. This is the reflection topography of Si 220. One end of the sample crystal plate was fixed and the crystal is bent by gravity. This topography was taken in the range of 16 to 20 mm from the free end of the plate, and the corresponding value of β is 0.15 mm^{-1} for $l = 16$ mm at the right end of the photograph and 0.23 mm^{-1} for $l = 20$ mm at the left end. Figure 9.31b shows the calculated result of interference fringes between multiple mirage diffracted waves and the reflected waves from the bottom surface, which is called MIFMD (Modified IFMD). Furthermore, Fig. 9.31c shows the calculated result of interference fringes between multiple mirage diffracted waves, that is, IFMD. Figure 9.31d shows the calculated result of IFMRB. In these figures, M_1 to M_7 are fringes belonging to IFMD in the first zone, and M'_1 to M'_3 are belonging to IFMD in the second zone. At the left end of Fig. 9.31a, six fine fringes appear among the wide fringe of M_1, which are IFMRB. Three fringes appear on M_2, two fringes appear on M_3, and one fringe appears on M_4 to M_7. Of the three fringes of IFMRB in M_2, the dark contrast at the top becomes somewhat thinner as it goes to the right, then darkens again at the right edge, where it connects to the bottom fringe in M_1. In addition, two fringes are visible in M_3 at the left end, and its upper fringe is thinner than the lower, but it becomes darker at the right end and connects to the lower fringe of M_2. This trend is consistent with

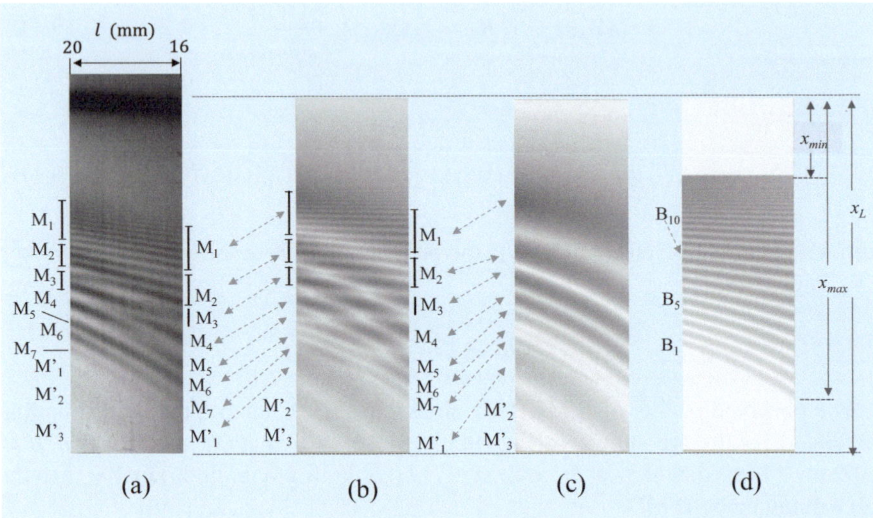

Fig. 9.31 Si 220 topography (adapted from Fig. 9.5 in [19]). (**a**) Experimental topography. (**b**), (**c**) and (**d**) show the calculated MIFMD, IFMD and IFMRB. The value of β is 0.15mm^{-1} at the right end and 0.23mm^{-1} at the left end. x_L is 8.4 mm

the calculated results of MIFMD in Fig. 9.31b. In contrast to the complex changes of observed and calculated fringes seen in Fig. 9.31a, b, the changes in IFMD in Fig. 9.31c and the IFMRB in Fig. 9.31d are regular and monotonous as a function of x as well as β.

9.6.7 Interference Fringes in the Transmitted Beam (IFTB)

When IFMRB are observed from the incident surface, interference fringes in the transmitted beam (IFTB) are also observed from the transmission surface. Figure 9.32 shows the beam geometry for IFTB. The observed results are shown in Fig. 9.30b. IFTB is caused by interference between $\boldsymbol{S}_0$ and $\boldsymbol{S}_3$. The former reaches the bottom surface directly from the incident point, while the latter reaches the bottom surface after once reflected from the bottom surface and once reflected from the entrance surface. As in the case of IFMRB, the fringe interval becomes wider as x increases. Since the differences in wave number as well as in the path length between the two beams for IFTB are both larger than those for IFMRB, the fringe interval is narrower than that of IFMRB.

We write the electric field component of the diffracted wave due to the pseudo-refracted wave $\boldsymbol{S}_0$ in Fig. 9.32 as E^a_{BT}, and that due to the pseudo-refracted wave $\boldsymbol{S}_3$ as $E^a_{\text{Bb}^2\text{T}}$. 'T' in the subscript means that the wave satisfies the boundary condition of the Bragg case on the transmission surface. As shown in Fig. 9.32, IFTB is observed

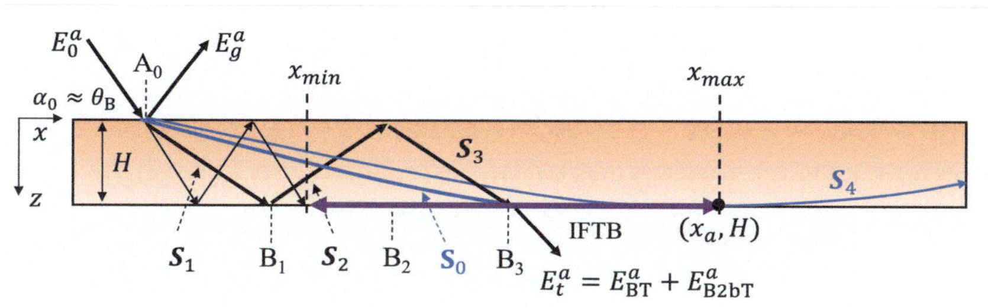

Fig. 9.32 Beam geometry for formation of IFTB

in the region from x_{min} to x_{max} which is less than half of the region where IFMRB is observed.

The electric field of the transmitted wave E_t^a for a non-absorbing crystal is given by

$$\left|E_t^a\right|^2 = \left|E_{\mathrm{BT}}^a + E_{\mathrm{Bb^2T}}^a\right|^2 = \left[T_1^2 + T_2^2 + 2T_1T_2\cos(\Delta\gamma_r)\right]\left|E_0^a\right|^2. \tag{9.119}$$

T_1 and T_2 are

$$T_1 = 1 - \frac{r_1(W_{s0})}{r_2(W_{e0})},\ T_2 = \frac{r_1(W_{s1})}{r_2(W_{e1})}\left[1 - \frac{r_1(W_{s1})}{r_2(W_{e1})}\right], \tag{9.120}$$

where W_{s0} and W_{e0} are the values of W_s at $\mathrm{A_0}$ and at $\mathrm{B_3}$ for $\boldsymbol{S}_0$, respectively, and W_{s1} and W_{e1} are those for $\boldsymbol{S}_1$. $\Delta\gamma_r$ in (9.119) is given by

$$\Delta\gamma_r = \frac{G}{2\beta}[3\{M(W_{s1}) - M(W_{e1})\} - M(W_{s0})]. \tag{9.121}$$

Figure 9.30d shows an example of calculated IFTB. The peaks of IFTB are numbered from the x_{max} side toward x_{min}, where n-th is given by $n = \Delta\gamma_{rn}/(2\pi)$. In the case of IFMD and IFMRB, a mirage diffracted wave is essential, but it is not in the case of IFTB. IFTB can be observed even when the crystal is not strained ($\beta = 0$) or there is no mirage diffracted wave when $\beta < 0$ as described next.

9.6.8 Interference Fringes When Strain Gradient is Negative (IFRB)

We consider the interference fringes when the strain gradient β is negative. Under the anomalous transmission condition, $W_s \leq -1$ so that $\beta W_s > 0$. The trajectory of the pseudo-refracted wave is shown in Fig. 9.33a, and the movement of the corresponding tie-point along the dispersion surface as the wave propagates is shown in Fig. 9.33b. The trajectory of the pseudo-refracted wave $\boldsymbol{S}_1$ becomes a hyperbola with an open

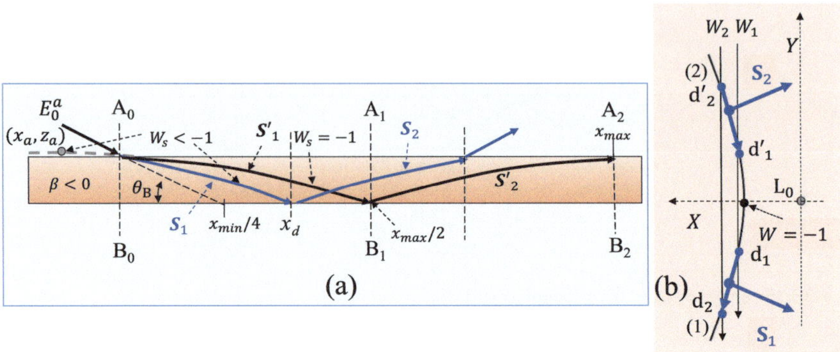

Fig. 9.33 Trajectories of pseudo-refracted waves in the strained crystal when $\beta < 0$ (**a**), and the movement of the corresponding tie-points on the dispersion surface (**b**)

bottom and reaches the bottom surface of the crystal. As seen in (9.66), the vertex (x_a, z_a) does not exist inside the crystal, because $x_a < 0$ and $z_a < 0$. The tie-point of the wave $\boldsymbol{S}_1$ moves from d_1 to d_2 on branch (1) as the wave propagates. When the wave $\boldsymbol{S}_1$ reaches x_d on the bottom surface, a part of the wave is reflected and becomes the wave $\boldsymbol{S}_2$, and the corresponding tie-point moves from d'_2 to d'_1 on branch (2) as the wave propagates.

When the deviation parameter at the incident point W_s is -1, the vertex of the hyperbolic trajectory of the pseudo-refracted wave is at the incident point and the direction of the trajectory is downward. Then the pseudo-refracted wave enters the crystal and propagate the maximum distance (x_{max}) in the x direction after once reflected from the bottom surface as shown in Fig. 9.33a.

As shown in Fig. 9.20, IFRB is caused by interference between the pseudo-refracted waves $\boldsymbol{S}_1$ and $\boldsymbol{S}_2$. $\boldsymbol{S}_1$ is the wave reflected once from the bottom surface and $\boldsymbol{S}_2$ is the wave reflected twice from the bottom surface and once from the top surface. On the other hand, IFTB is caused by interference between the pseudo-refracted waves $\boldsymbol{S}_1^0$ and $\boldsymbol{S}''_3$. The wave $\boldsymbol{S}_1^0$ reaches the bottom surface directly and the wave $\boldsymbol{S}''_3$ reaches the bottom surface after reflected once at the bottom and once at the top surface. The observed results of IFRB are shown in Fig. 9.34a, and those of IFTB are shown in Fig. 9.34b. The calculated results of IFRB are shown in Fig. 9.34c, and those of IFTB are shown in Fig. 9.34d. The fringe interval of both IFRB and IFTB increases as x increases.

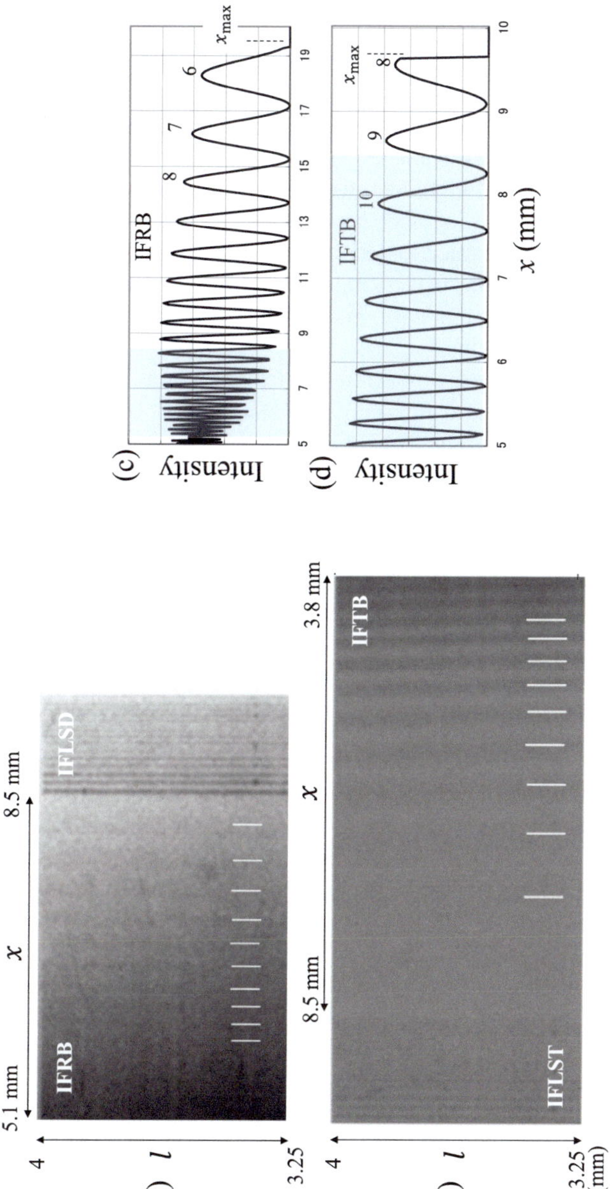

Fig. 9.34 Reflection (**a**) and transmission (**b**) topographies of Si 220 for $\beta < 0$. The calculated IFRB (**c**) and IFTB (**d**) under the corresponding conditions. The value of $\beta = -0.02\ \text{mm}^{-1}$ is used. Note that the direction of x is reversed in (b). The photographs are reproduced from Figs. 9.4a and 9.5a in Fukamachi [18]

9.7 Some Studies on Dynamical Diffraction for Crystals Without Sufficient Perfection

In the X-ray dynamical theory of diffraction for perfect crystals proposed by Ewald, the dispersion equation (dispersion surface) plays a major role. Among existing crystals, there are very few that can be called ideally perfect crystals, and most crystals may have some kind of defects that affect their physical properties. The degree of perfection of crystals has been evaluated using dynamical diffraction theory. For this purpose, it is necessary to elucidate the diffraction phenomenon from crystals containing defects, and various methods have been proposed. In this book, we have considered a local reciprocal lattice to derive a local dispersion surface, and studied diffraction phenomena in a deformed crystal with a strain gradient. We have focused on mirage diffraction in the Bragg case, and described various interference fringes observed in reflection topography.

Some related works on mirage interference fringes are as follows. IFMD was first observed by Zaumseid [8] using quartz crystals. Later, Bak-Misiuk et al. [9] grew a SiO_2 film on Si, observed IFMD under strain and analyzed it using the Takagi equation. In addition, Chukhovskii and Petrashen [10] predicted IFMD using spherical wave theory and estimated its peak position. Yan and Noyan [11] bent a Si flat plate crystal using a cantilever method and observed the shift of the peak of mirage diffraction by varying the strain. They analyzed the observed data using spherical wave theory in [12]. Smirnova et al. [13] observed IFMD in the Si 400 reflection topography from a $Si_{1-x}Ge_x$ heterostructure film by applying strain and analyzed it using the Takagi equation. They investigated the change in the period of IFMD when the curvature of the film was changed. Fukamachi et al. [14] and Jongsukuswat et al. [15] observed IFMD from a cantilever crystal. In the latter, strain gradients were determined from the interval of interference fringes. IFMRB was observed by Fukamachi et al. [16], IFTB was observed by Jongsukswat et al. [17], and IFRB was observed by Fukamachi et al. [18]. MIFMD was investigated by Fukamachi and Kawamura [19].

The Takagi equation [20, 21] is widely used to analyze strained crystals instead of using local dispersion surface. The Takagi equation is briefly described in Appendix D.

Dynamic diffraction theories for crystals that do not have sufficient perfection are classified into the following three types in order of decreasing degree of imperfection.

Type 1. When the strain is small and the concept of local dispersion surface can be applied.

Type 2. When Takagi's equation is applicable.

Type 3. When statistical dynamical diffraction theory is applicable [22, 23].

The monograph by Authier [3] provides a good introduction to these theories.

References

1. P. Penning, D. Polder, Philips Res. Report **16**, 419–440 (1961)
2. J. Gronkowski, C. Malgerange, Acta Cryst. **A40**, 507–514 (1984)
3. A. Authier, *Dynamical Theory of X-ray Diffraction* (Oxford University Press, 2001)
4. M. Hart, Z. Phys. **189**, 269–291 (1966)
5. T. Fukamachi, S. Jongsukswat, D. Ju, R. Negishi, K. Hirano, T. Kawamura, Appl. Cryst. **47**, 1267–1272 (2014)
6. T. Fukamachi, S. Jongsukswat, D. Ju, R. Negishi, K. Hirano, T. Kawamura, Appl. Cryst. **48**, 312 (2015)
7. B. D. Cullity, *Elements of X-Ray Diffraction*, 2nd edn, Addison-Wesley Publishing Co (1978)
8. P. Zaumseil, Krist. Tech. **13**, 983–990 (1978)
9. J. Bak-Misiuk, J. Gronkowski, J. Härtwing, W. Wierzehowski, Phys. Status Sodidi **A99**, 345–351 (1987)
10. F.N. Chukhovskii, P.V. Petrashen, Acta Cryst. **A44**, 8–14 (1988)
11. H. Yan, I.C. Noyan, J. Appl. Cryst. **39**, 320–325 (2006)
12. H. Yan, Ö. Kalenci, I.C. Noyan, *J. Appl. Cryst.* **40**, 322–331 (2007)
13. I.A. Smirnova, E.V. Suvorov, E.V. Shulakov, Phys. Solid State **53**, 35–40 (2011)
14. T. Fukamachi, M. Tohyama, K. Hirano, M. Yoshizawa, J.D. Negishi, K. Hirano, T. Kawamura, *Acta Cryst.* A**66**, 421–426 (2010)
15. S. Jongsukswat, T. Fukamachi, K. Hirano, D. Ju, R. Negishi, M. Shimojo, K. Hirano, T. Kawamura, Jpn. J. Appl. Phys. **51**, 076702 (2012)
16. T. Fukamachi, S. Jongsukswat, K. Hirano, R. Negishi, M. Shimojo, D. Ju, K. Hirano, T. Kawamura, *J. Phys. Soc. Jpn.* **80**, 083002(2011)
17. S. Jongsukswat, T. Fukamachi, K. Hirano, D. Ju, R. Negishi, M. Shimojo, K. Hirano, T. Kawamura, *J. Phys. Soc. Jpn.* **81**, 094804 (2012)
18. T. Fukamachi, S. Jongsukswat, D. Ju, R. Negishi, K. Hirano, T. Kawamura, Acta Cryst. **A75**, 842–850 (2019)
19. T. Fukamachi, T. Kawamura, Acta Cryst. **A78**, 422–429 (2022)
20. S. Takagi, *Acta Cryst.* **15**, 1311–1312 (1962)
21. S. Takagi, J. Phys. Soc. Japan **26**, 1239–1253 (1969)
22. N. Kato, Acta Cryst. **A36**, 763–769 (1980)
23. N. Kato, Acta Cryst. **A36**, 770–779 (1980)

Appendix A
Darwin's Method

Based on the reflection from a thin film obtained in the optical description in Chap. 4, we will explain Darwin's method, which describe diffraction from a crystal consisting of layers of lattice planes called mesh planes. In the Darwin's method, we do not use the reciprocal lattice space.

A.1 Reflectance from One Thin Mesh Film

Darwin's diffraction model treats the diffraction from a crystal as a stack of thin mesh planes. The reflectance from one thin film with thickness d is given by (4.81),

$$r_s = -i\frac{r_e \lambda \rho d}{\sin\theta_i}.$$

If the thin film consists of only one layer of unit cells (its thickness is d), the electron density is $\rho = |F|/V_c$ by using the crystal structure factor F and the unit cell volume V_c. By writing the Bragg condition as $m\lambda = 2d\sin\theta_i$, the scattering amplitude q for the scattering angle θ_i is given by

$$q = -\frac{r_e}{m}\frac{2d^2}{V_c}|F| = -\frac{\pi d|\chi_g|}{\lambda\sin\theta_i}. \tag{A.1}$$

$|q|$ is a very small quantity of about 10^{-5}. The scattering amplitude of forward scattering is written by

$$q_0 = q\frac{|F_0|}{|F|} = q\frac{|\chi_0|}{|\chi_g|}, \tag{A.2}$$

T. Fukamachi and T. Kawamura, *Resonant Dynamical X-ray Diffraction*, Springer Series in Solid-State Sciences 208, https://doi.org/10.1007/978-981-96-8406-9

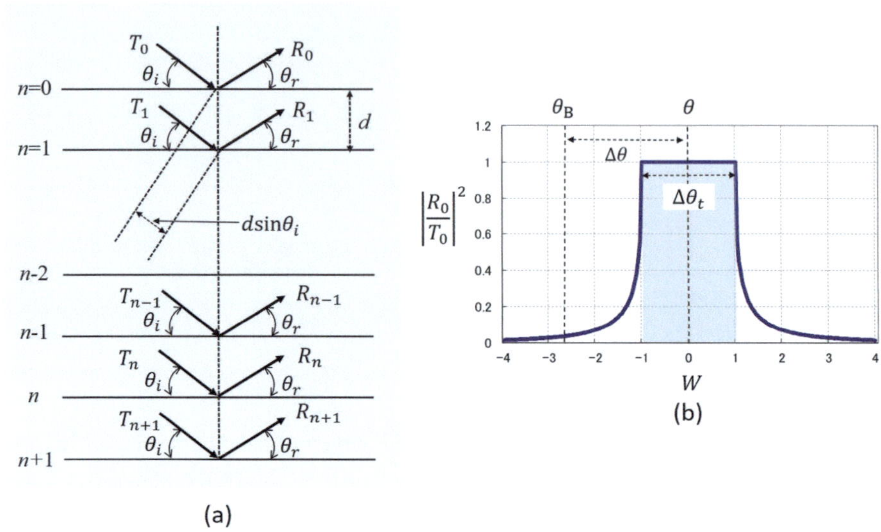

Fig. A.1 **a** A schematic diagram of Darwin's diffraction model. **b** Darwin curve

where F_0 is the crystal structure factor for forward scattering.

If the crystal is a stack of mesh planes as shown in Fig. A.1a, the optical path difference between two consecutive planes is $d\sin\theta_i$, and the phase difference ϕ is

$$\phi = kd\ \sin\theta_i, \tag{A.3}$$

using the relation $k = 2\pi/\lambda$. If we denote the amplitude of the transmitted wave coming to the n-th plane as T_n and that of the reflected wave as R_n, we have the following relations

$$T_n = (1 + iq_0)T_{n-1}\exp(-i\phi) + iqR_n\exp(-i2\phi), \tag{A.4}$$

$$R_n = iqT_n + (1 + iq_0)R_{n+1}\exp(-i\phi). \tag{A.5}$$

The first term in (A.4) represents the amplitude of transmitted wave from (n-1)-th plane T_{n-1}, and the second term represents that of reflected wave R_n which travels to (n-1)-th plane and is reflected from it, so the phase difference is 2 ϕ.

A.2 Diffraction from Semi-Infinite Crystal

By using (A.4) and (A.5), we can obtain an equation for only the amplitude of transmitted wave as

$$(1 + iq_0)\exp(-i\phi)(T_n + T_{n+1}) = T_n\left\{1 + q^2\exp(-2i\phi) + (1 + q_0)^2\exp(-2i\phi)\right\}. \tag{A.6}$$

Since (A.6) is an arithmetic equation, we can solve it by putting,

$$T_n = T_0 x^n. \tag{A.7}$$

Then we can write as

$$T_{n-1} = T_0 x^{n-1}, T_{n+1} = T_0 x^{n+1}.$$

By inserting these equations into (A.6), we get the quadratic equation with respect to x. Similarly, we can write as

$$R_n = R_0 x^n. \tag{A.8}$$

From the Bragg condition, we get

$$\frac{2d\sin\theta_{\mathrm{B}}}{\lambda} = m, \tag{A.9}$$

where m is a positive integer. We introduce a parameter which represents the difference between the incident angle and the Bragg angle as

$$v = kd(\sin\theta_i - \sin\theta_{\mathrm{B}}) = kd(\theta_i - \theta_{\mathrm{B}})\cos\theta_{\mathrm{B}} \tag{A.10}$$

and write

$$\phi = m\pi + v. \tag{A.11}$$

As q_0, q and v are all very small quantities, x can be obtained as

$$x = (-1)^m\{1 \pm \left\{q^2 - (v - q_0)^2\right]^{\frac{1}{2}}\right\}. \tag{A.12}$$

For the solution of x, we refer to James [1].

If we find the reflectance from a semi-infinite crystal, we set $n = 0$ and obtain $|R_0/T_0|^2$. From (A.5), we get

$$R_0 = iqT_0 + (1 + iq_0)R_0 x\exp(-i\phi). \tag{A.13}$$

Then, R_0/T_0 is given by

$$\frac{R_0}{T_0} = \frac{iq}{1 - x(1 + iq_0)\exp(-i\phi)}. \tag{A.14}$$

Since v is very small, we can put

$$\exp(-i\phi) = \exp[-i(m\pi + v)] = (-1)^m(1 - iv),$$

and using x in (A.12), we obtain

$$\frac{R_0}{T_0} = \frac{iq}{i(v - q_0) \pm \left\{q^2 - (v - q_0)^2\right\}^{1/2}}. \tag{A.15}$$

From the correspondence with the deviation parameter W in the Bragg case given by (6.76), we define

$$W = \frac{v - q_0}{|q|}, \tag{A.16}$$

following the treatment of Miyake [2]. Then for $|W| \geq 1$, (A.15) becomes

$$\frac{R_0}{T_0} = -\frac{q}{|q|}\left\{-W \pm \left(W^2 - 1\right)^{\frac{1}{2}}\right\}. \tag{A.17}$$

The plus sign (+) is applied for $W \geq 1$ and the minus sign for $W \leq 1$. For $|W| \leq 1$, (A.15) becomes

$$\frac{R_0}{T_0} = -\frac{q}{|q|}\left\{-W \pm i\left(1 - W^2\right)^{\frac{1}{2}}\right\}. \tag{A.18}$$

The reflectance is

$$\left|\frac{R_0}{T_0}\right|^2 = \left\{|W| - (W^2 - 1)^{\frac{1}{2}}\right\}^2 = \frac{1}{\left\{|W| + (W^2 - 1)^{\frac{1}{2}}\right\}^2} \tag{A.19}$$

for $|W| \geq 1$, and

$$\left|\frac{R_0}{T_0}\right|^2 = 1 \tag{A.20}$$

for $|W| \leq 1$. (A.19) and (A.20) are consistent with (6.93) and (6.94), which are the reflectance from a semi-infinite crystal obtained by the Ewald-Laue method. The reflectance for $-1 \leq W \leq 1$ is 100%. The calculated reflectance is shown in Fig. A.1b. This curve is called the Darwin curve. Since $q_0 < 0$ for $W = 0$, $-v = q_0$. By writing $\Delta\theta$ as

$$\Delta\theta = \theta_i - \theta_B, \tag{A.21}$$

and using (A.1) and (A.2), we get

$$v = -q_0 = kd\,\Delta\theta\cos\theta_{\mathrm{B}} = \Delta\theta\frac{2\pi d\cos\theta_{\mathrm{B}}}{\lambda} = \frac{\pi d|\chi_0|}{\lambda\sin\theta_{\mathrm{B}}}. \tag{A.22}$$

Then we obtain

$$\Delta\theta = \frac{|\chi_0|}{\sin 2\theta_{\mathrm{B}}} = \frac{2\delta}{\sin 2\theta_{\mathrm{B}}}. \tag{A.23}$$

This means that the Bragg angle θ in this case is larger than the Bragg angle θ_{B} in the kinematical theory of diffraction by $\Delta\theta$. This difference is the same as derived from the X-ray refraction effect and the same as (6.74) derived from the Ewald-Laue method. The width $\Delta\theta_t$ of the angular region of the total reflection for $-1 \le W \le 1$ is given by

$$\Delta\theta_t = \frac{2|\chi_g|}{\sin 2\theta_{\mathrm{B}}} = \frac{2|\chi_g|}{|\chi_0|}\Delta\theta. \tag{A.24}$$

A.3 Extinction Effect

We will find the intensity I_n $(= |T_n|^2)$ of the n-th plane. Expressing x in (A.12) as a function of W, we have

$$x = (-1)^m\left\{1 \pm |q|\left(1 - W^2\right)^{1/2}\right\} = (-1)^m\exp\left\{\pm|q|\left(1 - W^2\right)^{1/2}\right\} \tag{A.25}$$

by assuming $|q|$ is very small. Since $T_n = T_0 x^n$, we get

$$I_n = I_0\exp\left\{-2n|q|\left(1 - W^2\right)^{1/2}\right\} = I_0\exp(-2st), \tag{A.26}$$

where

$$s = \frac{k|\chi_g|}{2\sin\theta_i},\ t = nd. \tag{A.27}$$

At the exact Bragg condition ($W = 0$), the depth where I_n becomes $1/e$ of the incident intensity I_0 is given by

$$t_{\mathrm{B}} = \frac{1}{2s} = \frac{\sin\theta_i}{k|\chi_g|}. \tag{A.28}$$

t_B is called the extinction distance and the value is approximately $1\,\mu$m. In the total reflection region, X-rays are localized near the crystal surface and cannot penetrate deep into the crystal even though it is a non-absorbing crystal. (A.28) is consistent with (6.98) obtained by the Ewald-Laue method.

So far, we have been dealing with non-absorbing crystals, but by adding the scattering amplitude due to resonance scattering to q and making it a complex number, we can extend the above argument to absorbing crystals.

References

1. R.W. James, *The Optical Principles of The Diffraction of X-Rays.* (G. Bell and Sons Ltd, London, 1967)
2. S. Miyake, Solid State Physics. **8**, 547–555 (In Japanese) (1973)

Appendix B
Resonant Dynamical Diffraction Theory Using the Riemann Surface

By using the Riemann surface, Saka [1–6] has reformulated the dynamical theory of X-ray diffraction involving resonant scattering. To reveal the characteristic feature of dispersion surface, four parameters are used, which are the real and the imaginary parts of s and β given by

$$s = s_r + is_i = K_0(\theta - \theta_{\mathrm{B}}) + \frac{1}{2}K_0\chi_0\left(1 - \frac{\gamma_g}{\gamma_0}\right)\frac{1}{\sin 2\theta_{\mathrm{B}}}, \tag{B.1}$$

$$\beta = \beta_r + i\beta_i = K_0 P(\chi_g\chi_{-g})^{1/2}\left(\frac{\gamma_g}{\gamma_0}\right)^{1/2}\frac{1}{\sin 2\theta_{\mathrm{B}}}, \tag{B.2}$$

Here, s represents the degree of deviation from the exact Bragg condition, β the reflection strength, $\gamma_g = \cos\theta_2$, $\gamma_0 = \cos\theta_1$ and s/β is regarded as a deviation parameter. If the resonance errors in the direction of the incident wavevector and the diffracted wavevector are denoted by $\Delta\eta_0$ and $\Delta\eta_g$, they are given by

$$\Delta\eta_0 = \Delta\eta_{0r} + i\Delta\eta_{0i} = \alpha\gamma_0\left\{s \pm \left(s^2 + \beta^2\right)^{\frac{1}{2}}\right\}, \tag{B.3}$$

$$\Delta\eta_g = \Delta\eta_{gr} + i\Delta\eta_{gi} = \alpha\gamma_g\left\{-s \pm \left(s^2 + \beta^2\right)^{\frac{1}{2}}\right\}, \tag{B.4}$$

where α is expressed as

$$\alpha = \frac{\sin 2\theta_{\mathrm{B}}}{2\gamma_g}. \tag{B.5}$$

Then the real and imaginary parts of dispersion equations are obtained as

$$\Delta\eta_{0r}\Delta\eta_{gr} - \Delta\eta_{0i}\Delta\eta_{gi} = \frac{P^2K_0^2}{4}\mathrm{Re}(\chi_{-g}\chi_g), \tag{B.6}$$

T. Fukamachi and T. Kawamura, *Resonant Dynamical X-ray Diffraction*, Springer Series in Solid-State Sciences 208, https://doi.org/10.1007/978-981-96-8406-9

$$\Delta\eta_{0r}\Delta\eta_{gi} + \Delta\eta_{0i}\Delta\eta_{gr} = \frac{P^2 K_0^2}{4}\mathrm{Im}(\chi_{-g}\chi_g). \tag{B.7}$$

These equations are the same as (7–59) and (7–60). Visualizing the geometric relationship between the four parameters using the Riemann surface is useful for understanding the characteristics of rocking curves under special conditions due to resonant scattering. Using these equations, the phase shift of Pendellösung fringes in the diffracted and transmitted waves in the Laue case is investigated [5]. The results show that the phase of the Pendellösung fringes of the diffracted wave is opposite to that of the transmitted wave when $|\chi_{gr}| \gg |\chi_{gi}|$. As the ratio of $|\chi_{gi}|/|\chi_{gr}|$ increases, the phase difference becomes smaller. When $|\chi_{gr}| \ll |\chi_{gi}|$, the two phase are the same. This result shows good agreement with the calculated and experimental results by Negishi et al. [7–9]. In [6], the wave field of a spherical wave is obtained in the Laue case when the real part of χ_{gr} is zero or $|\chi_{gr}| \ll |\chi_{gi}|$. The results show that the wavefield in a crystal is the same as that given by the conventional theory based on a plane wave when $|\chi_{gr}| \gg |\chi_{gi}|$, but that it is necessary to modify it regarding the wavefield on the vacuum side.

References

1. T. Saka, Acta Cryst. A**72**, 338–348 (2016)
2. T. Saka, Acta Cryst. A**72**, 472–479 (2016)
3. T. Saka, Acta Cryst. A**74**, 578–585 (2018)
4. T. Saka, Acta Cryst. A**74**, 586–594 (2018)
5. T. Saka, Acta Cryst. A**76**, 132–136 (2020)
6. T. Saka, Acta Cryst. A**73**, 474–479 (2017)
7. R. Negishi, T. Fukamachi, M. Yoshizawa, K. Hirano, K. Hirano, T. Kawamura, J. Phys.Soc. Jpn. **77**, 023709 (2008)
8. R. Negishi, T. Fukamachi, M. Yoshizawa, K. Hirano, K. Hirano, T. Kawamura, Phys. Status Solidi. A**206**, 1865–1870 (2009)
9. R. Negishi, T. Fukamachi, M. Yoshizawa, K. Hirano, K. Hirano, T. Kawamura, Phys. Status Solidi. A**208**, 2567–2570 (2011)

Appendix C
Strain in an Elastically Bent Plane-Parallel Crystal

C.1 Pure Bent Crystal (Laue Case)

We describe the value of β given by (9.31) in the symmetric Laue case when a crystal is deformed due to pure bending where the strain gradient is constant. We assume that the size of the crystal is sufficiently smaller than the radius of curvature R in pure bending. Let P and P' be the positions before and after deformation in the crystal as shown in Fig. C.1. The displacement u due to the deformation is given by

$$u = \overline{\mathrm{PP}'} = -R\frac{\varphi^2}{2} = -\frac{x^2}{2R} \tag{C.1}$$

using $\overline{\mathrm{A_0P}} \approx R\varphi$ and $\overline{\mathrm{PP}'} = \overline{\mathrm{A_0P}}\varphi/2$. Here, x is the distance $\overline{\mathrm{A_0P}}$ and $x = (x_0 + x_g)\cos\theta_\mathrm{B}$, and R is the curvature radius. Since the reciprocal lattice vector $\boldsymbol{g}$ is parallel to the incident surface (symmetric Laue case) and can be considered parallel to the displacement vector $\boldsymbol{u}$, we get

$$\boldsymbol{g} \cdot \boldsymbol{u} = -\frac{gx^2}{2R}. \tag{C.2}$$

By inserting the Bragg gap $G_\mathrm{B} = 2\pi/\Lambda_0$ into (9.32), the parameter β becomes

$$\beta = \frac{-\Lambda_0}{2\pi\cos^2\theta_\mathrm{B}}\frac{\partial^2(\boldsymbol{g}\cdot\boldsymbol{u})}{\partial x_0 \partial x_g},$$

where Λ_0 is the spacing of Pendellösung fringes. From (C.2), we get

$$\frac{\partial^2(\boldsymbol{g}\cdot\boldsymbol{u})}{\partial x_0 \partial x_g} = -\frac{g\cos^2\theta_\mathrm{B}}{R}. \tag{C.3}$$

T. Fukamachi and T. Kawamura, *Resonant Dynamical X-ray Diffraction*, Springer Series in Solid-State Sciences 208, https://doi.org/10.1007/978-981-96-8406-9

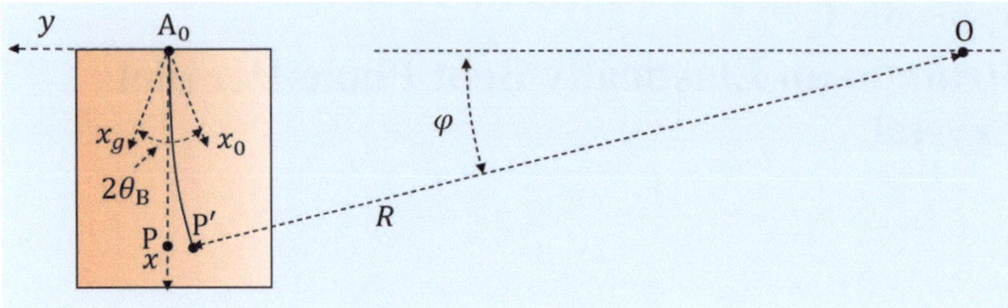

Fig. C.1 Deformation of a plane-parallel crystal due to pure bent

We can write β as

$$\beta = \frac{g\Lambda_0}{2\pi R} = \frac{\Lambda_0}{d_0}\varepsilon'. \tag{C.4}$$

Here, d_0 is the lattice spacing and ε' is the strain gradient given as

$$\varepsilon' = \frac{1}{R}. \tag{C.5}$$

If we assume that $\Lambda_0 = 22$ μm, $d_0 = 0.19$ nm and $R = 50$ m, we get $\beta = 2.3\ \text{mm}^{-1}$.

C.2 *Pure Bent Crystal (Bragg Case)*

Let the length of a pure bent crystal be ρ and its thickness be H. When the crystal is bent as shown in Fig. C.2a, tension is applied to the front surface and compression force is applied to the back surface. At $H/2$, there is a neutral layer without strain. We write the curvature radius of an arcuate crystal (from the origin O to the neutral layer) as R, and the increment of ρ shifted by an increment z_R from R as $\Delta\rho$. Then the longitudinal strain is given as

$$\varepsilon_1 = \frac{\rho + \Delta\rho - \rho}{\rho} = \frac{(R + z_R)\phi - R\phi}{R\phi} = \frac{z_R}{R}. \tag{C.6}$$

The transverse strain ε_t at a distance z from the front surface is given by

$$\varepsilon_t = \frac{d - d_0}{d} = -\upsilon\varepsilon_1 = -\frac{\upsilon z_R}{R} = -\varepsilon' z_R,\ \varepsilon' = \frac{\upsilon}{R}. \tag{C.7}$$

by writing Poisson's ratio (transverse strain/longitudinal strain) as υ. Here, d is the lattice spacing of the bent crystal, and d_0 is that of the crystal without deformation. ε' is the strain gradient. If we write the front surface of the crystal as $z = 0$ and the

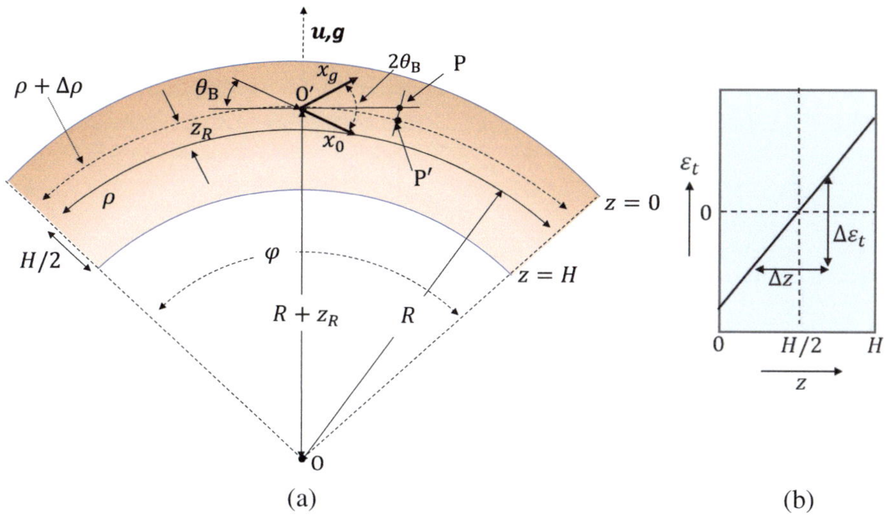

Fig. C.2 Strain in a curved plane-parallel crystal. **a** Curved crystal. **b** Strain in the crystal

bottom surface as $z = H$, the transverse strain is written by

$$\varepsilon_t = \varepsilon' \left(z - \frac{H}{2} \right). \tag{C.8}$$

Figure C.2b shows the strain distribution within the curved crystal. In a curved crystal as shown in Fig. C.2a, the strain gradient is constant, and the transverse strain is negative at the front surface and positive at the bottom.

C.3 Cantilever Crystal with End Load

An example of the end load to a cantilever crystal is shown in Fig. C.3. The bending moment M is given by the load applied to the free end of the crystal P as

$$M = -Pl, \tag{C.9}$$

where l is the distance from the load position to the measurement point. According to material mechanics, there is a relation

$$\frac{1}{R} = \frac{M}{EI_Z}. \tag{C.10}$$

Here, E is Young's modulus and I_Z is the second moment of area. Deflection D is given by

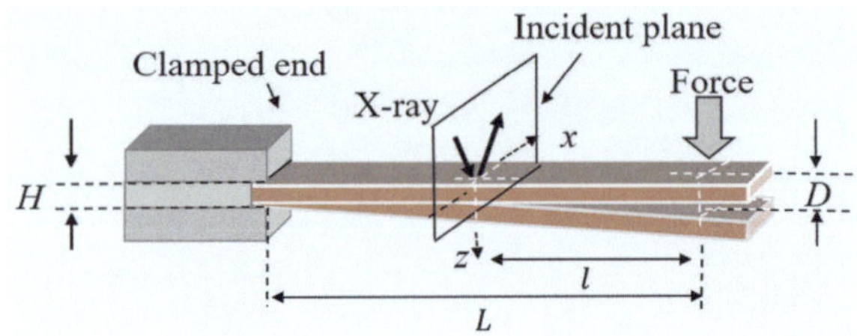

Fig. C.3 Deformation of cantilever crystal by applying force

$$D = \frac{PL^3}{3EI_Z} \tag{C.11}$$

by writing the distance from the load position to the fixed end as L. Substituting P in (C.11) into (C.9), we get

$$M = -3EI_Z \frac{Dl}{L^3}. \tag{C.12}$$

Further substituting this into (C.10), we get

$$\frac{1}{R} = -3\frac{Dl}{L^3}. \tag{C.13}$$

From this relation, we obtain

$$\varepsilon_\mathrm{t} = \varepsilon'\left(z - \frac{H}{2}\right) = \frac{\upsilon}{R}\left(z - \frac{H}{2}\right) = 3\upsilon\frac{Dl}{L^3}\left(z - \frac{H}{2}\right) \tag{C.14}$$

and

$$\varepsilon' = 3\upsilon\frac{Dl}{L^3}. \tag{C.15}$$

The strain gradient ε' is proportional to the deflection D and the distance l.

C.4 Cantilever Crystal with Uniformly Distributed Load

A cantilever crystal generally deforms due to gravity. Then, the uniformly distributed load is applied to the crystal. If the weight per unit length is w, the moment in this case is written as

$$M = -\frac{wl^2}{2}. \tag{C.16}$$

The deflection D is given by

$$D = -\frac{wL^4}{8EI_Z}, \tag{C.17}$$

where L is the length of the crystal. Using this relation, the lateral strain is written as in the case of the end load as

$$\varepsilon_t = \varepsilon'\left(z - \frac{H}{2}\right),$$

and the strain gradient is given by

$$\varepsilon' = 4\upsilon\frac{Dl^2}{L^4}. \tag{C.18}$$

ε' is proportional to the deflection D and l^2.

C.5 Relation Between Strain Gradient ε' and β (Bragg Case)

Using Fig. C.2a, we consider the relation between strain gradient ε' and β. When there is no strain in a crystal, we take a point P which is located at a distance x from point O′ in the direction perpendicular to the reciprocal lattice vector $\boldsymbol{g}$. When the crystal is bent, the point P moves to P' which is shifted by u in the z direction. From (C.1), the magnitude of the displacement vector $\boldsymbol{u}$ is written by

$$|\boldsymbol{u}| = \overline{\mathrm{PP'}} = \frac{x^2}{2R}.$$

Then we can write

$$\boldsymbol{u}\cdot\boldsymbol{g} = -\frac{gx^2}{2R} = -\frac{gx^2}{2}\varepsilon'.$$

By using the relation

$$x = (x_0 + x_g)\sin\theta_B, \tag{C.19}$$

we can get

$$\frac{\partial^2 \boldsymbol{g}\cdot\boldsymbol{u}}{\partial x_0 \partial x_g} = -\varepsilon' g \sin^2\theta_B = -\frac{2\pi\varepsilon'\sin^2\theta_B}{d_0}. \tag{C.20}$$

Using (9.31), β becomes

$$\beta = \frac{-\Lambda_0}{2\pi \cos^2\theta_B} \frac{\partial^2 \boldsymbol{g} \cdot \boldsymbol{u}}{\partial x_0 \partial x_g} = \frac{\Lambda_0 \tan^2\theta_B}{d_0} \varepsilon'. \tag{C.21}$$

β in the Bragg case in (C.21) differs from that in the Laue case in (C.4) by a factor of $\tan^2\theta_B$.

The right-hand side of (C.21) was derived by Yan et al. [1].

Reference

1. H. Yan, Ö. Kalenci, I. C. Noyan, *J. Appl. Cryst.* **40**, 322–331 (2007)

Appendix D
Takagi Equation

The Takagi theory is a dynamical theory of diffraction that can be applied to deformed crystals [1–3]. We will describe Takagi equation based on the review article by Miyake [4].

When there are two waves in the crystal, one wave with wavevector $\boldsymbol{k}_0$ connected to the incident wave and one diffracted wave with wavevector $\boldsymbol{k}_g$, the Bloch wave is expressed as

$$D(\boldsymbol{r}) = D_0 \exp(-i\boldsymbol{k}_0 \cdot \boldsymbol{r}) + D_g \exp\left(-i\boldsymbol{k}_g \cdot \boldsymbol{r}\right). \tag{D.1}$$

From the basic (5.42) and (5.43), we have

$$\frac{2\xi_0}{\kappa_0} D_0 - P\chi_{-g} D_g = 0, \tag{D.2}$$

$$P\chi_g D_0 - \frac{2\xi_g}{\kappa_0} D_g = 0. \tag{D.3}$$

In the dispersion surface in Fig. D.1, we write N at the intersection of normal line ν and asymptote T_0 of the dispersion surface.

By expressing

$$d_0(\boldsymbol{r}) = D_0 \exp\left\{-i(\boldsymbol{k}_0 - \boldsymbol{k}_0^N) \cdot \boldsymbol{r}\right\}, \tag{D.4}$$

$$d_g(\boldsymbol{r}) = D_g \exp\left\{-i(\boldsymbol{k}_g - \boldsymbol{k}_g^N) \cdot \boldsymbol{r}\right\}, \tag{D.5}$$

we can rewrite (D.1) as

$$D(\boldsymbol{r}) = d_0(\boldsymbol{r}) \exp\left(-i\boldsymbol{k}_0^N \cdot \boldsymbol{r}\right) + d_g(\boldsymbol{r}) \exp\left(-i\boldsymbol{k}_g^N \cdot \boldsymbol{r}\right). \tag{D.6}$$

T. Fukamachi and T. Kawamura, *Resonant Dynamical X-ray Diffraction*, Springer Series in Solid-State Sciences 208, https://doi.org/10.1007/978-981-96-8406-9

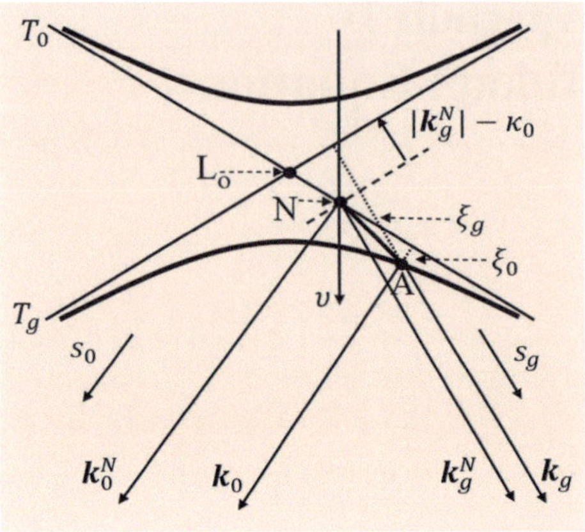

Fig. D.1 Dispersion surface in Takagi's method

Here, $d_0(\boldsymbol{r})$ and $d_g(\boldsymbol{r})$ are slowly varying function of $\boldsymbol{r}$. Equation (D.6) is the extended form of (D.1). From Fig. D.1, we get

$$\boldsymbol{k}_0 - \boldsymbol{k}_0^N = \boldsymbol{k}_g - \boldsymbol{k}_g^N = \overrightarrow{\mathrm{AN}}. \tag{D.7}$$

$\boldsymbol{s}_0$ and $\boldsymbol{s}_h$ components of $\overrightarrow{\mathrm{AN}}$ are respectively

$$\xi_0 (= |\boldsymbol{k}_0| - \kappa_0) \tag{D.8}$$

and

$$\xi_g - \left(\left| \boldsymbol{k}_g^N \right| - \kappa_0 \right). \tag{D.9}$$

Derivatives of the (D.4) and (D.5) give

$$\frac{\partial d_0(\boldsymbol{r})}{\partial s_0} = -i \xi_0 d_0(\boldsymbol{r}), \tag{D.10}$$

$$\frac{\partial d_g(\boldsymbol{r})}{\partial s_g} = -i(\xi_g + \zeta_N) d_g(\boldsymbol{r}). \tag{D.11}$$

Here, ζ_N is the resonance error given by

$$\zeta_N = \kappa_0 - \left| \boldsymbol{k}_g^N \right|. \tag{D.12}$$

Substituting the (D.2) into (D.10), and using the relations between (D.11) and (D.3) and (D.9), we get

$$\frac{\partial d_0(\boldsymbol{r})}{\partial s_0} = -\frac{i}{2}\kappa_0 P \chi_{-g} d_g(\boldsymbol{r}), \tag{D.13}$$

and

$$\frac{\partial d_g(\boldsymbol{r})}{\partial s_g} = -\frac{i}{2}\kappa_0 P \chi_g d_0(\boldsymbol{r}) + iKB'_g d_0(\boldsymbol{r}) \tag{D.14}$$

Here, B'_g is

$$B'_g = \frac{\left|\boldsymbol{k}_g^N\right| - \kappa_0}{K} = -\frac{\zeta_N}{K}, \tag{D.15}$$

which is the dimensionless resonance error for point N. Equations (D.13) and (D.14) are one form of the Takagi equation. In the above, we considered only one tie-point A on the dispersion surface, but we can apply the Takagi equation to any Bloch waves whose tie-points lie on the same dispersion surface, and to multiple superpositions of those Bloch waves. This is clear from the fact that (D.13) and (D.14) do not include ξ_0 or ξ_g, which are related to the reciprocal lattice.

D.1 Boundary Conditions

Since $d_0(\boldsymbol{r})$ and $d_g(\boldsymbol{r})$ are slowly varying functions of $\boldsymbol{r}$, the periods of the spatial changes are very large compared to that of the wavelength of the incident wave or the lattice constant of the crystal, and are about the same order of the extinction distance. For the incident wave, the electric field is expressed by

$$E(\boldsymbol{r}) = E^a(\boldsymbol{r})\exp(-i\boldsymbol{K}_0 \cdot \boldsymbol{r}), \tag{D.16}$$

with $E^a(\boldsymbol{r})$ being a slowly varying function of $\boldsymbol{r}$. Equation (D.16) can be used when the crystal surface is curved, as far as it can be regarded to be flat within the range of the extinction distance.

If we rewrite the incident wave as

$$E'(\boldsymbol{r}) = E^a(\boldsymbol{r})\exp\{-i(\boldsymbol{K}_0 - \boldsymbol{k}_0^N) \cdot \boldsymbol{r}\} \tag{D.17}$$

and a point on the surface as $\boldsymbol{r}_c$, we can obtain one of the boundary conditions as

$$d_0(\boldsymbol{r}_c) = E'(\boldsymbol{r}_c),\ d_g(\boldsymbol{r}_c) = 0, \tag{D.18}$$

which are relations between slowly varying functions of $\boldsymbol{r}_c$.

D.2 For a Deformed Crystal Due to Strain

We have considered local reciprocal vectors in Chap. 9, and the reciprocal lattice vector $\boldsymbol{g}$ is a function of position within the crystal and is written as $\boldsymbol{g}'(\boldsymbol{r})$. This is because T_0 is fixed in Fig. D.1, but T_g varies as a function of $\boldsymbol{r}$. The resonance error ζ_N when the standard point N of T_0 is fixed becomes a function of $\boldsymbol{r}$, that is, B_g' becomes a function of $\boldsymbol{r}$. From (9.8), we have

$$\boldsymbol{g}'(\boldsymbol{r}) = \boldsymbol{g} - \nabla(\boldsymbol{g} \cdot \boldsymbol{u}(\boldsymbol{r})).$$

In a strained crystal, B_g' in (D.15) is expressed as

$$B_g' = (B_g')_0 - \frac{1}{K}\frac{\partial}{\partial s_g}(\boldsymbol{g} \cdot \boldsymbol{u}(\boldsymbol{r})), \tag{D.19}$$

where $(B_g')_0$ is B_g' in (D.15). In this way, the Takagi equation can be applied to deformed crystals. The Takagi equation was originally proposed to extend the dynamical diffraction theory for deformed crystals.

D.3 For a Perfect Crystal

For a perfect crystal, B_g' is given by (D.15). If point N in Fig. D.1 is chosen as the Lorentz point L_0, B_g' is equal to zero. Then, (D.13) and (D.14) become

$$\frac{\partial d_0(\boldsymbol{r})}{\partial s_0} = -\frac{i}{2}\kappa_0 P \chi_{-g} d_g(\boldsymbol{r}), \tag{D.20}$$

and

$$\frac{\partial d_g(\boldsymbol{r})}{\partial s_g} = -\frac{i}{2}\kappa_0 P \chi_g d_0(\boldsymbol{r}). \tag{D.21}$$

By solving this system of partial differential equations under the boundary conditions of the wave fields at the entrance and exit surfaces of the crystal, it is possible to derive equations for rocking curves, intensity distributions of Pendellösung fringes, and so on.

By combining (D.20) and, (D.21) we can get second-order partial differential equations for $d_0(\boldsymbol{r})$ and $d_g(\boldsymbol{r})$. The standard procedure of solving the equations is to use the Riemman's method. Such analytical solutions are possible only in a small number of cases. For most deformed crystals, numerical calculation methods are widely used.

D.4 Solution for a Narrow Incident Flux

We will discuss the results of diffracted intensities from a perfect crystal in the symmetric Laue case using a very narrow slit as shown in Fig. D.2a. We assume that a very narrow beam is incident at point O on a plane-parallel crystal of thickness H, by placing a slit just above the incident surface. This incident beam excites quasi-spherical waves in the Borrmann fan of the triangular region OAB in the crystal, as already mentioned. $d_0(\boldsymbol{r})|_{z=H}$ and $d_g(\boldsymbol{r})|_{z=H}$ on the bottom surface are given by Riemann's method [5] as

$$d_0(\boldsymbol{r})|_{z=H} = \delta(x+l) - \frac{2\pi\sqrt{\chi_{-g}\chi_g}}{\sin 2\theta_{\mathrm{B}}}\frac{(l-x)}{\sqrt{l^2-x^2}}J_1\left(\pi G_{\mathrm{B}}H\sqrt{1-\left(\frac{x}{l}\right)^2}\right), \quad \text{(D.22)}$$

and

$$d_g(\boldsymbol{r})|_{z=H} = -\frac{i\pi\chi_{-g}}{\sin\theta_{\mathrm{B}}}J_0\left(\pi G_{\mathrm{B}}H\sqrt{1-\left(\frac{x}{l}\right)^2}\right). \quad \text{(D.23)}$$

Here, we define the coordinates of the origin O as $x = 0$ and $z = 0$. J_0 and J_1 are Bessel functions of the first kind and G_{B} is the Bragg gap. The calculated diffracted intensities obtained by changing the crystal thickness are shown in Fig. D.2b by using the relation $|d_g(\boldsymbol{r})|^2 \propto J_0^2$.

The diffracted intensities in Fig. D.2b show interference fringes obtained in 8.2.1. In deriving these diffracted intensities, it is not necessary to find the dispersion surface or to consider the movement of the tie-point of the pseudo-refracted wave, its trajectory and phase.

Kanamaru et al. [6] used the Takagi equation to derive the diffracted intensity formula from a semi-infinite crystal for the resonant dynamical diffraction in Chap. 7, and calculated the rocking curves at the K absorption edge of Ge.

The solution to the Takagi equation or the numerical calculation method for obtaining the diffracted intensity from a crystal with constant strain gradient is briefly described by Authier [7].

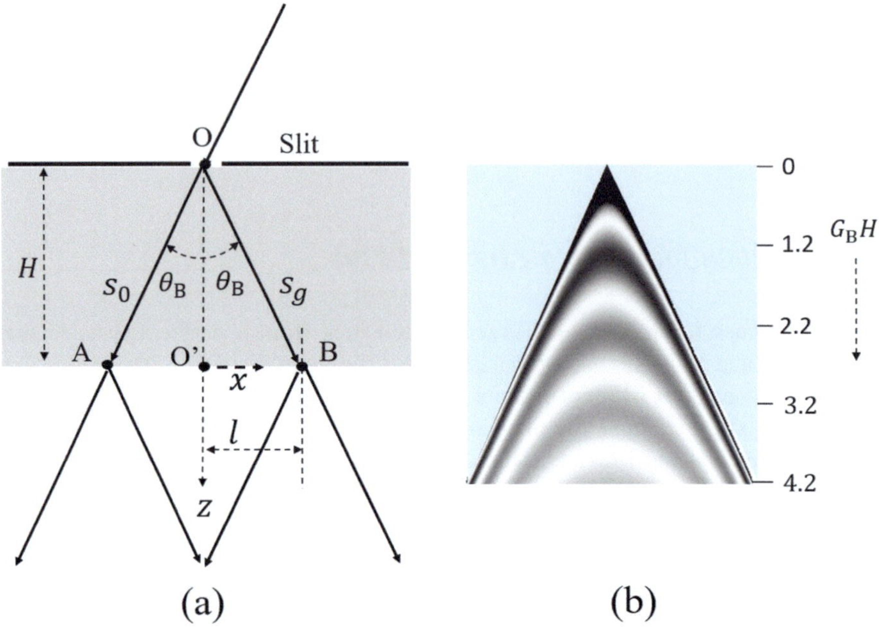

Fig. D.2 **a** Beam geometry of diffraction by a very narrow incident beam. **b** Intensitiy distributions ($|d_g(\boldsymbol{r})|^2$) in the Borrmann fan

References

1. S. Takagi, Acta Cryst. **15**, 1311–1312 (1962)
2. S. Takagi, J. Phys. Soc. Jpn. **26**, 1239–1253 (1969)
3. D. Taupin, Bull. Soc. Fr. Minėr. Crist. **87**, 469–511 (1964)
4. S. Miyake, Solid State Physics. **9**, 319–328 (In Japanese) (1974)
5. S. Takagi, J. Cryst. Soc. Japan. **19**, 248–272 (In Japanese) (1971)
6. T. Kanamaru, M. Kuribayashi, K. Ishida, J. Phys. Soc. Jpn. **64**, 3529–3536 (1995)
7. A. Authier, *Dynamical Theory of X-ray Diffraction.* (Oxford University Press, 2001)

Index

T. Fukamachi and T. Kawamura, *Resonant Dynamical X-ray Diffraction*, Springer Series in Solid-State Sciences 208, https://doi.org/10.1007/978-981-96-8406-9

R

S

MIX
Papier aus verantwortungsvollen Quellen
Paper from responsible sources
FSC® C105338

If you have any concerns about our products,
you can contact us on
ProductSafety@springernature.com

In case Publisher is established outside the EU,
the EU authorized representative is:
Springer Nature Customer Service Center GmbH
Europaplatz 3, 69115 Heidelberg, Germany

Printed by Libri Plureos GmbH
in Hamburg, Germany